Bruce Batchelor

February 1986

VLSI Image Processing

Edited by
R. J. Offen
Manager of the Computer Systems Research Laboratory,
GEC Research Laboratories, Hirst Research Centre,
Wembley, UK

COLLINS
8 Grafton Street, London W1

Collins Professional and Technical Books
William Collins Sons & Co. Ltd
8 Grafton Street, London W1X 3LA

First published in Great Britain by
Collins Professional and Technical Books 1985

British Library Cataloguing in Publication Data
VLSI image processing.
1. Image processing 2. Integrated circuits—
Very large scale integration
I. Offen, R.J.
621.38'0414 TA1632

ISBN 0-00-383055-1

Typeset by Hope Services, Abingdon, Oxon
Printed and bound in Great Britain by
Mackays of Chatham, Kent

Contents

Contributors' Biographies

B. Arambepola is Leader of the VLSI Signal Processing Group, Very High Performance Integrated Circuit Design Laboratory, GEC Research Laboratories, Hirst Research Centre, Wembley, UK.

S. R. Brooks is Manager of the Marconi Space and Defence Systems Research Laboratory, GEC Research Laboratories, Marconi Research Centre, Chelmsford, UK. He is involved in microwave surveillance and remote sensing, from airborne and spaceborne platforms, including synthetic aperture radar.

H. Buxton holds a lectureship in Information Technology at Queen Mary College, London, in the Department of Computer Science and Statistics. With a background in cognitive psychology, BSc (Bristol) and PhD (Cambridge), she has been researching temporal aspects of visual perception using an Artificial Intelligence approach; that is, developing a computational theory of both human and machine systems. This research involves parallel computer implementations of processing algorithms to interpret image sequences.

P. V. Collins was, until recently, Manager of the Marconi Systems Division, Computer Systems Research Laboratory, GEC Research Laboratories, Hirst Research Centre, Wembley, UK. He is currently with Marconi Underwater Systems Ltd.

M. M. McCabe is Manager of the Computer Science Research Division and Leader of the Computer Vision Group, Computer Systems Research Laboratory, GEC Research Laboratories, Hirst Research Centre, Wembley, UK.

R. J. Offen was previously Senior Lecturer in Physics at the University of Otago, Dunedin, New Zealand. He is currently Manager of the Computer Systems Research Laboratory, GEC Research Laboratories, Hirst Research Centre, Wembley, UK.

I. N. Parker is Manager of the Very High Performance Integrated Circuit Design Laboratory, GEC Research Laboratories, Hirst Research Centre, Wembley, UK.

P. N. Scharbach is Leader of the Formal Methods Group, Computer Systems Research Laboratory, GEC Research Laboratories, Hirst Research Centre, Wembley, UK. The research interests of the Group include the development and application of formal techniques for specifying and verifying highly concurrent systems.

J. S. Wiejak. After obtaining a BA in Mathematics from Christ's College, Cambridge, followed by a PhD in Physics from Imperial College, he joined the Computer Systems Research Laboratory, GEC Research Laboratories, Hirst Research Centre, Wembley, UK. He is currently working at Meta Machines, Oxford, on sensor-based robotics systems.

Preface

Image processing is a large, complex and diverse area of endeavour, and is becoming more so. As practised at present, image processing relies to a great extent on two major, effectively distinct and self-contained domains of activity, computational algorithms and processor architectures – especially those architectures facilitated by VLSI technology. This book has grown out of our belief, based on actual experience, that future progress in image processing, especially in the context of highly concurrent processing, is going to be increasingly dependent on the establishment of productive formal links between these two important domains. This book has much to say about both algorithms and advanced computers, often in the context of specific image processing applications, but the underlying, sometimes implicit, theme is the importance of understanding the potential mappings which exist between the two domains.

This book is not a primer in image processing, in that a certain amount of prior knowledge is assumed. Many excellent introductory and review texts already exist, and it was not our intention to produce yet another. Rather, we have tried to present some new and novel perspectives of image processing in the hope that they are of interest and eventual utility. As a consequence, both the choice of material and its organisation represent something of an experiment. I hope that you, as the reader, feel that in some measure it succeeds. Chapter 1 covers some introductory background material on image processing techniques and computer architectures, as well as setting the scene for what is to follow. Chapters 2 and 3 cover algorithms and architectures respectively, while Chapter 4 attempts an initial synthesis between these two domains. In order to provide some concrete illustrative examples, Chapters 5 and 6 cover two very different image processing applications areas – machine vision and synthetic aperture radar. Finally, Chapter 7 presents one view of what the future might hold.

Many people, in one way or another, have contributed directly and indirectly to the preparation of this book. In particular, various members of the research and support staff of the GEC Research Laboratories, Wembley, UK, and especially Mrs Nell Peters, who has borne much of the brunt of manuscript preparation. To them all, my heartfelt thanks.

R. J. Offen

Chapter 1

Whence Image Processing?

R. J. OFFEN

1.1 INTRODUCTION

An increasingly accepted contemporary view of image processing is to regard it, in a broad and generic sense, as being nothing more than one aspect of 'information processing' (Machlup and Mansfield, 1983), as opposed to the more *ad hoc* traditional view which tended to embrace one or more, depending on individual preference or prejudice, of various technological niches including, *inter alia,* picture processing, image processing, pattern recognition, scene analysis, image interpretation, optical processing, video processing and image understanding. These niches overlap to greater or lesser extents and each has its own history, peculiarities and ardent supporters.

There have been a number of attempts to produce some measure of unification and demystification by categorising these niches into more reasonably circumscribed areas, such as 'signal processing', 'classification' and 'understanding' (Cohen and Feigenbaum, 1982). Signal processors transform an input image into another image that has desirable properties. For example, the output image may have a better signal-to-noise ratio or have certain features enhanced in some way. Image processing and picture processing are the most common terms for this class of processing. Both digital and optical techniques can be used. Classification techniques classify images into predetermined categories. Character recognition and signature verification are typical examples. Pattern recognition and pattern classification are the most commonly used terms, though the word 'recognition' has mainly a historical context. Image understanding has a definite semantic context, requiring both knowledge about the specific problem domain, often complex and three-dimensional, as well as sophisticated image processing techniques. The inherent complexity of image understanding has resulted in the fact that decomposition of processing into 'early', 'intermediate' and 'late' stages is almost axiomatic.

We will use the term 'image processing' as a catch all for all of these activities, with the implicit understanding that the fundamental underlying activity is that of 'information processing', performed on two-dimensional projections of some property of the physical world.

The most important factors in bringing together the disparate aspects of these diverse yet strongly related fields has been the burgeoning availability and use of advanced computing, storage and display technologies, especially those incorporating VLSI design and fabrication techniques. This intrusion of radically new technology has caused some fundamental rethinking of many aspects of image processing. In fact, we shall adopt the premise that VLSI has become the natural implementation route for realising many of the essentially parallel algorithms which are characteristic of image processing.

Figure 1.1 illustrates, in a simple schematic form, the linkages which exist between the key conceptual aspects of implementing image processing systems. The algorithms–architectures linkage is fundamental, yet it is still not clearly understood. Currently it represents the weak link in the design pathway. It is the exploration of this link, in the context of the very considerable potential of VLSI design and fabrication technology, which provides the underlying theme of this book. This lack of understanding is especially critical in the context of systems with high levels of concurrency. The current state of system implementation is very much a case of machine architectures dictating algorithmic design. The classical von Neumann architecture is so central to the intellectual ethos of modern computing that the architecture–algorithm duality is firmly skewed towards this type of architecture. It is only with the developing interest in highly concurrent non-von Neumann architectures, coupled with the high degree of versatility

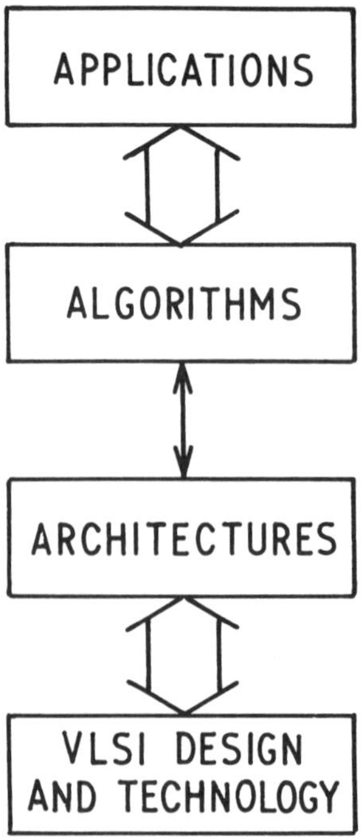

Fig. 1.1 The interrelationships which exist between the key stages which underlie the design and implementation of VLSI-based image processing systems. The width of the arrows represents the extent of our understanding of the implicit mappings.

inherent in VLSI design and fabrication, that researchers have started to question some of the conventional routes for implementing algorithms on computers.

In studying, in a general sense, the interrelationship between computational algorithm and implementation architecture some fundamental issues soon become apparent. These include questions relating to representation, description, data structures, mappings, data and control flow, and performance. These are all issues which will be explored in later chapters. At an even more fundamental level, in the context of this book, is the issue of what one is doing in attempting to process images, especially at the level of image understanding. The central paradox of much of image processing, especially image understanding, is that it is in general underconstrained in the sense that given one or more images of a scene the set of possible real-world scenes is very large indeed. In this sense it is reasonable to regard any image of the real world as a two-dimensional projection of the physical universe. An intrinsic characteristic of image formation is the information loss inherent in the process of capturing the image. The intensity values represented by the individual pixel values are the result of the interaction of many factors, including the intensity, colour, location and nature of the light sources, the position, reflectance, transparency and opaqueness of the objects in the scene, the transmission, refractance, absorption and scattering properties of the transmission media, the optical and electrical properties of the imaging device, etc., etc. The difficulty in interpreting images occurs not because we do not understand these diverse phenomena but because of the confounding of complexities inherent in the mappings of these various processes onto the resulting image. While a given scene gives rise to a unique image in any given direction, that image (and knowledge of the direction of projection) does not uniquely specify the three-dimensional scene that it resulted from. There is, for example, no way to fix the depth of any point seen in the image. Mathematically speaking, a projection operator does not have a unique inverse. The question is how properly to represent and sensibly to resolve these uncertainties? Mackworth (1983a) has proposed a set theoretic approach to this problem, where the imaging process is represented, in the simplest case, by mappings between the 'world', 'projection' and 'image' domains. This is illustrated in Fig. 1.2. This, at least in principle, makes possible the representation of all the solutions, potentially infinite in number, to the image interpretation problem in a finite manner. This is a representation, which though it may be of considerable theoretical interest, needs further exploration if it is to be of any practical use.

As implied above, an 'image' in the context of this book is a two-dimensional, almost invariably Cartesian, array of data resulting from sampling the projected instantiation of a local variable, the scene brightness function, obtained via a sensing device. The function values are either brightness values or vectors of brightness values sensed in different spectral bands, e.g. colour images. In the black-and-white case these values are

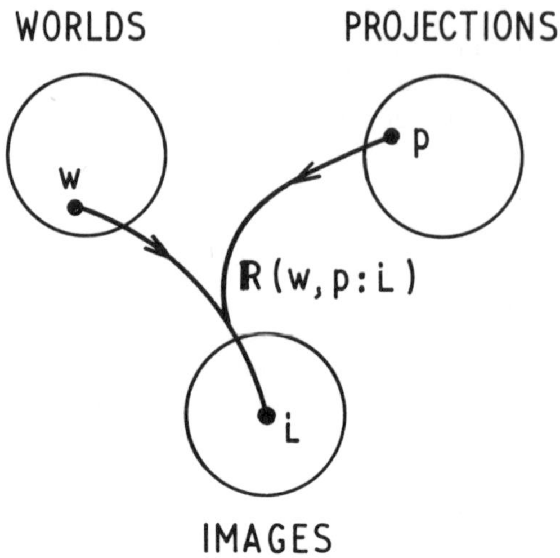

Fig. 1.2 Symbolic interrelationships between the world, projection and image domains (*ibid.*). **R** represents the 'mapping function'.

usually called grey levels. The array values are usually real, non-negative, bounded and implicitly zero outside the field of view bounded by the array dimensions. Digital image arrays can be very large. For example, digitising a fairly standard black and white photographic image can involve a storage requirement of 8 Mbits for an array of 1024 × 1024 elements. In many applications even finer resolution is required, with the attendant increase in storage overheads.

Image sensors as such will not be discussed here but in essence such sensors may be regarded either as 'array' devices or 'scanning' devices. An array sensor provides a discrete array of samples of the scene brightness function, while a scanning sensor provides a set of cross-sections of this function along the path of the scanning pattern. We will be concerned in this book only with the situation where the sample array is digitised subsequent to being sensed. These digitised array elements are often called 'picture elements' or 'pixels'. Obviously, in areas such as optical processing the samples are captured and stored on photographic film, or other image transducers, in an analogue form. The images and the processing tasks used to illustrate the various themes and concepts raised in this book will be drawn from areas as diverse as medical imaging, synthetic aperture radar and stereoscopy.

In the sense that much contemporary computer programming is concerned with the manipulation of relatively simple word-oriented data structures, the situation is starting to arise where it makes more sense to move towards programming with actual images, as opposed to their decompositions. This will require the development of new techniques, environments and computing facilities so that users can easily understand, specify and solve a wide range of image-specific problems. Programming, be it with very simple data or with complex images, is essentially a modelling process involving abstraction, decomposition and simplification. It is already clear that many image processing tasks are going to require comprehensive yet tractable computational models if much in the nature of practical applications are to be realised. In this context it is easy and sensible to regard image processing as a generic activity in the sense that much of its application is going to be dependent on common representations, models and underlying theory. This book has a lot to say about this theoretical support and the consequent design and implementation tools, as well as the underlying technologies which are necessary to support such a unifying concept.

In subsequent chapters various different types of image processing operation will be discussed in the context of a range of different applications. Many of them can be classified as either *point, local* or *geometric* operations. Point operations involve transforming single pixels in a way that does not depend on any neighbouring pixels. Local operations involve transformations on pixels so that the new value of each pixel depends also on the values of pixels in some neighbourhood. Geometric operations involve pixel values at some other point, defined by a geometric transformation, or in a neighbourhood of that point.

The following sections of this chapter briefly introduce some of the central concepts necessary for a proper understanding of subsequent chapters. The reader well versed in the field of image processing should skip Section 1.2 and resume at Section 1.3.

1.2 THEMES AND TECHNIQUES: SOME BASIC CONCEPTS

The history of the processing of images by digital methods is relatively circumscribed, in that it is essentially no older than the digital computer itself.

Computer vision began in the 1950s with statistical pattern recognition, the goal of which was to assign an input image to one of a small number of predetermined classes. Optical character recognition is a representative application. Digital image processing technology for the enhancement, restoration, coding and transmission of images soon started to appear at about the same time, and now constitutes a large and sophisticated field that incorporates many recent computer vision techniques. True computer vision, with the goal of 'understanding' images of complex three-dimensional scenes, was first attempted in the early 1960s. The immense computational complexity inherent in mimicking the human visual process soon became apparent. Intuitively appealing detectors for visual features, such as object boundaries and schemes to control processing proved unreliable and inadequate. Devoting massive amounts of processing at the early stages of vision was technically and economically impossible, so in the 1970s a cognitive approach to computer vision emerged which conveniently minimised image-level computation and emphasised the symbolic manipulations to which computers are well adapted. In such 'knowledge directed' vision, computational effort is directed by processes that use facts about such phenomena as gravity, support, occlusion, or the likely spatial relations between objects in the scene. Research turned toward representing and manipulating facts about particular domains, such as polyhedral blocks or terrain features, and exploiting the domain-specific knowledge in vision. The representation and application of knowledge is itself, however, a very difficult branch of Artificial Intelligence, and the available techniques proved inadequate to bridge the gap between the input image and the desired symbolic descriptions of it. In the 1980s the consensus of the computer vision community is that the gap is best bridged by a varied and redundant set of visual data representations arranged in a hierarchy of increasing abstraction. The production of many intermediate representations requires a huge amount of computation, but the human vision system seems to do it, albeit with neural structures that operate quite differently from today's digital computers.

Certainly, the most explosive growth has been during the past decade, with the rapidly increasing use of image processing in a multiplicity of new applications, some of which are discussed later in the book. This growth has been the concomitant of considerable improvements in the speed, size, versatility and cost-effectiveness of digital computers and their peripherals. This, in turn, has been the consequence of significant advances in the areas of LSI and VLSI technology. This is a topic to which we will return in later sections. Certain other key events and activities have contributed in a seminal manner to this explosive growth of the application of image processing techniques.

Without the popularisation of the fast Fourier transform algorithm in the mid sixties much of the current activity in the transform processing of images would, in all probability, still be in the realm of optics.

The space programmes of a number of countries have provided a major impetus for computer picture processing. Television coverage of the lunar landings involved major programmes in image coding, transmission, correction and restoration. The capture, transmission and analysis of images captured during planetary 'fly-pasts' imposed even more stringent conditions and so further extended the 'state of the art'.

Attempts to automate the extraction of data from complex images arising, *inter alia,* in medicine, nuclear physics, industry, remote sensing and astronomy have also contributed significantly to our knowledge of image processing. Some of these are discussed later.

Inevitably, the endless search for military superiority by the major powers has provided an engine for progress in such areas as coding, restoration, target recognition, ranging and scene understanding, especially in non-visible portions of the electromagnetic spectrum.

The development and commercialisation of computers more architecturally suited to image processing than the traditional von Neumann machine has been a crucial energising process. Specifically, a number of parallel-array machines and very heavily pipelined special purpose conventional machines have been developed which have performance figures in the Giga operations per second range. It is only by using machines with this level of performance that significant real-time image processing becomes a realistic proposition.

The emergence of Artificial Intelligence (AI) as a discipline in its own right has been a significant factor. The majority of researchers in AI tend to view perception as a form of problem solving, at least at some stages of visual processing and perception. Thus, computer vision and its many potential applications have become major topics of research within the AI community, world wide. An early and highly fruitful consequence of this interest was the decomposition of visual processing into passive, domain-independent, data-driven image analysis (or low level vision) and domain-dependent goal-driven scene analysis (or high level vision): a highly pragmatic decomposition which has persisted. It should be pointed out that since the early sixties there has been a marked divergence between the pattern recognition and AI approaches to the machine understanding of images. The former approach has continued to stress the use of *ad hoc* image features in combination with statistical classification techniques. More recently, use has been made of 'syntactic' methods, where images are 'parsed' in terms of a hierarchical set of primitive constructs. In contrast, the AI approach has continued vigorously to pursue and expand the problem solving metaphor.

The rapid growth of computer graphics and digital scene simulation techniques has provided a tangential yet important metaphor for image processing.

Much of current research centres around the production of intermediate representations formed before object recognition is attempted. These image-like representations are registered with the input image and contain values of physical parameters of scene points such as the distance from a sensor to the point, the albedo of surfaces, the direction of motion of objects, the orientation of surfaces, and so forth. It is usual to assume that the processes that produce these images are part of 'early' or low level vision in that they do not require domain-dependent facts. In fact, these processes cannot be regarded as completely general and reliable, since so much information is projected away in the two-dimensional input image. The fact that they so often work correctly in animal vision seems to imply that they rely on natural constraints or assumptions about the world to derive unambiguous output. A goal of modern computer vision research is the identification and use of such constraints. This, in turn, calls for seeking out properties of the physical world that could help a visual process do useful work, making mathematical models of their interaction with visual phenomena, and implementing the mathematics in computer programs. Currently, attention is centred on the designing of processes that can be implemented on parallel computational architectures, since only in this way is the sophisticated performance of human vision remotely attainable.

The remaining sections of this chapter are concerned primarily with briefly expanding on some of the ideas and terminology of the essentially classical areas of image processing and computing, and in doing so provide an introduction for what is to follow in subsequent chapters. Much of what follows in Sections 1.2.1 to 1.2.8 can be found in considerably more detail in standard works, such as those by Rosenfeld and Kak (1982), Hall (1979) and Pratt (1978).

1.2.1 Enhancement and restoration

The formation of an image, or its conversion from one form into another, or its transmission from one place to another, often involves some degradation of image 'quality', with the result that the image then requires subsequent improvement, 'enhancement' or 'restoration'. In practice there are many different sources of image degradation. Some affect only the grey levels of individual pixels, without producing any spatial distortion. These are often referred to as 'point degradations'. Those that do involve spatial distortions are often called 'spatial degradations'. Temporal and spectral effects can introduce further degradation.

Image restoration has commonly been defined as the reconstruction or estimation of an image to correct for image degradation and approximate an ideal degradation-free image as closely as possible. Image enhancement involves operations that improve the appearance of an image to a human viewer, or operations to convert an image to a format better suited to machine processing. The basic distinction between enhancement and restoration is that with the former no attempt is made to establish the underlying physical degradation processes while with the latter, specific known or estimated degradation processes are assumed. The former, more heuristic approach, embraces such techniques as contrast modification, deblurring, smoothing and noise removal, while restoration tends to revolve around the formalism of filter theory.

The quality of an image can depend very much on the use to which it is put. The relative 'fidelity' of an image to its progenitor is often expressed as a simple point-wise measure in an appropriate space. Measures such as mean-square and mean-absolute deviations are often used. If the image mapping process is to be evaluated an average measure over an ensemble of input images is usually considered. Classically, image quality has often been measured in terms of resolution, or resolving power, consequent on the imaging of single-step, bar and point patterns.

The enhancement techniques used in practice are many and varied. Some of the more common are briefly explained below.

Contrast modification involves two basic types of operation: 'grey scale correction' and 'grey scale transformation'. The former attempts to compensate for uneven contrast, usually the consequence of non-uniform exposure, by modifying the grey levels of individual pixels. The latter is concerned with increasing the dynamic range of the image contrast and includes the well known technique of 'histogram modification'.

An ideal imaging system should monotonically map the object brightness distribution into image grey scale levels in such a way that the mapping is uniform, i.e. independent of pixel location. In practice the exposure mapping is often non-uniform and requires some form of grey scale correction. If it proves possible to establish the non-uniform mapping that forms the image then correction is straightforward. The non-uniform mapping can be represented by the expression $g(x,y) = e(x,y)f(x,y)$, where f is the ideal grey scale intensity function resulting from a uniform mapping, g the actual grey scale image and e a correction function. The correction function can be determined by calibrating the imaging system with a known brightness distribution, usually constant. If g' is the image of such a constant distribution, such that $f' = c$, then $e = g'/c$ and any image captured by the system can be corrected.

Grey scale transformation is concerned with commonly used grey scale operations such as 'contrast stretching' and histogram modification, and involves transformations that are independent of pixel location. Such transformations can be expressed in the form $s' = t(s)$ where s and s' are respectively the original and the transformed grey scale values. In practice, of course, s and s' are quantised variables. Depending on the form of the function t, the grey scale of an image can be expanded or compressed. If required, part of the grey scale range can be stretched, at the cost of compressing another part, in order to emphasise detail. The inverse transformation, $s = t^{-1}(s')$, can in principle be used to correct for distortions and non-linearities in the grey scales of imaging devices.

Histogram modification is a technique which transforms an image's grey scale in such a way as to yield a given frequency distribution $f(s)$ of grey levels. There are a number of reasons for wanting to do this. These include the need to minimise quantisation error or to normalise images before comparison or classification. The former involves a preferential allocation of small quantisation steps where $f(s)$ is large, while the latter is concerned with removing sensitivity to ambient lighting conditions.

One of the most common image degradation processes is that of blurring. As a consequence it is often expedient to seek a means to enhance the blurred image by 'sharpening' up the delimiting contours between its various regions. Blurring is essentially an averaging, or integration, process in that high spatial frequencies are preferentially attenuated. The logical inference from this is that blurring can be corrected by the use of differential operations. If an image is noisy as well as being blurred, sharpening techniques have to be handled with care since noise often contains a considerable high spatial frequency component. Sharpening techniques should be restricted to images where the signal-to-noise ratio is large or where some attempt to remove the noise has already been made.

In that we wish to sharpen blurred features with arbitrary spatial orientations, an isotropic, i.e. rotation invariant, differential operator is mandated. This constraint can be used to derive some of the properties that such operators must obey. One of the most commonly used isotropic differential operators is some form of Laplacian. In the current context the Laplacian of an image f can be represented as $\nabla^2 f = \partial^2 f/\partial x^2 + \partial^2 f/\partial y^2$. In order to appreciate the significance of this operator, assume that the blurring is the consequence of a process which can be represented by the underlying diffusion process $\partial g/\partial t = k \nabla g^2$, where g is a function of x, y, and t, and $k > 0$. At $t = 0$, $g(x,y,0)$ is the unblurred image $f(x,y)$. At a time t (> 0) we can represent the blurred image $g(x,y,t)$ as the Taylor series $g(x,y,t) = g(x,y,0) + t\partial g(x,y,t)/\partial t + (t^2/2)\, \partial^2 g(x,y,t)/\partial t^2 + \ldots\ldots$ If we retain only the linear terms, replacing $g(x,y,0)$ with f and $\partial g/\partial t$ with $k\nabla^2 g$, then we obtain $f = g - kt\nabla^2 g$. Hence we have a mechanism, correct to the first order, to restore the unblurred image f. Obviously a diffusion process is not going to underlie all blurring phenomena but at least we have demonstrated the credibility of an approach involving a Laplacian operator, obviously in practice expressed in a discrete form.

Another way of viewing the effect of differential operators on images is to regard them as spatial frequency filtering operations, emphasising the higher spatial frequencies as opposed to the lower ones. Typically, a 'high-pass filter' has a transfer function which increases with increasing spatial frequency up to a frequency where noise starts to dominate and then falls fairly abruptly to zero. This spatial filtering can be performed either in the frequency domain or as a convolution operation.

Smoothing is essentially concerned with noise removal. A fairly basic problem is that most linearly based methods for smoothing noisy images usually result in some degrees of image degradation by blurring. This is particularly significant with images which have distinct sharp edge features. Consequently, the underlying aim in the search for effective smoothing techniques is the minimisation of this concomitant blurring. This implies the possibility of applying global optimisation techniques to some appropriately defined 'cost function'. In practice local operations are almost always used, owing to the very large number of degrees of freedom which exist in most practical problems.

An image may be affected by noise from a wide variety of different sources, including sensor noise, quantisation noise, 'grain' noise and channel noise. The noise may be systematic or random, the former being the easier to deal with in practice – for example with Fourier techniques. Noise in an image usually posesses significant high spatial frequency components in its spectrum relative to the image under consideration. As a consequence simple low-pass spatial filtering can be quite effective, even though it does not in any real sense distinguish between image and noise. In general, noise removal requires some knowledge of the underlying noise source mechanisms if it is to be at all effective.

If the underlying noise mechanism can be identified then it is, in principle, possible to remove most of the image noise and then rebuild the image by an interpolation process. If it proves impossible to identify, or make use of, the underlying noise mechanism then somewhat more gross techniques, such as averaging, must be resorted to. The price which must be paid for this generalising is that some degree of image blurring inevitably occurs. Consequently, a compromise between the amount of noise removed and the concomitant blurring must usually be sought.

It is easy to see why averaging techniques do remove noise from noisy images. If $s(i)$ represents the noise-free pixel grey scale signal and $n(i)$ represents the corresponding noise signal then in the average $\{[s(1) + \ldots + s(i) + \ldots + s(N)]/N + [n(1) + \ldots + n(i) + \ldots n(N)]/N\}$ the second term is effectively an instantiation of a random variable with mean zero and standard deviation $s/\sqrt{N}$, where s is the rms noise for a single pixel. Consequently, if we average over a number of independent copies of a given image, with independent noise sources, noise reduction without blurring can be achieved.

If only a single image is available then noise can be reduced by local averaging, but only if the image noise is more fine grained than the image detail. The spatial size of the local averaging process depends on the amount of noise reduction that is required. A crude indication is given by the result above, which indicates that to reduce the noise by a factor of N then an average over N^2 pixels must be performed.

An extension of the above ideas is the use of weighted averages, with the weights chosen in such a way as to minimise the averaging over image details such as lines and edges. For this approach to be of any use there must be a means of locating such features within the image, either explicitly or implicitly.

One technique which does remove noise without blurring is *median filtering*. This involves replacing a given pixel with a value equal to the median value of the grey scale values of a suitably chosen surrounding neighbourhood of that pixel. This technique is very effective at reducing noise spikes from an image, but is less effective in dealing with coarser grained noise. This absence of blurring means that median filtering can be iterated without degrading the image. For two-dimensional median filtering, as required for image processing, the neighbourhood over which the operation is performed must be carefully chosen so as to ensure that required features are not inadvertently atenuated or removed. For example, it is easy to remove thin lines and isolated points as well as clipping corners.

Restoration techniques usually involve some form of *a priori* insight into the degradation mechanism(s), which is then used to place constraints on the use of the appropriate restoration algorithm. In the majority of cases the underlying physical cause of the degradation is too complex to be amenable to analysis or is just not identifiable. Here some form of *a posteriori* knowledge, derived from the degraded image, is often used. Many restoration techniques revolve around the use of a 'point-spread function'.

The point-spread function of an imaging system is essentially the two-dimensional impulse response of the system. In the simplest case, i.e. a linear space-invariant system, the degraded response $g(x, y)$ of the system corresponding to the ideal image $f(x,y)$ is given by the two-dimensional convolution function $f(x,y)*h(x,y)$ where $h(x,y)$ is the point-spread function. The assumption of space invariance implies that the image of a point source in the focal plane will change only in location, not in functional form, as the point source moves in the object plane. This obviously has its dual, and highly useful, representation in Fourier space. From a mathematical viewpoint the aim of image restoration is to obtain as good an estimate of $f(x,y)$ as is possible in the light of the knowledge available regarding the underlying distortion mechanisms. In some cases it is possible to determine $h(x,y)$ exactly from an *a priori* knowledge of the imaging process while in others it may be necessary to try to deduce it *a posteriori* from the distorted image alone.

In addition to some insight into the point-spread function one needs to know something about the statistics of the image noise. The most common and straightforward assumption is a 'white' noise source, completely uncorrelated with the image itself. This is a convenient and tractable notion, though in practice the latter assumption is unfortunately too often false.

Inverse filtering and 'least-squares' filtering are two representative and commonly used restoration techniques. Using the notation above, the Fourier space representation of the degraded image $G(u,v)$ is given by $G(u,v) = H(u,v)\ F(u,v)$, where $F(u,v)$ is the undistorted image and $H(u,v)$ the point-spread function. In principle $F(u,v)$, and hence $f(x,y)$, can be established by evaluating $G(u,v)/H(u,v)$. The transfer function of this 'inverse' filter is $1/H(u,v)$. In practice the actual evaluation of $F(u,v)$ can be difficult if $H(u,v)$ has zeros, as is often the case. This usually results in the effective filter placing very little weight on the regions of the spectrum near the zeros of $H(u,v)$ yielding, somewhat arbitrarily, a correspondingly low value of signal-to-noise ratio.

An approach which has a lot of intuitive appeal is to seek to minimise the differences between the original image and a best, in some as yet undefined sense, restoration of it. This usually involves the minimising of an appropriately defined cost-function. On account of its mathematical simplicity, a widely used cost-function is the mean squared error. This gives rise to the so called non-recursive 'Wiener' and recursive 'Kalman' filters. In the use of Wiener and Kalman 'least-squares' filters it is implicitly assumed that both the original image and the noise belong to homogeneous random fields and that their power spectra are known. This is often not the case. If, apart from the underlying noise mechanism, only the noise variance is known then a restoration method known as 'constrained deconvolution' can be usefully applied. While a Wiener or Kalman filter gives the best restoration in an average sense for a class of pictures, constrained deconvolution gives an optimum restoration for a particular degraded image. An extension of the least-squares approach is one where the image is represented as a Markov process corrupted by white noise. The resulting filter is essentially recursive in nature. Restoration is discussed again, in considerably more detail, in Chapter 2.

1.2.2 Segmentation

Many techniques, some of which have been discussed previously, such as restoration and enhancement, involve an explicit transformation of the input image into an output image. Another group of techniques, including segmentation, are concerned with the situation where the output is a description of the content, or some part of it, of the input image. Since most physical objects do not produce images which can be characterised by a single or even a few measurable characteristics, the problem is basically a difficult one. *Segmentation* is a generic term for those techniques which involve taking an image and extracting information relevant to specific picture 'segments', such as lines, regions and objects, and their interrelationships. It is basically a process of data compression by pixel classification, in that an image is segmented into subsets by assigning individual pixels to particular classes. Obviously this is an area of considerable practical concern. Some representative techniques are mentioned below.

Clustering and pattern classification involve the classification of a set of pixels according to the value(s) of some image property which can be measured or evaluated for each relevant set of pixels. If the probability density of the property values for each class is known, along with the class occurrence probabilities, then a minimum error classification criterion can be established. This is a standard problem in pattern recognition (Rosenfeld and Kak, 1982; Hall, 1979).

Grey level thresholding is a segmentation technique which classifies pixels into classes on the basis of their grey levels. In its simplest form it involves classification into two classes, 'black' and 'white', on the basis of a chosen threshold. The most useful threshold, and one that is amenable to automation and dynamic feedback, is one which separates the peaks on a bimodal grey level histogram. Other approaches involve iterative improvement, by varying the threshold on the basis of some acceptance criterion such as a minimum error threshold, and modelling the grey level histogram with gaussians. If the grey level histogram is definitely multi-modal it is often convenient to classify pixels into a number of classes, effectively yielding a highly 'reduced' grey scale.

The above methods are essentially global in context. There are also a number of local operations which can be used to detect certain local image features which can then be used as segmentation data. These techniques are usually concerned with detecting abrupt changes in grey level, such as occur at edges and similar image discontinuities. We have already seen (Section 1.2.1) how differential operators respond to rapid changes in grey level. They are obvious candidates as edge detectors. The more commonly used local edge detection techniques include the Laplacian, Roberts, Sobel and Prewitt operators (Rosenfeld and Kak, 1982; Hall, 1979).

A further level of sophistication often required of an image processing system is the detection of features more subtle than straight edges: for example lines, curves and simple shapes. There exists a plethora of diverse techniques for extracting various image features (Rosenfeld and Kak, 1982; Hall, 1979), some of which are discussed in subsequent chapters.

1.2.3 Representation

Once an image has been segmented in some way there is a need to establish a list of properties of the appropriate image subsets extracted by the segmentation algorithms. This list of properties should specify and interrelate the segmented regions in the context of a representation, implicitly or explicitly, appropriate to the task in hand. Representations used in image processing typically involve relational structures such as directed graphs where the graph nodes represent 'parts' and the graph edges represent 'relations'. Each node is labelled with a list of associated property 'values'. Various properties are used in practice. In general one attempts to use linear, normalised, transform-invariant properties. These properties provide a description of the image in what is essentially a particular *ad hoc* representation. Properties used in practice can be local, such as shape, or global, such as projections and transforms. If image texture is of significance then grey scale statistical properties are often used. Instances of some of these properties are discussed in later chapters.

The representations used in practice must be judged in terms of their practical 'adequacy'. This must not be merely a mathematical adequacy. It must include enough elements of epistemological and heuristic (procedural and declarative) adequacy to enable all the requisite information to be available to the task in hand in such a way as to facilitate computation in a reasonable time.

A model, in the sense of a description of image classes, appropriate to the given image and image processing task, is constructed which is then used for establishing correspondences between the model and the image description so that the image can be interpreted in some appropriate way. Obviously the notion of representation, and how to choose it, is basic to this process.

Two general types of image model have been invoked, *declarative* and *procedural.* A declarative model consists of a set of constraints on the properties of, or relationships among, the parts of an image at the given level of description. The images whose descriptions satisfy these constraints constitute the classes defined by the model. A procedural model, on the other hand, is any process that recognises images or image features. The class that it defines consists of the pictures that it accepts. This is obviously quite arbitrary. In practice, the most commonly proposed class of procedural models is that of 'syntactical' models, involving string, array, cyclic, tree and graph grammars. Picture models are often 'fuzzy' or underconstrained. Declarative models often involve probabilistic constraints, while in procedural models rule application may well be controlled probabilistically.

Various levels of description and modelling can be identified. Representations of properties at the 'array' level, the 'region' or 'geometric' level and the 'relational' level constitute three potential but increasingly abstract image models. The models used in practice usually exhibit a number of intermediate stages bridging the gap between computational theory and physical implementation, each with its own explicit representation. Such intermediate representations must have certain important characteristics (Mackworth, 1983b). They must be capable of finitely representing all and only the configurations in the intermediate domains that could have produced the image. They must be capable of being refined as additional information is available in the form of additional imagery of the assumed common underlying scene. They must be efficiently computable from the information available in the image and the other intermediate representations. They must be capable of representing the trade-offs between intermediate domains that allow one factor to be varied while simultaneously varying another to maintain a constant image. Finally, they must be capable of representing and enabling access to model constraints and their logical consequences.

The importance of models and their implicit representations are discussed further in subsequent sections of this chapter and again in Chapter 4.

1.2.4 Matching and recognition

Image matching, particularly involving the use of correlation, has a long history. It has been used extensively for tasks such as the detection of change, character recognition, target recognition, aerial navigation and stereo mapping.

The requirement to match or register two images or 'frames' in the context of real-time imagery can arise in a number of different situations. Two images or sub-images may need to be brought into registration, temporal changes between successive images may be required directly or to derive the 'optical flow' of a scene, two spatially displaced images of the same scene may be used to derive depth information (stereopsis), or we may be seeking a particular pattern within the image.

If multiple images, involving spatial or temporal relative displacements, are used to derive depth information then detailed knowledge of the imaging geometry and the implicit coordinate transformations becomes important.

Registration is a process fundamental to many matching tasks. Registration, in practice, is complicated by various geometric imaging distortions, such as 'perspective', 'barrel' and 'pin cushion' distortions. In general, two images of the same scene, taken from different viewpoints, cannot be registered, or the distortion in a single image corrected, by applying a simple transformation to the entire image(s), since objects at different distances transform differently under the imaging process. Rather, it is necessary to proceed on a piecewise basis. Matching a pattern, or 'template', to its occurrences in an image is an important generic image processing task. There are many potential methods available for effecting this matching. The most common means of selecting the best of these, in the sense of providing the most efficient measure of matching, is to apply some form of 'distance measure' metric to the matching process. On this basis, one of the best linear matching operators is the 'normalised cross-correlation coefficient', operating with the template directly on the image binary or normalised grey scale values. It may in some instances be more efficient to perform this operation in the corresponding Fourier transform space. In this context it is possible to make use of a matched filter to optimise performance further. A linear operation such as cross-correlation enables one to find exact counterparts of the template within the image but does not readily cope with rotated, scaled or perspective transformed versions of the template, unless a large number of appropriate templates are used. Another matching technique, capable of coping with this situation, involves a search through a space of possible distortions and transformations of the original template. This, obviously non-linear, approach to matching is often referred to as 'relaxation labelling' and has been the subject of much investigation, especially in terms of the underlying control strategies. One persistent problem is the relatively large computational overheads inherent in this approach. Relaxation labelling is discussed again in Section 2.4.2.

1.2.5 Understanding

Image understanding stands preeminent as the still essentially unachieved long term goal of those researchers who work in the general area of computer-based image processing. It occupies a major niche within the subject of Artificial Intelligence (AI). It is often referred to by names such as image analysis and computer vision. As far as we are concerned these all mean essentially the same thing.

There is an enormous discrepancy between the tasks which can be performed by existing image understanding systems and the effortless variety, richness and sophistication of the human visual process. This is not merely a quantitative difference which might be overcome with increasing technological sophistication, such as vastly more processing power or seminally new processing architectures. Rather, it is rooted in some very fundamental shortcomings in our theoretical understanding of the computational paradigms underlying the visual process. It is in the area of image understanding that this fundamental lack of knowledge becomes most glaring.

Chapter 5 will discuss in some detail the particular problems that confront those who would attempt to design systems to understand automatically the content of single static, stereoscopically derived or time-varying two-dimensional images.

Historically, there have been many diverse approaches to the attempts to achieve acceptable image understanding systems. We do not have the space to go into them in detail here. Figure 1.3 illustrates, in summary form, some of the central concepts and techniques which have emerged to date (Marr, 1982).

It is in the early or low level parts of the classical image understanding problem of achieving the perception of three-dimensional shape that most progress has been made. It has been possible to identify the underlying assumptions that make a specific problem solvable and then to implement those assumptions as algorithms. Currently available implementations include those which perform stereopsis, structure from motion, shape from shading and photometric stereo.

In an attempt to establish tractable computational models, some vision researchers have been attempting to use the techniques of theoretical physics in order to establish a means to put all, or almost all, of the early

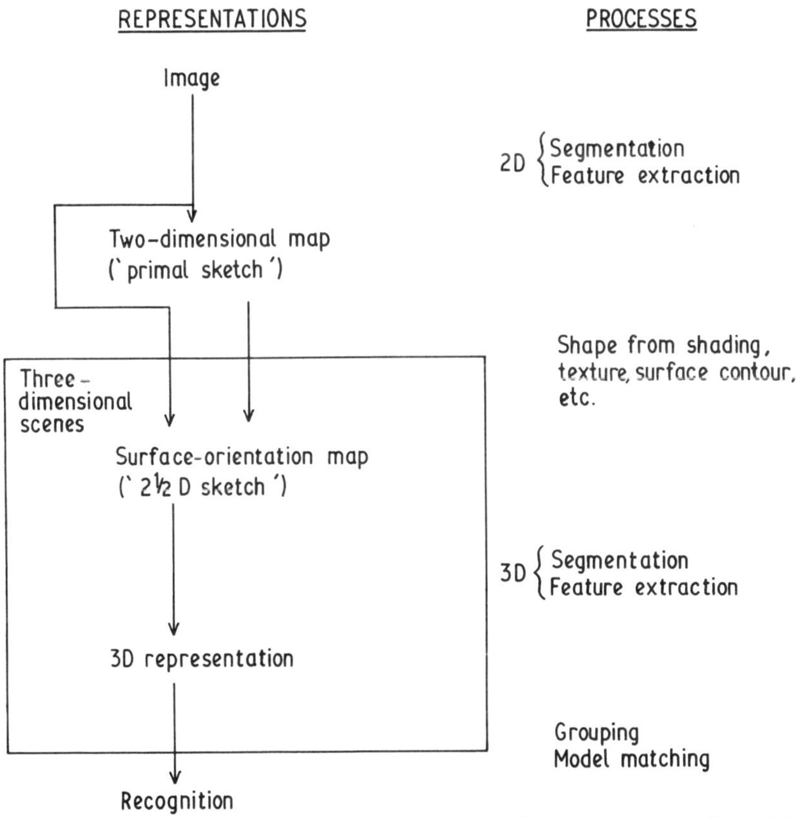

Fig. 1.3 Simplified diagram of an image understanding system. Based on Marr (1982).

vision problems into a common framework. The hope is that it will be possible to formulate the constraints inherent in an image processing problem as a special type of problem analogous to those in physics solved by using variational principles.

Beyond these early vision successes matters get rapidly worse. Even at the fairly low level of a two-dimensional image it is far from obvious what kind of surface information the eye is extracting from a transparent window pane or from the reflections in a rippled pool of water, let alone 'surfaceless' entities like fog or a wisp of smoke.

As one proceeds towards higher levels of visual processing, such as the 2½-D sketch, matters get rapidly worse. At this point the algorithms must start to include real-world knowledge, which means that all the problems of knowledge representation and inference mechanisms suddenly appear. The problem is further compounded by the fact that the image understanding task is highly underconstrained or 'ill-posed' and that the human brain has multiple representations for just about everything. In processes such as stereopsis and structure from motion only the depth and orientation values at discrete points are computed, typically where the image intensity changes abruptly. In the Marr–Hildreth approach, discussed in later chapters, these points correspond to the zero crossings of a so-called 'difference of Gaussians' or 'DOG' operator when it is applied to an image. Human perception involves complete, piecewise-smooth surfaces as an important aid to cognition. In the situation of only having discrete boundary data, as is the case in stereopsis, the number of potential surfaces which can be fitted to these boundaries is infinite and must in practice be constrained by further information relevant to the image or the imaging process. Care must be taken here to avoid any 'over commitment' which might jeopardise later stages of processing. The tension between top-down driven hallucination and bottom-up driven oversight is very strong here.

The twin concepts of representation and model are of fundamental importance here. Marr (1982) has strongly emphasised the importance of establishing an underlying abstract computational model, where the problem and its intrinsic constraints are characterised, before attempting to formulate image understanding algorithms. It is the topmost of three distinguishable levels at which an image understanding system can be specified. The bottom level is that of the physical computational system. The middle level, concerned with the algorithmic aspects of the problem, is the hardest to define precisely in that it is the bridge between an abstract model and its physical realisation. This question is taken up again in later sections.

1.2.6 Computer graphics and image synthesis

Digital image synthesis and computer graphics are closely related and rapidly growing fields, very closely allied to many of the central themes of image processing. In a very real sense they are activities reciprocal to image processing in its most general sense. Computer graphics is concerned primarily with sophisticated graphical representations of computational data structures, while digital image synthesis is concerned more with the production of computer-animated scenes. In practice the two disciplines are inextricably interdependent. Advances in one area usually feed through to the other. Computer aided simulation is one area where both concepts are very much relevant at the one time. In both it is the lack of computational power which is often a fundamental inhibitor of innovation, not the imagination of the potential user. A major impediment to extensive use of graphics has been, until recently, its cost, resulting from the extensive processing, storage and communications requirements and the need for expensive displays and specialised hardware. In both cases VLSI is already making significant contributions to the sophistication and performance of the new hardware coming on to the market. Indeed, it is only the rapid technology-driven improvements in the price-to-performance ratio of certain key components in recent years which have made possible the current generation of sophisticated graphics systems. In particular, the now widespread use of raster graphics systems is very much the consequence of advances in VLSI memory chip design and fabrication.

A factor in the rapid assimilation of graphics techniques in computing has been, and still is, the search for 'user friendly' operator–machine interfaces. This is a reflection of the fact that computer science is moving from a situation where people learned to adapt to machines to one where machines are increasingly expected to adapt to people. The almost explosive growth in the graphics-facilitated use of 'icons', 'windows', 'drawing' or 'painting' facilities and a large choice of fonts on personal machines has been a key factor in satisfying the development of ergonomically motivated 'what you see is what you get' user interfaces. This in its turn has put increasing demands on graphics technology. The Apple Lisa and Macintosh, Xerox Star and the various LISP machines are good examples of this trend.

A further, and powerful, market driven force accelerating this search for increased performance is the demand for advanced graphics, synthesis and animation facilities by vendors of computer games. Here the distinctions between graphics and synthesis are truly blurred. Related areas are the production of interactive training simulators and the film, TV and advertising industries.

Computer graphics can be classified into two fundamental processes. In one, an object is geometrically transformed or manipulated in some form of coordinate space, and that arbitrary coordinate space is then mapped back onto the screen. The other creates a rendering, or visual representation of the object that has been described geometrically in the previous step. Visual cues are added to indicate the object's true physical structure. In this context 'fractals' have proved a valuable means for generating surface texture (Pentland, 1984).

Most of the more traditional applications of computer graphics have tended to be two-dimensional. In recent years there has been increasing interest in three-dimensional applications, especially those involving user interaction. This has been the consequence of advances in modelling methods and computer hardware. Solid modelling packages are now being used as important and integral components of computer aided design and simulation facilities within the computer integrated manufacturing process. The automatic or semi-automatic use of computer vision and other sensing modalities within the manufacturing process depends critically on the use of sophisticated three-dimensional geometric parts modelling. This is to ensure that design and manufacturing performance can be 'seen' as well as analysed by more traditional methods. This is further discussed in Section 7.3.2.

In concept, three-dimensional graphics is relatively simple to implement. First define the three-dimensional geometry, manipulate it into a world coordinate system, and then draw it rapidly on the screen. The fundamental operations involve rotation, translation, scaling, and perspective. In rotation, an object defined in its coordinate space is turned on at least one axis. In translation, the object is moved to a different position in the coordinate space in its same orientation. In scaling, it is simply made larger or smaller. In perspective, the effect of objects receding into the distance is simulated in a way similar to that in which a human eye sees an object.

To display a realistic rendering of a fairly simple mechanical part may require up to a half megabyte of data. Manipulations on the graphics database realistically can pose a formidable computational load. To display the part on a graphics display may require up to 1 million floating-point operations and 5 million integer operations, repeated 30 times a second to move it dynamically. Although this capability is beyond the reach of reasonably priced current technology, the ultimate goal of computer graphics is to generate realistic moving pictures.

High performance graphics systems present a real challenge because they are, in a sense, a moving target. Special skills are required to create algorithms, VLSI chips and analogue circuitry to produce state of the art graphics systems. Although graphics algorithms are well described in the literature, the issue is not the design of the algorithm but its implementation – making the algorithm fast and efficient within reasonable memory size and systems cost. The high performance graphics designer must have a feel for which of the algorithms should be contained within the VLSI chip and which should be left in software or firmware, since they could be changed or superseded by improved algorithms.

Some actual graphics 'engines' are briefly described in Section 1.4.3.

1.2.7 Communication

Minimising the communication bandwidths and/or storage requirements of digital imaging systems is an important design consideration. For example, a standard television signal requires a bandwidth of the order of 10 Mbytes/s, and high resolution images can multiply this rate by as much as an order of magnitude. The easiest, and often the best, means of reducing the bandwidth and storage requirements is simply to reduce the spatial, temporal and grey scale resolution. In general such a compression coding alters the image quality. Consequently, to be meaningful any statements on bandwidth or storage overhead reduction must be predicated with a statement or measurement of image quality. The aim of many compression schemes is to reduce channel bandwidth without reducing the subjective image quality, as perceived under normal viewing conditions.

Assume a certain number of bits is needed to store, transmit or display an image in a distortion-free form. If the use to which the image is put implies spatial and grey scale resolutions inferior to those of the original image, then it is possible to compromise on such issues as sampling and quantisation and so represent the image with a reduced number of bits. This of course implies that a certain amount of distortion, between the original image and any subsequently reproduced versions, is tolerable. This is often an acceptable procedure, in that many images tend intrinsically to contain more information than is actually needed in practice. More generally one is led to ask how, in a given context, one can represent an image by the smallest possible number of bits? For a given implied distortion it would be nice to know if there are optimum coding and transmission strategies, in the sense of minimising the number of bits used to encode and transmit the image.

Coding is concerned with efficient distortion-free representation of image elements, be they pixels or features, using the minimum number of bits. The Huffman codes are a case in point, where the fewest bits are allocated to specifying the most commonly occurring features. This elegantly simple procedure (Huffman, 1952) ensures that one can obtain a uniquely decodable code with the minimum average number of bits per message. Another simple, reversible coding technique is 'run length' coding, which takes advantage of pixel to pixel correlation. In a digitally encoded image a 'run' is defined as a sequence of consecutive pixels of identical values along a specified direction. For simplicity a horizontal scan line is often used. The bit rate reduction is achieved by simply encoding the start and length of run, rather than the individual pixels. Obviously, this approach works best with images that have a relatively small high spatial frequency component. These and other similar techniques can be readily applied to single-image frames as well as to image sequences, such as video signals. For the latter, various inter-frame techniques, such as 'conditional replenishment', which encode statistically significant changes between successive frames, are an obvious extension of the simpler spatial, or inter-frame, techniques mentioned above (Hall, 1979). Inter-frame coding of typical video signals can result in a compression to less than one bit per pixel, which is considerably better than spatial (intra-frame) coding techniques such as DPCM (typically 2 to 3 bits per pixel) and transform coding (1 to 2 bits per pixel). Compression is concerned with encoding, for a given 'distortion', an image using the minimum number of bits. The ultimate goal in picture compression is to represent the encoded image with uncorrelated data. This is because if this is achieved then every data element representing the image is unique in the sense that it cannot be predicted from the rest of the data, at least by linear 'mean-square' estimators. In practice, even for uncorrelated data, the number of bits may still be unacceptably large. If this is the case, the image elements are ranked in order of their importance. It is then possible to delete terms which have only second order effects on image quality, e.g. terms representing high spatial frequencies, in the context of a particular fidelity requirement. It is possible to find an appropriate uncorrelated set only in very constrained circumstances. For images of any real interest, the problem of such a representation becomes a lot more intractable.

In practice, image compression is commonly effected by using reversible linear transformations, such as the Fourier, Hadamard, cosine and Karhunen–Loeve transforms. Here an image is expanded in terms of a set of orthonormal functions, with the appropriate, and uncorrelated, expansion coefficients playing the role of the requisite image samples (Hall, 1979). This is discussed further in Chapter 2.

If one is not restricted to the use of linear transform techniques then an easily implemented approach is that of predictive, in the mean-square estimator sense, compression. This has its origins in Wiener's theory of optimum linear prediction (Bozic, 1979). These are commonly known as differential pulse code modulation (DPCM) techniques (Hall, 1979).

Both transform and predictive compression techniques have their drawbacks as well as their attractions. Attempts have been made to bring together the best aspects of both, such as using generalised orthogonal transform kernels to yield generalised DPCM systems.

1.2.8 Reconstruction

The advent of digital computers has facilitated the reconstruction of two-dimensional images from a set of their one-dimensional projections in a variety of diverse fields. The area of medical imaging has probably developed these techniques to the greatest extent, where they underpin computerised X-ray tomography (CAT), computerised emission tomography (CET), ultrasound computerised tomography (UCT) and computerised nuclear magnetic resonance tomography (CNMRT), often referred to as magnetic resonance imaging (MRI). Other applications include radio-astronomy, synthetic aperture radar (SAR) (see Chapter 6), optical interferometry and electron microscopy.

The basic problem which underlies many of these diverse applications is deceptively simple. Given a subset of all possible projections of some property of an object, which varies with position within the object and characterises it, estimate the spatial distribution of this property by means of an appropriate mathematical algorithm. The most commonly used approach achieves reconstruction by back projecting the transmission projection data convolved with a spatial frequency filter. An equivalent process is to take the two-dimensional Fourier transform of a function which is built up from the one-dimensional Fourier transforms of the projections. Referring to Fig. 1.4(a), if $f(x,y)$ represents a two-dimensional image function then a line running through $f(x,y)$ is called a 'ray'. The path integral of $f(x,y)$ along a ray is called a 'ray integral' and a bundle of such rays is called a 'projection'. Mathematically this can be represented as follows. The line AA′ has equation $x \cos \theta + y \sin \theta = d$, where d is the perpendicular distance of this line from the origin 0. The ray integral then becomes

$$P(\theta,d) = \int_{AA'} f(x,y)\, ds$$

where ds is taken along the ray. The function P then defines the parallel projection of $f(x,y)$ for the angle θ. An example of three such projections is shown in Fig. 1.4(b). The basic technique then involves the reconstruction of the 'image' $f(x,y)$ from its projections taken at a number of different angles. If the projections are generated by integrating along a fan of rays emanating from a point source then a set of fan beam projections result, as shown in Fig. 1.4(c). In practice the use of fan beams results in the very much more rapid generation of the ray integrals, compared with the parallel beam method. A point source is used, along with an array of detectors covering the whole spread of the fan. Successive rotations then quickly build up a set of fan projections.

A result which is of fundamental significance for a number of reconstruction techniques is the so called 'Fourier slice', or 'projection slice', theorem. It provides a relation between the one-dimensional Fourier transform of a projection of the image function $f(x,y)$ and the two-dimensional Fourier transform of the same function. This theorem dictates that by taking the projection of an image at a number of different angles and Fourier transforming each of these projections, one can then determine the values of the two-dimensional transform $F(u,v)$ along a set of radial lines. The image function $f(x,y)$ can then be recovered by performing the inverse transform. In practice only a finite number of projections can be taken with the result that $F(u,v)$ is known only along a finite number of radial lines. Consequently, one has to resort to interpolation, with attendant image degradation, especially in the higher spatial frequency components.

Computerised X-ray tomography is perhaps the best known practical application of image construction techniques. Referring to Fig. 1.4(a), a parallel beam of X-rays is shown propagating in the direction AA′. Photons are continually scattered or absorbed out of this beam. If N is the number of monoenergetic photons emitted from the source into the beam at A during the exposure time, then the number N' exiting at A' is given by the ray integral $N' = N\exp(-\int_{AA'}\mu(x,y)ds)$, where $\mu(x,y)$ is the linear attenuation coefficient and ds is an element of ray length. This yields $\int_{AA'}\mu(x,y)ds = \ln(N/N')$, where the left-hand side is simply a ray integral for a projection of the function $\mu(x,y)$. In practice the basic assumption of monoenergetic photons is always violated, but is necessary because of the dependence of $\mu(x,y)$ on the photon energy. The consequence of this

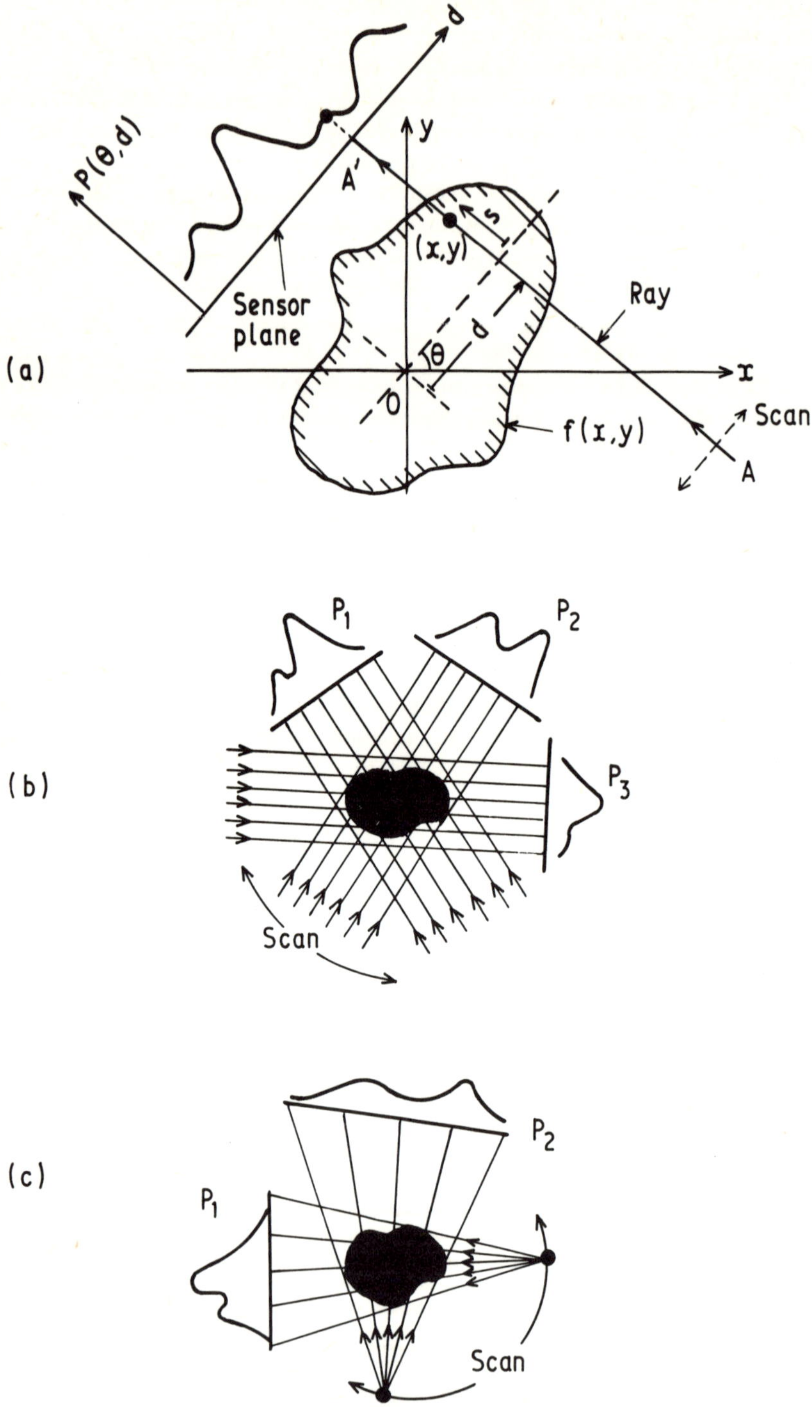

Fig. 1.4 (a) Schematic diagram of a typical parallel beam projection geometry; (b) three projections so produced; (c) typical projections obtained if a fan beam is used.

assumption in generating reconstructed images is the appearance of what are called 'polychromaticity artifacts'. The algorithm which is most commonly used for CAT reconstruction employs the so called 'filtered back-projection' method. It has the advantages of being very accurate and comparatively fast. It is used with both parallel and fan projections.

Computerised emission tomography is concerned with generating cross-sectional images of the distribution of radio-isotopes in the human body. Here one is concerned with the collimation and detection of gamma rays resulting from a radioactive decay process. The decay, mechanical and biochemical processes which affect the distribution of radio-tracer as a function of time mean that the projection data for reconstructing a cross-sectional image must be captured in a time that is short compared with the relaxation times of these background processes. If this can be achieved it does have the considerable advantage that it makes possible the generation of image sequences depicting dynamic processes in the body. With ultrasound computer tomography one can perform either time-of-flight or attenuation tomography. For the former the reconstructed image is that of the refractive index distribution of a slice, while the latter gives reconstructions more akin to those discussed above. Neither of these techniques has established the widespread useage achieved by CAT techniques.

Nuclear magnetic resonance (NMR) imaging has gained attention mainly because it has some important diagnostic advantages over the popular computerised X-ray tomography and in addition does not involve the use of ionising radiation. These advantages include a wide choice of imaging techniques exploiting different properties of tissue, the intrinsic three-dimensional nature of data collection and the non-hazardous nature of the magnetic fields used in MRI. Whilst CAT relies on rotating gantries and absorption of photons, the MRI procedure is entirely electronic and based on properties of the magnetic resonance spectra of hydrogen nuclei in soft tissues of the body. The property which CAT images is the spatial distribution of attenuation whilst MRI ideally images the spatial distribution of three physical properties of the body, namely the density of hydrogen nuclei and two relaxation times. Other physical properties of the tissue, such as flow and diffusion, also affect the output significantly. A combination of the three properties and the variety of excitation methods allow a diversity of image formation, data collection and image reconstruction unparalleled by other techniques. This is the basic strength of the MRI method. In CAT, the inputs are ray pulses aimed at different parts of the object and the output a spatially coded signal. In MRI a complicated four-component signal, all of which are functions of time, forms the input and a complex signal emerges that is temporally coded. In MRI, unlike CAT, the composite input can be tailored to control the interaction with the object to produce diverse images.

NMR is applicable to materials with nuclei containing odd numbers of protons or neutrons or an odd combination of the two, which exhibit spin. Tissues contain a measure of such nuclei in the form of hydrogen, calcium, phosphorus, etc., so that NMR may be applicable to most solid phase and liquid phase materials. In the absence of an external field, the magnetic moments possessed by spinning nuclei are randomly oriented and the object does not have a net magnetisation. However, when the object is placed in an external magnetic field, the nuclei experience a magnetic torque which aligns nuclei in the direction of the magnetic field. The fraction of the nuclei which are magnetised is dependent on the strength of the applied magnetic field (B) and the temperature of the object (T), i.e.

$$M \propto \frac{N\ B}{T}$$

where N = number of spins per elemental volume observed
M = net magnetisation

The spinning nucleus responds to the external field with motion like a gyroscope precessing about the direction of the field. The precessional angular frequency (ω), called the 'Larmor frequency', of the spin is related to the magnetic field as follows:

$$\omega = -\gamma B$$

where γ is the gyromagnetic ratio, a physical property of the nucleus. Different elements exhibit different gyromagnetic ratios so that discrimination between elements is possible.

If the external magnetic field is varied in a controlled manner such that each point in the object experiences a different field strength, then protons at different positions will precess at different frequencies, thus providing spatial information. For signals to be emitted the nuclei need to be excited, by a radio frequency pulse, to a higher energy state and then allowed to decay back to the equilibrium state. This return or relaxation of the nuclei to equilibrium produces the 'free induction decay' signal that is the source signal obtainable from an NMR system.

Two relaxation parameters are used to describe the decay process. The spin lattice or longitudinal relaxation time (T_1) describes the physical process of reorientation of spins into alignment with the applied magnetic field. The spin-spin or transverse relaxation time (T_2) describes the physical processes which cause a dephasing or randomising of individual spins, so that the sum of their transverse fields decays to zero. The physical phenomenon giving rise to the longitudinal decay also contributes to the transverse decay so that the parameters are not entirely independent. The relaxation time constants associated with these processes are important imaging parameters. Being dependent on molecular structures, they often vary between normal and pathological tissues, which can often be indistinguishable in an X-ray image.

It is the spin density and the relaxation parameters T_1 and T_2 that form the most commonly used imaging parameters. Ideally, images may be obtained that are a spatial map of any one parameter or a combination of them. This choice of imaging parameters is the principal reason making MRI versatile and powerful. However, this ideal applies to static images only and real images are affected, sometimes drastically, by flow, diffusion and chemical shifts. Even for the ideal static image, the relaxation parameters themselves may not correctly represent the decay phenomenon of real tissues and often tissues are reconstructed using a linear combination of a number of T_1 values. These effects add to the inaccuracies arising from field inhomogeneities and other sources of noise so that MRI is beset with a large number of unwanted parameters that have not been sufficiently incorporated into the imaging model. This, and the availability of a large number of modalities make the task of obtaining a good representation very much a matter of the skill of the radiographer.

A key concept in NMR imaging relates to the behaviour of the emission frequency with magnetic field intensity. Because the Larmor frequency is proportional to field intensity, spectral techniques can be employed to derive a spatial mapping once a variable magnetic field has been established. With a linear spatial gradient in the magnetic field, the resonant frequency will vary linearly in space and can be separated using Fourier spectral analysis techniques. This solves the problem for one dimension and reduces the imaging to a slice. Other methods involving oscillating field gradients can be applied to isolate a point or line in a patient. With the oscillating field gradient approach, NMR signals arise only from emissions originating at the null point of the field. Hence, the use of three separate gradients limits the emitted signals to a single physical point, which must then be scanned through the patient to form an image. In practice, the scanning is done in an interlaced fashion, with spectral techniques used to carry out a multiplicity of spatial measurements in parallel.

Mathematically, MRI is capable of very elegant formulation, the imaging process being regarded as a mapping of the spatial frequency domain. The image may then be recovered from this data, usually through a multi-dimensional Fourier transform. Current results compare favourably, in terms of image quality, with X-ray computerised tomography, but in the relaxation times T_1 and T_2 an entirely new range of tissue information is available.

As we have indicated, one of the great strengths of NMR imaging is its enormous flexibility in terms of scanning and image reconstruction schemes. It is possible under software control selectively to image points, lines, planes and even volumes of the object of interest. Each of the methods has its own advantages and disadvantages in terms of imaging time, computing requirements and signal-to-noise, and at present not all possible variants have been demonstrated. Indeed, although in several respects volume scanning is very attractive, its computing requirements have so far precluded its widespread application. Medical imaging is discussed again in Chapter 7.

Synthetic aperture radar (SAR) is a radar imaging modality which embodies a reconstruction technique. Synthetic array radars are airborne or spaceborne radars which obtain terrain 'images' of the earth's surface through the transmission and reception of electromagnetic energy in a manner analogous to more conventional radars (Hovanessian, 1980). SAR makes use of the phase of Doppler information obtained during the movement of the aircraft or satellite on which the radar is mounted. The radar is usually a sideways-looking radar although it is not necessary that the antenna beam should be at right angles to the direction of motion. The technique has been used by, and was pioneered by, radio astronomers. Radar pictures of Venus made use of similar techniques, employing the rotation of the planet which gives rise to a Doppler shift depending on the position of the scatterer on the planet surface.

The increasing use of SAR techniques has been brought about by improved technology, particularly in digital signal processing, and also by a growing need to provide a rapid surveillance system with an 'all-weather' capability. It should be acknowledged, however, that although SAR images can provide very good map information, the sensor frequency band is much lower than that of optical and near-optical bands. The information gathered is therefore much coarser than that for near-optical bands and needs a more specialised interpretation than is required for photographic images. In the SAR image the information lies in the amplitude of the processed image but this is not necessarily directly related to the appearance of the target. Radar shadowing can provide information on terrain height and contour. Polarisation of the transmitted signal provides added information. For a relatively new surveillance technique the means and techniques for image interpretation have not yet been fully developed, mainly because of the lack of adequate data on which to experiment and the very considerable computing overheads involved.

Chapter 6 discusses the processing of SAR image data in considerably more detail.

1.3 HUMAN VISUAL PERCEPTION: PARADIGM OR RED HERRING?

The most sophisticated image processing system with which we are immediately familiar is our own human visual system. This familiarity is both a great advantage and a frustrating restriction for workers in machine vision in particular and image processing in general. Consequently, it is worth digressing briefly in order to mention some of the genuine and persistent controversies relating to theories of human visual perception. The hope is to indicate how appropriate, or otherwise, the human visual system might be as a paradigm for the various manifestations of image processing which are discussed throughout the book. A very readable introduction to the basics of human visual perception can be found in Frisby (1979).

The subject of visual perception is eclectic, complex, diverse, contentious and difficult. A real drawback is the fact that we are easily tempted to overlook these factors by the effortless ease with which we see, recognise and comprehend our environment. All too often we are led quite naturally to believe that we can easily understand, and consequently mechanise, this facility. It is probably fair to say that, at a superficial level at least, some knowledge about human perception can be very useful to the designer and user of image processing techniques, but it may be a great distraction and hindrance to the machine architect. Obviously, one must know something about psychophysical parameters such as quality, brightness, contrast, spatial and temporal acuity, colorimetry and photometry, *inter alia,* when processing images for human viewing. But to design a piece of image processing hardware along the lines of a putative retinal-cortical map is rather like insisting that aeroplane wings should flap.

Perception can, along with most problems, be made more tractable by using structured models as a framework for both understanding and prediction. Figure 1.5, which owes primary allegiance to a fairly orthodox 'constructivist' viewpoint, illustrates such a model. This generic model has a considerable following in the AI community because the questions of representation and computation can be addressed quite explicitly and hence, in principle, become manageable. It permits both a top-down, hypothesis driven but hallucination fraught, viewpoint and also a bottom-up, data driven but oversight hindered one. A composite up-down control strategy is then, in principle, readily implementable. This general model, viewed primarily with a top-down emphasis and explicitly incorporating the metaphor of mental processes as computations defined over internal

CONCEPTUAL LEVEL	CONCERNS
Abstract and functional	Preconditions Goals Strategies Underlying computational theories
Representational and algorithmic	Realisations of particular computational theories Representations used for input and output domains Data structures Mapping functions (i.e. algorithms)
Mechanisms	Physical realisations of processing systems Virtual machines

Fig. 1.5 A structured model suitable for providing a workable framework to underpin the development of scene-understanding systems.

representations, is central to much of contemporary cognitive science (Ullman, 1980). This model has considerable intuitive appeal to those workers who regard the digital computer as a suitable metaphor for the computational aspects of visual perception. Digital computers have influenced theories about human perception and cognition in a unique manner. First, by providing a means for building working models, they have forced theorists to be explicit. Secondly, and more significantly, they have provided an information processing metaphor in which computation is performed by the sequential application of explicit stored rules (the program) to explicit symbol structures (the data).

It is this information processing metaphor that the proponents of 'direct' perception take exception to. J. J. Gibson (1979) has proposed a significantly different model, with a strong emphasis on a bottom-up viewpoint. His so called 'direct' or 'ecological' approach to visual perception and cognition does not explicitly involve computations of any sort. To Gibson and his followers, perception is the result of the direct assimilation of available information impinging on the organism from its environment. According to Gibson's theory of 'ecological optics', the ambient light is reflected off the stable and moving surfaces in the environment and is structured by them, in some part invariantly. Consequently, at every point of observation there is an ambient 'optic array' which contains invariant information about both the persisting structure of the environment and the events which are taking place in the environment. Perception is the direct detection of function-relevant information invariants, 'affordances' in Gibson's terminology, in the optic array. In a very real sense the proponents of the ecological approach to visual perception would claim that 'cognition is parasitic to perception'. Thus in the absence of adequate structure in the immediate stimulus it becomes necessary to go beyond the information given, and so cognition invades the domain of perception in order to try to make sense of the situation. It is none the less a serious attempt to provide an alternative to a computational theory, regarding it essentially as one of *deus ex machina*.

Much of the current controversy over perception and cognition revolves around fundamental questions about deeply embedded problems relating to causality, models, representations, metaphors, granularity, and meaningful decomposition. For the worker in image processing there is much food for thought but also much misleading and fanciful speculation to lead him/her astray. In essence, we effectively have to choose between various rival paradigms, or metatheories, of perception which to all intents and purposes are distributed along a continuum. At one extreme lies the Gibsonian perspective of direct visual perception, or ecological optics, while at the other extreme lies the constructivist viewpoint as proposed originally by Helmholtz, and currently championed by R. L. Gregory. With constructivism perceptions are inferences from stored and currently available data. With this paradigm there is no direct relation between neural activity and perceptual phenomena. Rather, perception is mapped into the paradigm summarised in Fig. 1.5, where the sense organs are regarded as transducers, providing the data for consequent computation in the context of a particular representation and computational model.

A knowledge of the extensive intellectual heritage which lies behind the many-faceted pursuit of the machine processing and perception of images is important. Beyond this, the engineer or computer scientist working on image processing should realise that they are still in a position to make real, potentially important contributions to our knowledge of the deep and varied processes which underlie human visual perception.

1.4 THE IMPACT OF VLSI

One means of assessing the significance of a new technology to a particular area of endeavour is to establish how well it satisfies the following two criteria. Does it do a number of existing tasks considerably better than existing methods, and does it catalyse a whole lot more new and previously unattainable applications. By both of these criteria, the application of VLSI design and fabrication technology to the processing of image data is proving to be enormously successful. Rapid advances in semiconductor design and fabrication technology in recent years have stimulated revolutionary advances in digital integrated circuit architectures. This in turn is making possible the development of high speed, architecturally sophisticated device designs. The extreme computational demands inherent in most non-trivial image processing tasks means that these advances are of considerable significance.

It is worth noting that the major performance gains arising from the use of VLSI are not solely the consequence of greater packing densities and faster switching times. Some of the most significant gains in performance and sophistication are emerging as a consequence of the use of structured design techniques and the adoption of novel computational algorithms and processing architectures which make full use of the large amounts of concurrency which VLSI can support. A central theme of this book is the close interdependencies which exist, and should be formally exploited, between algorithm, architecture, concurrency and design routes in implementing, *inter alia,* VLSI image processing structures. Design complexity is a problem in its own right. As the number of transistors per chip moves inexorably beyond the million mark the VLSI designer is facing the formidable task of specifying and connecting together these myriad devices to form a complex signal processing or computing system on a single chip.

As the level of integration steadily increases it is not only design complexity which needs to be addressed. Increasing chip size, relative to the area occupied by a single transistor, means that concurrency and communication related problems such as data transfer latency, pin-out and clock skew loom large. Architectural regularity and short range communication, both intrinsically appropriate for image processing, can do much to facilitate minimisation of these problems. The 'systolic array' is an excellent example of a special purpose signal processing or computing system architecture which is highly amenable to VLSI design, while in addition incorporating a number of features or properties which lead to a circuit of high computational efficiency and performance. A systolic array is normally a special purpose device optimised for a specific range of applications. It is not as versatile as a general purpose programmable computer, but for its specialised function provides improvements in performance of an order of magnitude or more over a more general machine.

In the initial stages of many image processing applications it is essential to perform local operations on every pixel in the image. High degrees of concurrency and high performance data transfer mechanisms are essential if such operations are to be carried out in useful times. At a more sophisticated level, establishing stereo 'correspondences' and utilising 'blackboard'-based image understanding principles will require novel computationally intensive implementations if commercially useful image processing systems are to result from existing research programmes. Later chapters of the book will continue and elaborate upon these themes in greater detail. In the meantime the following sections will explore the growing significance of VLSI technology, and outline some of the progress to date in applying advanced digital computers to image processing.

1.4.1 The potential for VLSI

The traditional view of computer architecture tends to be that of a stored-program machine with sequential control-flow program organisation and a machine architecture consisting of a single processing unit physically separated from a single linear memory space by a communications unit. This is the so called 'von Neumann' computer architecture. VLSI technology and design techniques are clearly improving the cost effectiveness of the traditional computer in terms of its component cost, performance and reliability, together with its power and size. However, the properties of VLSI implicitly encourage different types of architecture and thus enable specialised hardware to be employed in many diverse computer system applications. Image processing is very much a case in point. In this context it is perhaps worth pointing out that a particular computer architecture can best be 'measured' by its effectiveness as a host for applications and by the performance levels achievable by implementations of that architecture. A related but important point is how well a particular architecture supports an appropriate high level applications programming language.

There are a number of intrinsic properties of VLSI implemented systems which need to be borne well in mind in developing new processor architectures. Design complexity is a critical constraint. The rapidly increasing complexity of designing and testing VLSI circuits requires very careful management. Thus a design philosophy of simplicity, regularity and replication is highly appropriate. In the end complexity must be regarded as a cost. Communication is expensive in chip area. Much of the space on a typical integrated circuit tends to be occupied by the several levels of wires that carry control and data between the functional blocks. Thus in the design effort equal attention must be given to wiring and component placement. Communication assumes a progressively dominant and limiting role in VLSI because, as chip size increases, it becomes relatively more expensive in chip area, signal energy and propagation time. Thus the incentive to design integrated circuits with localised communication pathways is strong. VLSI greatly favours on-chip communication as opposed to off-chip communication. VLSI circuit components based on low power silicon technologies such as CMOS are unlikely to be significantly faster than their LSI counterparts of ten years ago. Thus major advances in computational performance must come from the concurrent operation of a number of components, rather than from fast components. Input and output must be balanced with internal computation for maximum performance. Thus trade-offs between pin bandwidth and data memory size, both of which affect input-output costs, must be carefully optimised.

On the basis of these intrinsic properties it is possible to make some fairly general statements about preferred on-chip organisations for VLSI-implemented systems. Data and control paths should be simple and regular. Cells should be connected by a network with local and regular interconnections, thus avoiding or minimising long distance or irregular communication. Regular interconnection implies that the design can be made modular and extensible, so a large computer system can be implemented by combining the designs of simple cells and simple chips. Communication is best localised. Ideally, there should be no global communication, only communication between adjacent cells and chips. This may be viewed as locality of reference applied to VLSI. VLSI-based architectures can make extensive use of concurrency for increased performance. High performance is obtained in VLSI architectures from the concurrent operation of a large number of simple processors rather than from a single large processor. This concurrency may be obtained either by pipelining the stages involved in a computation or by multiprocessing independent computations in parallel. The architecture should make multiple use of each input data item. To balance input-output bandwidth with computation, it is necessary to make multiple use of input data. This can be achieved either by broadcasting the input data to all cells, or by having the data traverse a regular structure of cells.

In summary, VLSI is best suited to computer architectures composed of a collection of simple processors connected together in a regular structure. These general purpose and/or special purpose processors should to the greatest possible extent operate concurrently. Ideally there should be no global communication, only communication with adjacent or nearby processors. Such an architecture should then give rise to integrated circuits having regular interconnection, local communication wall structured data and control flows, and utilising many repetitions of a few 'canonical' designs.

Chapter 3 explores the ideas underlying the general concept of VLSI and VLSI-based architectures, and their implicit potential, in more detail.

1.4.2 Computational structures for processing images

The existence of the appropriate computational support for the vast majority of image processing activities is fundamental to the extent that without this presumed support most image processing would be effectively impossible. Only those techniques relying on optical methods would persist and be of any practical use.

The basic computational requirements, even for straightforward image processing tasks, are fairly imposing. A typical relatively high resolution image, such as that obtained from a LANDSAT earth resources satellite, is about 1000 pixels square. Low level operations, such as filtering or enhancement, typically require of the order of some tens of machine instructions. This is a computational requirement of some tens of millions of instructions per image, before any substantive higher level processing is performed. Higher level processing will increase this figure by several orders of magnitude. If such a simple low level operation is to be performed at a video rate, say 25 or 30 frames per second, this imposes a computational requirement of about a billion instructions per second, with any image input and output overheads to be added on top of this. For this very simple reason much of the image processing performed on standard serial machines has been either rather straightforward or very time consuming. The cost has either been in time or money. It is this very fundamental constraint which has provided the impetus towards the development of new high performance, cost effective computer architectures more capable of supporting the computational burden imposed by sophisticated image processing tasks. In the context of this book, it is those machines whose underlying architecture reflects the capabilities inherent in VLSI which are of most interest. Implementation details and discussion of specific machines are contained in subsequent chapters.

Computer architectures have traditionally been classified according to the uniqueness or multiplicity of their instruction and data streams. By considering the cartesian product

$$(\text{Single Instruction, Multiple Instruction}) \times (\text{Single Data, Multiple Data})$$

abbreviated as (SI, MI) $\times$ (SD, MD), we obtain four generic architectural 'styles': SISD, SIMD, MISD, and MIMD. This taxonomy is still very much in use, though some subdivision of these categories has been found necessary in practice. Most conventional uniprocessor architectures can be unambiguously classified as SISD, which in effect categorises the classical von Neumann architecture. SIMD architectures involve a single control unit fetching and decoding instructions which are then executed in the control unit itself, such as a 'jump' instruction, or are broadcast to a number of identical processing elements (PEs) interconnected by a switching network. These processing elements operate synchronously, but their local memory spaces will generally contain different items. A number of inter-PE switching networks can be envisaged. The role of such a network is to allow the transfer of data, in a single operation, from processor i to processor j according to some mapping function $j = f(i)$ with all processors being in an active state. If this function is such that we can perform all possible one-to-one mappings from source PEs to destination PEs, we have a 'permutation' network. If we have one-to-many mappings of sources to destinations, we have a 'broadcast' network – if only a subset of the destinations can be reached then we have a 'selective broadcast' network. Permutation networks are expensive to implement in practice. Often only a subset of the connections are wired in. In this case there are two important metrics to consider: the number of switches in the network and the time taken to perform any given mapping, allowing that several iterations through the network may be needed before an intended transfer is realised. Within this superficially simple SIMD scheme, depending on PE complexity, PE processing power, PE addressing scheme and inter-PE communication facilities, one can distinguish between 'pipeline (vector) processors', 'parallel-array processors', 'processing ensembles' and 'associative processors'.

Pipeline processors are difficult to classify. They can, depending on implementation details, be considered to be either SIMD, SISD or MIMD architectures. In simple terms, pipelined processing is analogous to an assembly line organisation. The machine's computational structure is segmented into consecutive units, while the program processes are decomposed into temporally overlapping subprocesses which have to pass sequentially through each unit. In a parallel-array processor the PEs operate synchronously in a 'lockstep' mode and communicate with neighbouring processors via the switching network. Ensemble processing is characterised by having a control unit which is a complete computer and in order to communicate the PEs have to pass all their messages via this control unit. Associative processors generally consist of several processing elements organised in the parallel-array processor style but with special memories which enable data to be retrieved associatively, that is by content rather than by address. Despite their attractive properties, associative processors are rarely used. They tend to be expensive, since they usually need extra gates for the logic needed in the memory itself in order to test its contents for matches. VLSI implementations may well overcome this stumbling block in the near future. Finally, MIMD architectures involve several processors operating concurrently but asynchronously, usually with shared access to a common memory space. Designs differ in terms of processor homogeneity and the switching strategies used for interconnecting processors and memory. This is discussed further in Chapter 4.

From the specific point of view of manipulating image data, especially at the lower levels of processing, the main focus to date for high speed machines has been the parallel-array processor and its close relative the 'systolic-array processor'. The regularity inherent in their intrinsic pixel to PE mappings means that such architectures are prime candidates for efficient VLSI implementation. Consequently, it has been possible to develop compact high performance machines whose performance, for a typical 64 × 64 single-bit processor array, is of the order of 50 GBOPS (BOP = Binary Operations per Second). This is adequate for many sophisticated real-time image processing tasks.

A typical parallel-array processor, shown in schematic form in Fig. 1.6, is built up from a number of relatively simple bit-serial PEs which are interconnected, usually to nearest neighbours, in a two-dimensional pattern. Apart from its single-bit internal registers, each PE may have access to its own private bit-organised PE memory. There is a single control unit to supply the sequence of PE instructions and PE memory addresses. During each instruction cycle, all the PEs execute the same instruction, using the same PE memory address. Instructions are provided to activate only those PEs where a program-specified condition is satisfied and to deactivate those PEs where this condition is not satisfied.

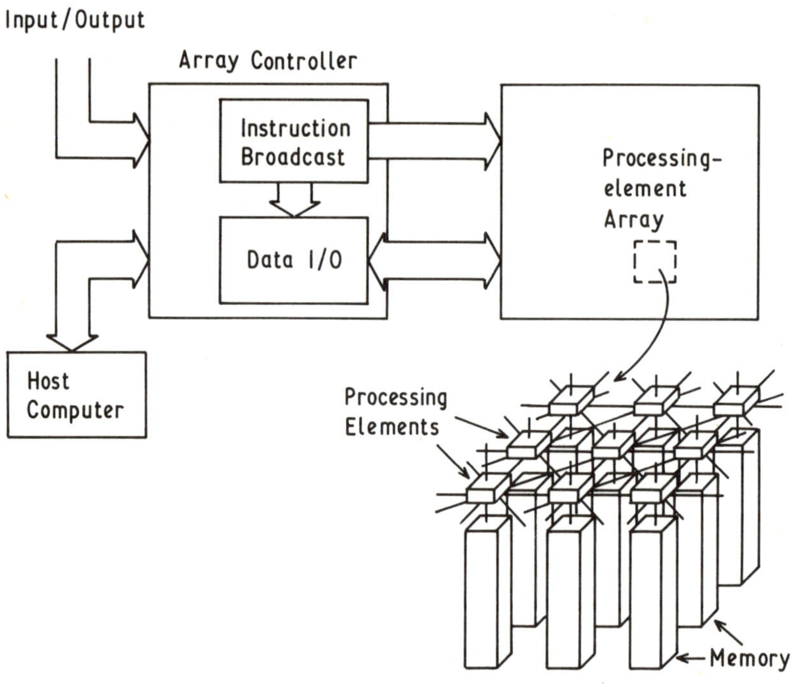

Fig. 1.6 A schematic block diagram of a typical parallel-array processor. In general, such machines are attached as a peripheral to a host computer.

In practice it is not necessary to have a one-to-one pixel to PE mapping. The control unit can, in practice, take care of the requisite many-to-one mappings automatically.

In a systolic array the PEs tend to be much more specialised and the interconnectivity more limited. As with the parallel-array organisation, a systolic array utilises common control and common timing flows, so that all the identical PEs perform the same function simultaneously, but on different data items. Each module is connected only to its nearest neighbours for the purpose of data transfer, and several data streams move at constant velocities over fixed paths in the network and interact within a PE whenever they meet. This is shown schematically in Fig. 1.7. Multiple use is therefore made of each data item, which results in a high computational throughput without the need for high bandwidth memory links.

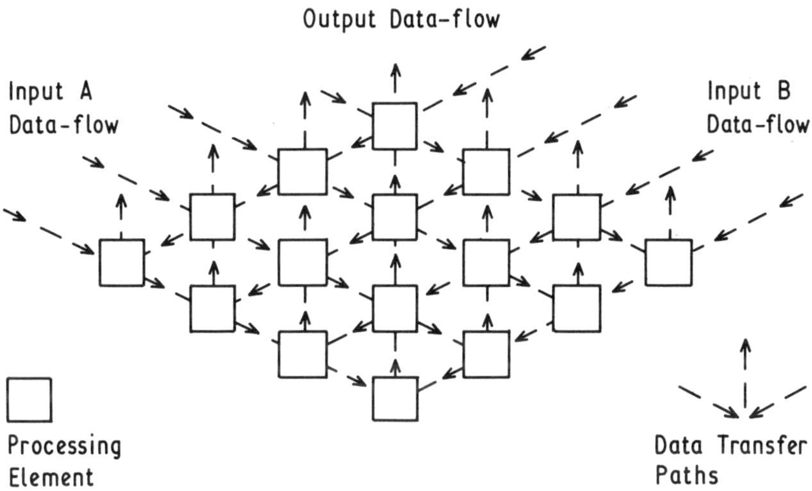

Fig. 1.7 Schematic diagram of the organisation of a systolic array.

In a particular implementation, the precise function of the PEs and their interconnection patterns is determined by the class of problem to be solved. Each PE could perform, for example, a multiply and accumulate function and could consist of a digital multiplier, adder and latches. For other applications, a much simpler PE containing a few basic logic gates might be more appropriate. The systolic-array architecture is obviously well suited to VLSI implementation. There are a number of fundamental reasons for this. First, the inherent modularity eases the design and layout problem, since the circuit designer can concentrate his effort on optimising one or two simple cells. Complex circuits can then be implemented by connecting these cells together in a completely regular fashion. Many basic image processing functions can, in fact, be implemented in this way from a small library of simple cells. Extensive functional parallelism is employed which exploits the potential for concurrency in VLSI circuits to increase functional throughput and improve system performance. The short and regular interconnections between cells lead to a small on-chip interconnection area, and to short data propagation delays. Interconnections between cascaded chips are also intrinsically short and high speed data transfer is possible using low power on-chip drivers.

Another important feature of the systolic architecture is that data transfer to and from the array is minimised. Individual data items are input to the array once only, whereupon they are utilised as often as necessary in combination with other data items before finally being discarded. Thus the systolic array is able to operate with a high functional throughput, without requiring excessively high bandwidth memory links. Because data flows at a uniform rate along fixed paths in the network the control overhead of the systolic array is small, and for some fixed function arrays almost non-existent. Other advantages are the reduced system memory requirements that arise because of the distributed processing and memory within the array, the ease with which the architecture may be scaled with the technology, and the ease with which the array may be expanded to handle larger problems.

There are a number of other styles of machine which have very high intrinsic performances, the so called 'super-computer' for example, but few have the specific utility for image processing, especially at the data-intensive stages, that the parallel-array and systolic processors possess.

1.4.3 Some dedicated image processors

Many image processing tasks typically involve an intrinsic trade-off between a system's speed and its flexibility. A typical minicomputer, if used for image storage and processing, will usually prove to be both too bulky and too slow for many dedicated real-time applications. If special purpose hardware is developed for a particular application then very high processing speeds can be achieved, but only at the cost of forfeiting easy reprogrammability. Both of these alternatives can prove to be uneconomic if versatile or low cost systems are being sought, for whatever reason. A sensible middle ground has emerged in recent years in that a number of dedicated, programmable image processors have appeared on the market. These devices are special purpose LSI or VLSI based processors, possessing architectural features which efficiently support a wide variety of image processing algorithms and applications. Some representative examples are briefly discussed below.

One of the first digital signal processing devices designed to support a wide range of high speed or computationally intensive applications, including image, speech, sonar and radar processing, is the Texas Instruments TMS320. This 16-bit external, 32-bit internal, single-chip, highly pipelined microprocessor, shown in schematic form in Fig. 1.8, combines the flexibility of a high speed controller with the numerical capability of a vector processor, and as such offers an inexpensive alternative to more conventional bit-slice processors. The TMS320 contains a 32-bit ALU and accumulator that supports double-precision arithmetic as

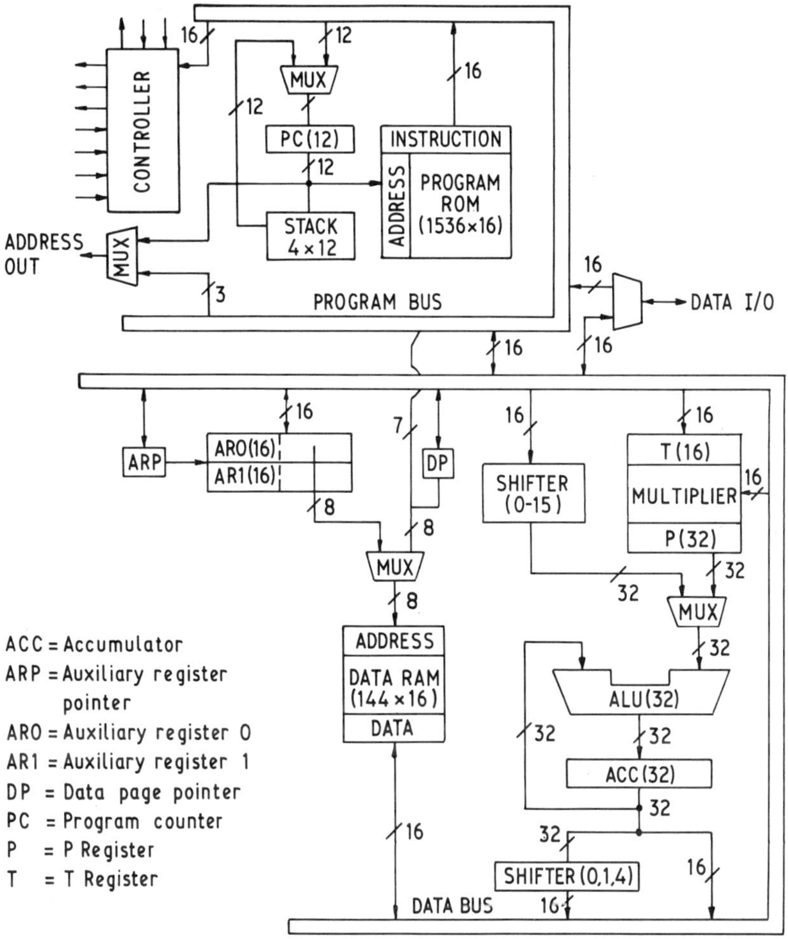

Fig. 1.8 A schematic diagram of the TMS320.

well as logical operations. Two barrel shifters provide powerful facilities for scaling and bit extraction. A high speed multiplier performs a 16 × 16 two's-complement multiplication and stores the 32-bit result in a single 200 ns instruction cycle, providing the ability to perform such basic operations as convolution, correlation and filtering at effective rates in excess of 2 million data samples per second. Further, pipelined multiply and accumulate operations can be performed at a rate of 400 ns each. The 16-bit parallel data bus is capable of supporting burst input/output rates of 40 MHz.

Similar devices are manufactured by NEC, Intel and AMI. A full 32-bit 'second generation' device is due from Texas Instruments late in 1985.

The recently announced NEC μPD7281 'data flow' (see Section 4.7.2) image processor incorporates a novel internal circular pipeline architecture and a powerful optimised image processing oriented instruction set. It is the first commercial VLSI device specifically to exploit a data flow architecture. As a data flow processor it does not fetch instructions in the conventional manner. Rather, it processes 'tokens', or packets, which represent operand data. The particular data flow architecture implemented in the μPD7281 allows it to maximise efficiency in a variety of image-specific multiprocessing applications. The internal circular pipeline architecture, shown in schematic form in Fig. 1.9, allows the processing unit to operate at a continuous 5 MHz rate. The link table and the function table store object code, while the data memory temporarily stores the data and instruction tokens that need to be matched. The address generator and flow controller attempts to match two tokens, and the queue temporarily stores the matched tokens before they are processed by the processing unit. The processing unit contains a 17 bit × 17 bit multiplier, and an ALU which performs normal arithmetic, logical, barrel shift, comparison and bit manipulation operations. As a consequence the device is well suited for image processing applications using algorithms involving two-dimensional convolution, enlarging, shrinking and rotation. For example, it can rotate a 640 × 400 pixel image through an arbitrary angle in 1.4 s. The output queue temporarily stores outgoing tokens, and the input controller examines the incoming token's processor address fields in

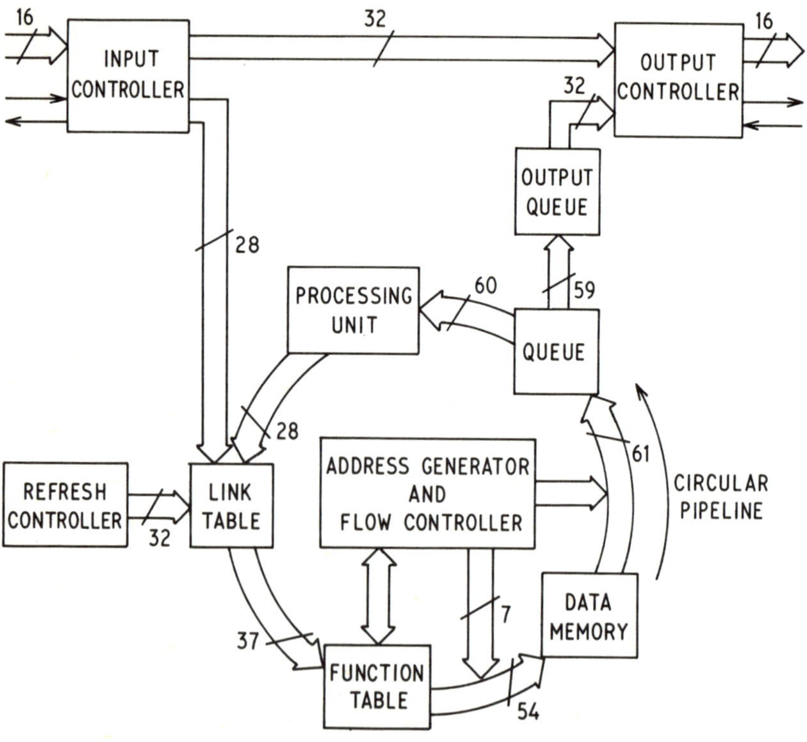

Fig. 1.9 Schematic organisation of the NEC μPD7281 data flow image processor.

order to ascertain the token's subsequent destination. The refresh controller generates refresh tokens for the on-chip dynamic RAM.

The μPD7281 can be used in two basic multiprocessor configurations, the 'cascade' and the 'ring'. The cascade involves a pipeline of processors, while with a ring configuration there is an image memory, a host processor, and hardware logic to multiplex data from the host processor and the image memory.

A two-dimensional systolic-array based device, the Geometric and Arithmetic Parallel Processor (GAPP), specifically targetted at high speed image processing applications, has recently been announced by NCR (*Electronics,* 1984b). Developed jointly by NCR and Martin Marietta Aerospace for use in real-time target recognition systems, the device integrates 72 individual single-bit processors laid out as a 6 × 12 array, each with 128 bits of memory, onto a single CMOS chip. When running at a 10 MHz clock rate a GAPP chip can typically perform 8-bit integer additions at a rate of 28 million operations per second. This rate increases linearly as GAPP chips are cascaded together. Each of the single-bit processors contains an ALU, various registers and latches, and bidirectional communications lines connected to its four nearest neighbours. Instructions can be globally broadcast to each processor in the array during every 100 ns cycle over a number of control and memory address lines. In practice the GAPP chip can be used in a purely systolic mode or more like a conventional bus-connected array, depending on the application requirements.

Though not an image processor in the strict sense, General Instrument Corporation's AGILE (Animated Graphics Imaging through List Execution) graphics and display system, which simulates full motion video, exhibits some novel features which are of more than passing interest to image processing practitioners (*Electronics,* 1984a). The primary market currently envisaged for the AGILE system, shown in schematic form in Fig. 1.10, is sophisticated video games animation but it embodies features which render it of far greater utility. The idea behind AGILE – the creation of primitive list structures that are linked to create pictures – is one which underlies the AI language LISP. The efficient manipulation of these archetypal structures provides a kind of video shorthand and thus has the potential to free implementers from the need to create every individual motion.

The high performance memory requirements of an AGILE system can be kept reasonably low, depending on the complexity of the graphics. The image-writing hardware uses only two lines worth of memory at any time, so that if the lines contain 512 pixels, each 12 bits deep, only 12 kbits of memory are required. In contrast, an equivalent interactive bit-mapped system would require 480 times as much. The basic image storage requirements in both types of systems would be equivalent, however. As well as displaying graphics programs, AGILE can facilitate developing them. From images entered manually, through a terminal, or from a camera or video disc, the host system creates a linked list that is a chronological 'script', or sequence of events, describing the three-dimensional staging of a game. A hardware

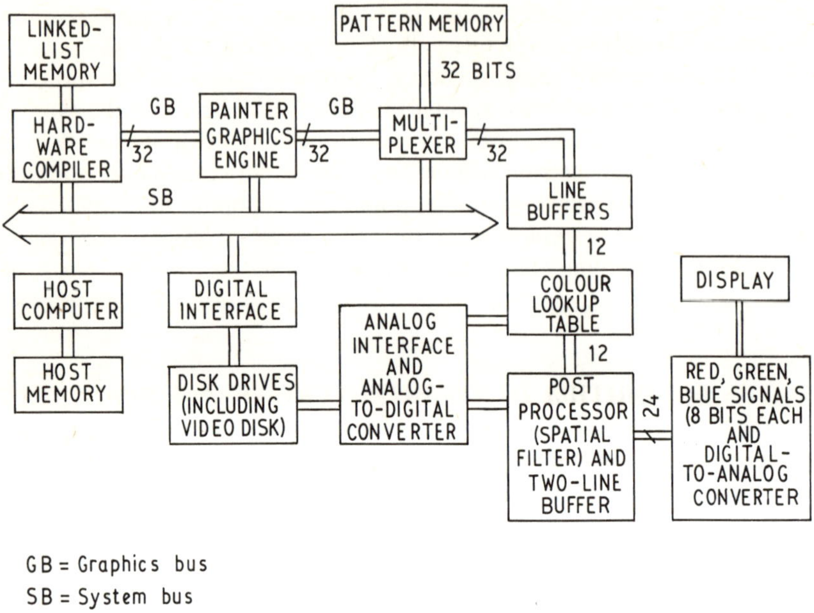

Fig. 1.10 Schematic diagram of the AGILE system.

compiler moves this list into a dedicated memory read 60 times a second by a special purpose graphics processor, called a 'painter' graphics engine. A 32-bit high speed (450 Mbit/s) graphics bus links the graphics side of the system. A different memory stores bit patterns representing primitive forms of each object in the animation. These patterns, whose diameters range in size from a single pixel to many screens, are like 'micro bit-maps', which describe discrete objects independently of their position. These patterns, each at this point 12 bits long, include information about an object's luminosity, expressed in 256 shades of gray. Corresponding gradations of colour, considered an attribute of this general form, are painted in later, according to the script. Up to 8 bits of the pattern can describe the object's shape; the rest are for colour. At the same time, the painter, also working from the script, compiles another, shorter list of 'picture hooks' for each frame – actually, addresses designating where the objects are stored in pattern memory and where they should be 'hung' in the frame's three-dimensional space. The painter hangs the objects on the appropriate hooks, merging information through the multiplexer. The linked data is loaded a line at a time, from the top to the bottom of the display, into one of two line buffers, from which it is sent to a colour lookup table containing 4096 hues. From there, the data emerges on a 12-bit bus organised as 4 bits each for red, green and blue. A fast postprocessor for spatial filtering then takes over. Operating at some 100 million equivalent arithmetic operations per second, it filters an image in two dimensions, line by line. Holding the previous two lines in a buffer for interpretation, it generates intermediate hues to create crisp images with no jagged edges. The 12-bit pixel that entered the buffer is then sent to the display characterised by 8 bits each for red, green, and blue; defined, in effect, with 24-bit accuracy, or 16 million colour variations.

Another graphics oriented special purpose VLSI device is the Stanford 'geometry engine' (Clarke, 1982). This is a four-component vector, floating-point processor for accomplishing three basic operations in computer graphics, matrix transformations, clipping and mapping to output device coordinates. A full system is composed of a number of geometry engines cascaded into a pipeline configuration.

1.5 THE ALGORITHM TO ARCHITECTURE TRANSFORMATION

In practice it soon becomes apparent that the efficient translation of an *ad hoc* computational algorithm into an actual piece of physical equipment is a far from trivial task. Such a transformation involves at least four clear and distinct conceptual stages: computational theory, the corresponding algorithm and its implicit or explicit representation(s), the abstract computational architecture and finally the actual physical implementation. This essentially sequential set of representations and transformations is shown schematically in Fig. 1.11. Such a simple picture leaves much unsaid. What representations are appropriate, or – in some sense – optimal, for the various domains? What sorts of mappings are involved? The representation

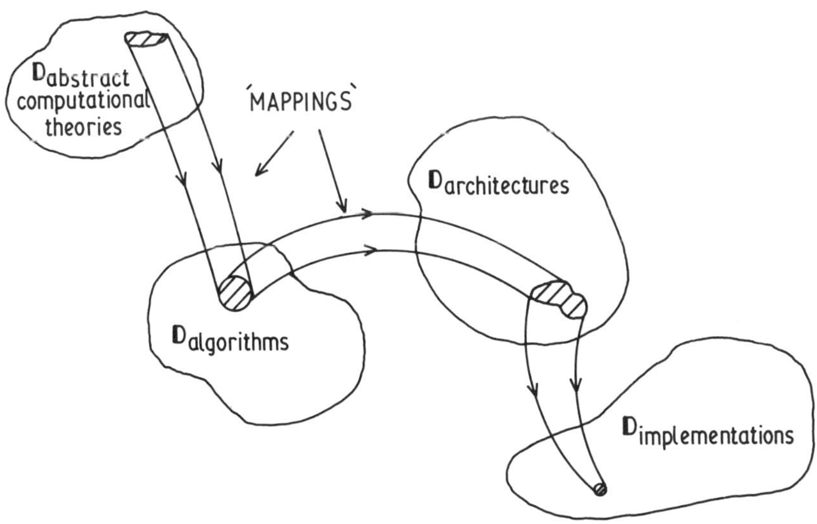

Fig. 1.11 An abstract picture of the pathways between computational theory and physical implementation. The symbol **D** represents an abstract 'domain' of possible entities.

of both algorithms and architectures by directed graphs, or their equivalent, but more formal, matrix representations seems appropriate, but how do we decide? Are the mappings to be simple isomorphisms or are they going to be more esoteric? The only way sensibly to address these problems in practice is to have some way of establishing metrics whereby these and related questions can be quantitatively answered. As well as providing a path through the mapping and representation methodologies, appropriate metrics should also be able to provide performance related measures which take the design process, the fabrication technology and the cost effectiveness of the implementation into account.

A computational theory is concerned primarily with formally elucidating questions about the specifiable goals of an abstract computation, why it is appropriate and what is the logic underlying the strategy for carrying it out.

An important aspect of a computational theory is the precise specification of the form and content of inputs and ancillary information available for the process(es) under investigation. Similarly the form and content of the outputs must be precisely specified. Finally, the logical, mathematical, symbolic, or other transformations which are to produce the output from the input must be explicitly specified, as must the intermediate data representations in the process(es). If the computational theory is indeed precise enough to be implemented, the resulting algorithm, and hence the resulting abstract architecture and final implementation, is a finite, formal realisable artifact whose characteristics and performance can be thoroughly and quantitatively evaluated.

The relevance of computational theories to computer vision was proposed, developed and strongly championed by the late David Marr (1982). Much current thinking in this area owes an explicit or implicit debt to this pioneering work which sought to elevate image analysis from a stereotyped copying of ill understood biology into a computational science.

At the level of algorithms and representations the questions are about how a computational theory can be instantiated in a particular situation. In particular, what representations are to be used for input, intermediate internal states and output, and what algorithms are going to be used to effect the transformation?

There is a strong duality between processes and representation. This is an important point in scene understanding because of the way in which a particular image of a given scene underconstrains that scene. Proper representation of the internal states of the computational model, and proper choice of the corresponding algorithms, is essential if hallucination in top-down driven systems and oversight in bottom-up driven systems is to be avoided. Some typical internal representations generic to most scene understanding problems include the image itself, representing raw intensity; the 'primal sketch', which makes explicit important low level information about the two-dimensional image; the '2½-D sketch', which makes explicit depth and orientation cues in a viewer centred framework; and the 3D model representation, which describes shape and spatial organisation in an object centred coordinate frame (*ibid*).

A computational architecture, in the context of this discussion, should be regarded as an abstract type of a system object, as opposed to an instantiation of that object. There is a sense in which this abstract architecture need not ever be fully defined. Figure 1.12 shows diagramatically the region, of an abstract domain, which represents a particular abstract architecture. The architecture is in fact specified in terms of various operational 'projections' onto, or 'windows' into, or 'views' of, this region in such a way that the abstract architecture equals an appropriate sum over these projections. This sum could in some circumstances underconstrain the abstract architecture, which will consequently not be uniquely defined. This lack of completeness could be a definite advantage in that a complete architectural definition does not need to be maintained, only the projections relevant to the problem in hand. It is still an open question as to which windows are the most appropriate to choose in practice. In order to make

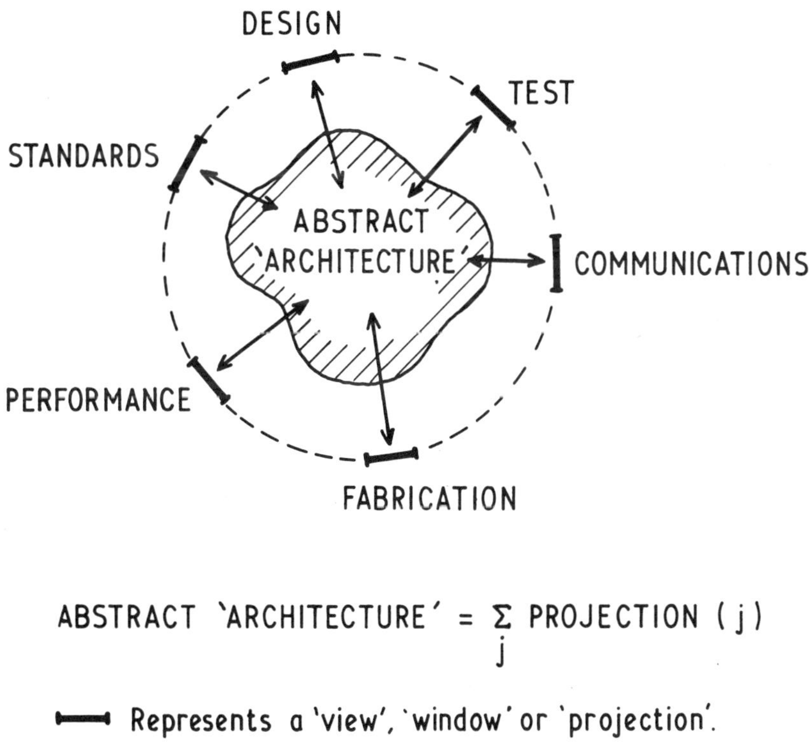

Fig. 1.12 A projection or 'view' based model of an abstract system architecture. The strength of this model is that the architecture need never be fully specified in its abstract form.

the implied computational model more tractable it is convenient to consider the windows to contain embedded layers, which effect a stepwise transformation from the abstract to the particular. This is illustrated schematically in Fig. 1.13.

The preceding discussion is concerned essentially with a static view of the connectivity and data structuring inherent in an abstract architecture. A dynamic, or control flow, picture could be obtained by considering an appropriately defined 'flux' through the relevant windows.

As far as any physical implementation is concerned, the basic question is how can the algorithm and its abstract implementation architecture be physically realised? Obviously questions relating to technology, manufacturing

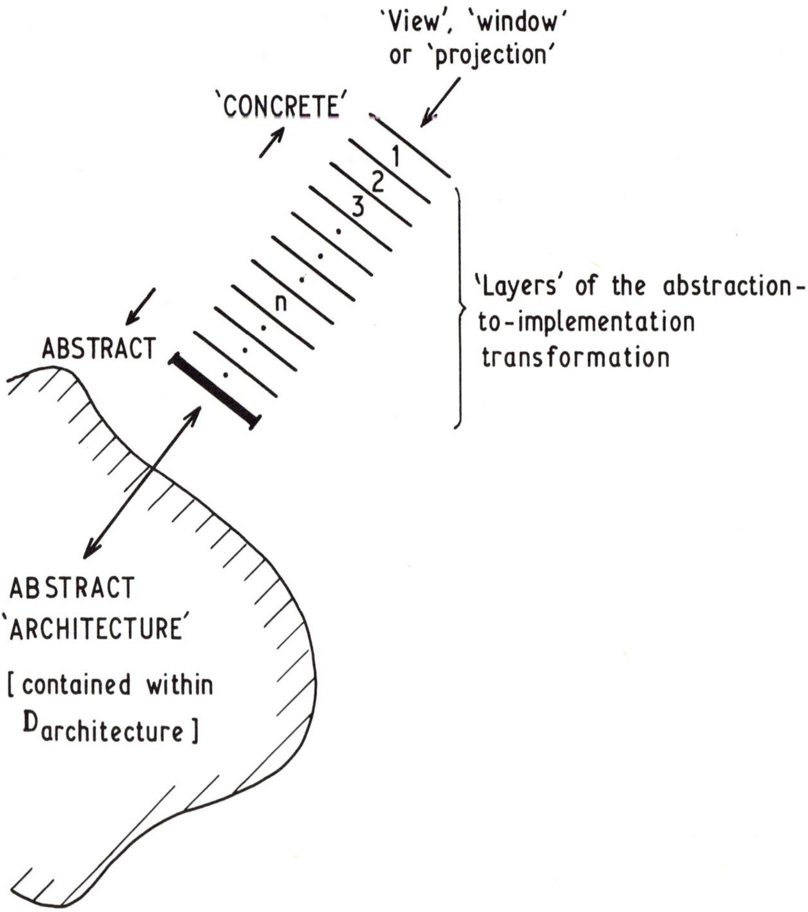

Fig. 1.13 A layered projection or view of an abstract system architecture of the type depicted in Fig. 1.12.

and efficiency will become highly germane and in general will be very implementation specific. Here, theory must give way to practice.

The above elements are all complex, yet essential, elements of the intellectual, and eventually practical, route from an abstract idea about an image processing task to its physical implementation. What is absent today, and is sorely needed, is an underlying formalism for linking all these elements together in a clear and tractable manner.

1.6 THE NEED FOR SOME THEORETICAL UNDERSTANDING

It is not merely that the points discussed in the previous section need to be given a useable formal basis if a sensible design route is to be established. The possibility of having the system hardware implemented as custom VLSI devices has completely changed how we need to think about the way in which one moves from the instantiation of a computational theory to its physical implementation. The crux of this difference in the way in which we should think about implementing image processing systems lies in the fact that we no longer have just one more or less standard abstract architecture on which to base our implementation thinking – the von Neumann machine. One of the prime consequences of recent advances in VLSI design methods and fabrication technology is the sheer variety of computational structures which are, in principle, now readily available. It is this element of choice which significantly compounds the already difficult design task, of working efficiently from computational theory to physical implementation, outlined in the previous section. In the present context 'efficiently' means that various appropriately chosen parameters relating to size, weight, thermal behaviour, power requirements, performance, modularity, fault tolerance, cost, generality, etc., etc., are optimised to the designer's satisfaction. Figure 1.14 attempts to illustrate this point schematically. The transformations from abstract computational theory to physical implementation are shown flowing from top to bottom. The width of this flow represents the number of potential choices that the designer has as he moves along the flow. To the right of (a) the conceptual stages occurring along the flow are shown. These relate directly to the design stages discussed in the previous section. Note that this representation can be regarded recursively in that if the implementation is in fact some form of virtual machine then this can in turn be regarded as an embedded instantiation of an abstract computational theory, and so on. At the algorithmic level, with appropriate representations, there is a very wide set of choices with only very weak measures available for evaluating these choices. With the traditional implementation route, it is only the very constrained target architecture – von Neumann – which makes the process more or less tractable.

Advances in VLSI technology and the increasingly sophisticated design environments now available to chip designers have completely changed this picture. In (b) the choice is still very wide at the implementation level as a

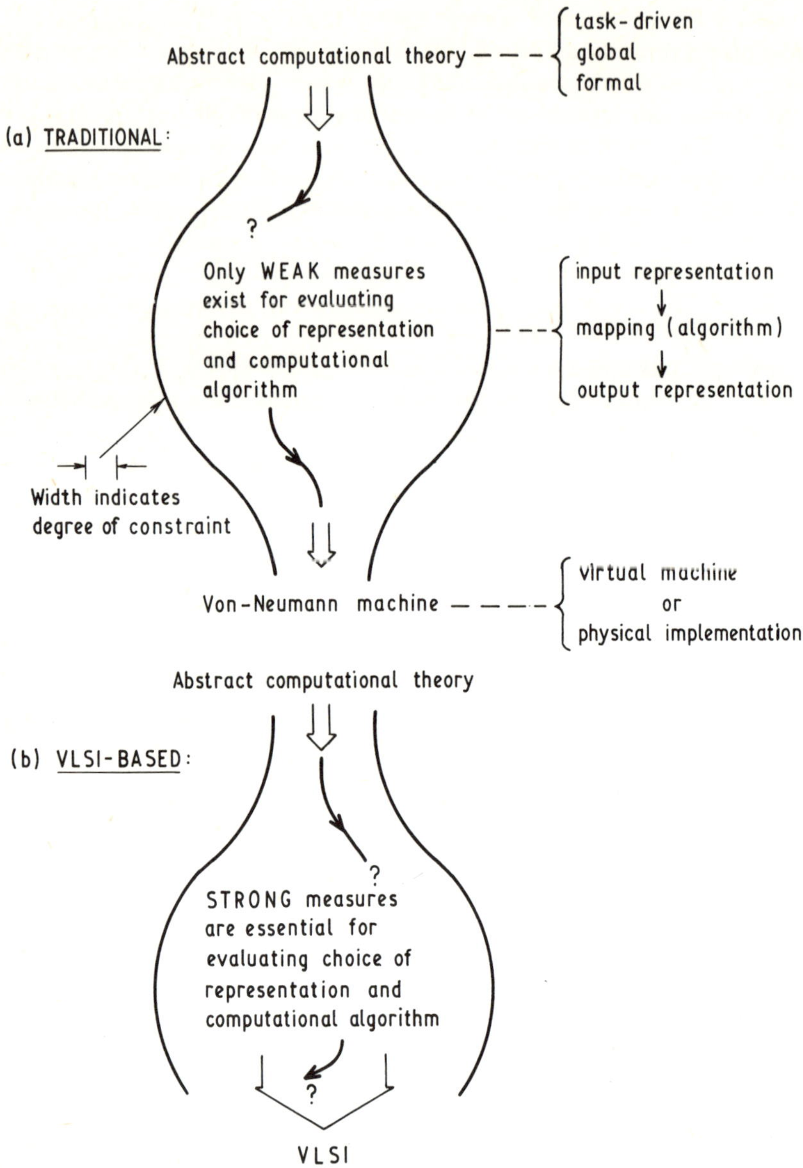

Fig. 1.14 Schematic representations of the 'flow' from an abstract computational theory, actual or implied, to a physical implementation for (a) a conventional von Neumann machine and (b) an *ad hoc* custom VLSI device.

consequence of the fact that it is now possible, following a custom VLSI route, to implement efficiently what previously would have been regarded as highly exotic, and hence impracticable, system architectures. The problem of efficiently, in the sense discussed above, matching an algorithm, within the context of a particular computational theory, to an implementation architecture is now almost impossible without the existence of some strong measures for evaluating and guiding the design process. It is this extra degree of freedom introduced by potential custom VLSI implementations which forces us into the situation of attempting to seek some methods or tools to guide us through the design process. As well as looking for means to assess the efficiency of the overall design process, as mentioned above, we also need some appropriate metrics with which to assess the choices of embedded concepts such as representations and mappings – the question of how one makes the initial choices is also one which must be resolved. Ideally, these metrics, or meta-parameters, should arise naturally out of the formalism used to underpin the design process.

Figure 1.15 shows conceptually how one might formalise, and hence automate, the processes described above. In a sense this can be regarded as combining the characteristics of Figs. 1.11 and 1.13. The highly complex mapping between abstract specification and physical implementation is reduced to a much more straightforward mapping by layering the problem and solution spaces. Within each space a set of manageable and formalisable transformations moves towards achieving tractably mappable representations. It should be clear by now that the whole question of what are suitable representations and how to choose and evaluate them is central to

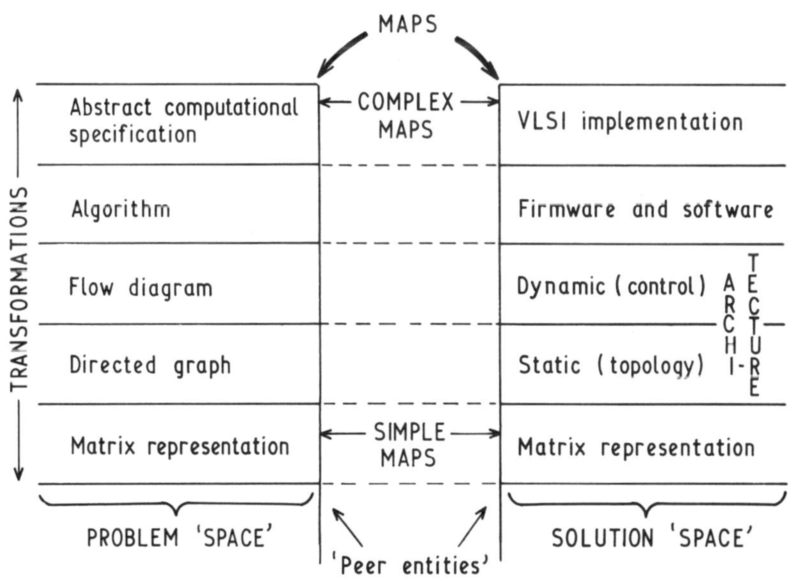

Fig. 1.15 A schematic diagram of the hierarchy of potential mappings which exist between the various levels of representation of the problem and solution spaces for a particular computational requirement. In the context of this figure the simplicity or complexity of the mappings refers to the ease or otherwise of performing the mappings.

establishing and supporting the formally underpinned design model outlined above. This point, and the other questions raised in this and the previous section, will be taken up again in Chapter 4. Prior to this Chapters 2 and 3 will elaborate on more of the relevant background by covering the general areas of image processing algorithms and VLSI architectures. Subsequently Chapters 5 and 6 will discuss two representative application areas, computer vision and synthetic aperture radar, while Chapter 7 will round off the book with a look at some aspects of the future of image processing.

BIBLIOGRAPHY

Baer, J.L. (1980) *Computer Systems Architecture,* Computer Science Press.

Brady, M. (1981) 'Computational Approaches to Image Understanding', *AI Memo No. 653,* MIT.

Duff, M.J.B. (ed.) (1983) *Computing Structures for Image Processing,* Academic Press.

Fu, K.S. and Ichikawa, T. (eds.) (1982) *Special Computer Architectures for Pattern Processing,* CRS Press.

Hall, E.L. (1979) *Computer Image Processing and Recognition,* Academic Press.

Haralick, R.M. and Shapiro, L.G. (1985) 'Image Segmentation Techniques', *Comp. Vis. Graphics and Image Proc.,* **29**, 100–132.

Hennessy, J.L. (1984) 'VLSI Processor Architecture', *IEEE Trans. on Computers,* **C–33**, No. 12, 1221–1246.

Hwang, K. and Briggs, F.A. (1984) *Computer Architecture and Parallel Processing,* McGraw-Hill.

Pratt, W.K. (1978) *Digital Image Processing,* J. Wiley.

Rosenfeld, A. and Kak, A.C. (1982) *Digital Picture Processing* (2nd edition, Volumes 1 and 2), Academic Press.

Seitz, C.L. (1984) 'Concurrent VLSI Architectures', *IEEE Trans. on Computers,* **C–33**, No. 12, 1247–1265.

Uhr, L. (1984) *Algorithm-Structured Computer-Arrays and Networks,* Academic Press.

REFERENCES

Bozic, S.M. (1979) *Digital and Kalman Filtering,* Edward Arnold.

Cohen, P.R. and Feigenbaum, E.A. (1982) *The Handbook of Artificial Intelligence* (Volume 3), Pitman.

Clarke, J.H. (1982) 'The Geometry Engine: A VLSI Geometry System for Graphics', *Comp. Graphics,* **16**, No. 3, 127–133.

Electronics (1984a) Jan. 12, 47–48.

Electronics (1984b) Nov. 12, 17–18.

Frisby, J.P. (1979) *Seeing: Illusion, Brain and Mind,* Oxford University Press.

Gibson, J.J. (1979) *The Ecological Approach To Visual Perception,* Houghton-Mifflin, Boston.

Hall, E.L. (1979) *Computer Image Processing and Recognition,* Academic Press.

Hovanessian, S.A. (1980) *Introduction to Synthetic Array and Imaging Radars,* Hughes and UCLA.

Huffman, D.A. (1952) 'A Method for the Construction of Minimum-Redundancy Codes', *Proc. IRE,* **40**, No. 9, 1098–1101.

Mackworth, A.K. (1983a) 'Constraints, Descriptions and Domain Mappings in Computational Vision', in O.J. Braddick and A.C. Sleigh (eds.), *Physical and Biological Processing of Images,* 33–40, Springer-Verlag.

Mackworth, A. (1983b) 'On Seeing Things, Again', *Proc. of the Eighth Int. Conf. on Art. Int.* (IJCAI–83), 1187–1189.

Machlup, F. and Mansfield, U. (eds.) (1983) *The Study of Information: Interdisciplinary Messages,* J. Wiley.

Marr, D.C. (1982) *Vision,* Freeman.

Pentland, A.P. (1984) 'Fractal-Based Description of Natural Scenes', *IEEE Trans. on Pattern Anal. and Mach. Int.,* PAMI–6, No. 6, 661–664.

Pratt, W.K. (1978) *Digital Image Processing,* J. Wiley.

Rosenfeld, A. and Kak, A.C. (1982) *Digital Picture Processing* (2nd ed, Volumes 1 and 2), Academic Press.

Ullman, S. (1980) 'Against Direct Perception', *The Behavioural And Brain Sciences,* 3, 373–415.

Chapter 2

Image Processing Algorithms

M. M. McCABE and P. V. COLLINS

2.1 INTRODUCTION

It is not the purpose of this chapter to provide an exhaustive description of known image processing algorithms and techniques; there are already many books in existence or preparation which cover the field. Rather our aim is to explore and discuss the types of image algorithm, means of implementing the algorithms, and the use of the algorithms in image processing systems. Particular emphasis will be placed on the mappings and connections between algorithms and architectures. The discussion will proceed by examining different classes of algorithm/technique and their usage. Examples will be extensively used, but these will be illustrative rather than exhaustive.

Throughout this chapter the general development will be made in the context of two-dimensional, grey scale image frames. However, the principles apply equally well to most other cases of image processing, for example colour, image sequence and motion analysis, or three-dimensional images. Further, although only enough examples have been chosen to illustrate the basic principles, these principles are intended to be applicable to all algorithms and applications.

There are in fact three classes of image processing technique that we consider here. At the most basic level there are what might be classified as tools, components or modules – essentially the basic building blocks for constructing image processing systems. Examples of such components are filters, smoothing operators or edge-detection operators. It is at this component level that the link between algorithms and architectures can be made most precise. Thus the degree and nature of any parallelism inherent in an algorithmic procedure can be determined, and the optimum means of implementation specified. Of course this optimum may be different for different applications, depending, for example, on whether cost or real-time implementation are the most important factors.

A second class of technique that we consider is generic in nature. In other words we are dealing not with a specific building block, but with a general method. Such methods may be suitable not only for image processing, but for other areas as well. Examples of generic techniques that are particularly relevant for image processing are dynamic programming, spectral analysis and the use of variational principles. In many cases a generic technique has associated with it a type of architecture particularly well matched to its implementation. Thus again we have examples of a mapping between algorithm and architecture.

The third class of technique we consider is at the system or application level. In fact we shall be more general in that we discuss the systems and applications *per se.* In terms of algorithms, the critical, and often very difficult, factor is the choice and configuration of the basic building blocks in order to provide a paradigm for solving the system application. Success requires a thorough understanding of the performance and efficiency of the component algorithms, and the choice cannot be divorced from consideration of architectures and implementations. One may envisage automatic configuration procedures, driven either by direct measures of image properties, for example the density of edges or characteristic correlation lengths, or by a model of the application and its constituent parts. The choice of architecture and implementation is itself of fundamental importance, as it affects the efficiency and cost of the resulting system.

In the next section we discuss a number of different aspects in general terms. This discussion is intended to be controversial and subjective, in that it represents the authors' own views. Section 2.3 treats algorithms and architectures from the point of view of basic building blocks. The algorithms themselves are relatively well known, so the discussion will concentrate on connections between algorithms, and on mappings between algorithm and architecture. Section 2.4 treats a number of generic techniques. Here the discussion of the algorithms themselves will be more detailed, and some recent research will be described. The use of variational principles as a potential unifying theme for much of early visual processing will be discussed, and architectural implications will be explored. Section 2.5 contains a treatment of systems and applications. Here we choose a relatively small number of specific applications and examine them in detail with a view to illustrating how algorithms are chosen and configured, and how this is affected by system constraints. Finally, some concluding remarks are contained in Section 2.6.

2.2 IMAGE PROCESSING ALGORITHMS

This section contains a discussion of various aspects of image processing algorithms in general terms, before the more detailed treatments in subsequent sections. A unifying theme of this discussion is the concept of hierarchies of processing levels, including the *processing cone.*

At the lowest processing level, which includes many conventional image processing algorithms, the image is treated essentially as a set of samples or picture elements (pixels), without reference to the structures or objects contained within the image. Thus the operations act on the image on a pixel-by-pixel basis, as though it is a sampled waveform, with the samples organised in a two-dimensional grid. Examples of such operations are simple filters, smoothing operators, and coding operators such as differential pulse code modulation (DPCM) or discrete cosine transform codes. The image is treated as raw data rather than as a representation of some scene containing objects. While such techniques are very useful, either directly, for example for enhancement or coding of television or medical images, or as a preprocessing stage for more complex image processing tasks, they are limited due to the unrealistic manner in which they treat the image. No account is taken of structural information, although this is directly related to the structure and disposition of objects in the imaged scene. Moreover, the criteria used for optimisation or assessment are frequently inappropriate: often some form of minimum mean-square-error distortion criterion is used, as opposed to a measure related to the information content of the image or to a model of the human receiver characteristics.

For more sophisticated tasks and applications it is necessary to go beyond treating the image merely as sampled data. The next level of processing consists of attempting to extract and manipulate the information which is contained within the image. This information may be described in terms of structures such as edges, boundaries, shapes or textures which form the constituent components of objects. Alternatively it may be described in terms of elements obtained by a statistical decomposition of the image; this decomposition can be carried out with the aid of a decision function incorporating a model of relevant image properties.

The difference between processing at the data and at the information levels may be illustrated by means of two examples. Consider first the problem of extracting edges in an image. At the data level it is relatively easy to locate pixels which are candidates for being edge points, for example by using the 3×3 Sobel or Laplacian operators and then applying a suitable threshold. However, it is much harder to form edge segments and bounding contours. If the threshold is set low, then many spurious candidate edge points are produced due to noise and textural fluctuations, whereas if it is set high, then insufficient candidates are produced to provide a contiguous set of points that may be identified as an edge segment. In general, in order to proceed from a set of candidate edge points to a set of edge segments or contours, a very complex procedure involving sorting, interpolating and perhaps adaptive thresholding is required. Often these procedures are empirical and their performance unsatisfactory. The problem is that one is trying to extract information from the image using mainly data level processing. In contrast there exist a number of algorithms and procedures which seek to extract edge elements and/or boundary contours directly. An example, which will be discussed later, is provided by the Marr–Hildreth operator. Of course, these techniques also have their problems: for example it is difficult to define an edge between some types of texture regions.

Our second example is the coding of images for data rate reduction. Conventional coding techniques operate strictly at the data level. Thus DPCM predicts the value of a pixel from the values of its neighbours and codes the error in this prediction, while block transform codes apply a given transform, typically a discrete cosine transform, to blocks of pixels and then code the transform coefficients. The codes are related to the pixel values and not to the information in the image. Often the 'codebooks' are designed to minimise the mean square distortion. These, and other related coding techniques, give good results for those situations where the quality of the coded images is specified in terms of sampled waveform features, for example spatial frequency constraints and mean-square error. Such a situation is the coding of broadcast television images. Typically, compression ratios of 3–10 can be achieved with high quality coded images. The exact value of the achievable compression ratio depends on both the nature of the image and the specification of coded image quality. These coding techniques are, however, inefficient for situations where it is the information within the image, rather than the image itself, which is important. In these cases much greater compression ratios can, at least in theory, be obtained by adopting an analysis/synthesis approach, whereby the information is extracted and described, and the coded image obtained by synthesis from this information. Of course, in terms of individual pixel values the discrepancies between original and coded images can be large. What is important is that, in some appropriate sense, the coded image should closely resemble the original. The improved compression ratio is achieved by matching the coding technique to the information rather than to the data provided by the image. We may note the close parallel with developments in the coding of speech signals, with waveform coding methods such as pulse code modulation (PCM) or DPCM corresponding to the data level and analysis/synthesis methods such as channel vocoding and linear predictive coding corresponding to the information level. As an example of information level coding, we may imagine first segmenting the image into its constituent regions, and then applying shape analysis to the region boundaries and texture analysis to the interiors of the regions.

For yet more complex tasks and applications, where some degree of interpretation and understanding is required, higher level operations that include and utilise knowledge about the domain will need to be used. This will involve the use of intelligent knowledge based systems (IKBS) techniques. Typical operations involve object recognition, scene description or 3-D modelling. Thus, for example, given a model of the objects and the environment and a high level description of a task, an ideal rule based system should be able to configure algorithms into a working procedure completely automatically. The architectural implications of this highest or knowledge level of processing are far from clear. In this chapter we shall not be concerned explicitly with knowledge level processing. However, even for those tasks that do need to include a significant amount of knowledge processing, the lower levels of processing are still essential. This is because the knowledge based system needs a set of reliable features with which to work, and any information that is lost in the data and/or information processing stages cannot subsequently be recovered. Of course, for some tasks knowledge based techniques may be used to drive the lower levels of processing.

The above discussion has been in terms of a processing hierarchy consisting of three levels: data, information and knowledge. As such it provides a crude classification scheme for algorithms and techniques. Although very useful in many cases, it should not be taken too rigidly. Thus for certain algorithms it may be difficult to specify that they operate at one particular level. Indeed, some techniques involve elements from all levels, for example some approaches to knowledge driven feature extraction. Also, for image processing systems, the interactions and feedback between different processing levels are often of fundamental importance. The value of the classification lies, first, in that a large number of algorithms can be classified accurately, secondly in the close connection between this hierarchy and the concept of the processing cone, which has significant implications for architectures, and thirdly in the relationship between this hierarchy and others that may be defined in system terms. (The 'processing cone' refers to the often used analogy between a cone and the rapidly increasing computational requirements inherent in processing images at the knowledge, information and data levels of representation.)

Algorithms that operate at the data level act, in general, on a large data set (say a 256×256 pixel array) but with a high degree of parallelism since, except possibly near the image boundaries, the action of the algorithm should be independent of pixel position. In addition, for many applications processing speed is of critical importance. For these types of algorithm conventional von Neumann architectures, which execute a series of instructions sequentially on the data, are not well matched. Rather, one requires architectures with a high degree of parallelism but with relatively simple individual processing elements. Ideally this parallelism should be two-dimensional, so as to map onto the image data itself. The structures may be programmable, as in single instruction multiple data (SIMD) machines, or dedicated modules such as systolic arrays or special purpose discrete Fourier transform (DFT) circuits. The inherent parallelism within the algorithms may be local in nature, i.e. expressible in terms of near neighbour operations, or more global. For these cases different types of parallel architecture may be suitable. The different architecture possibilities are discussed more fully in subsequent chapters.

At the information level the degree of parallelism is considerably reduced. However, even here one frequently has some parallelism, although not matched directly to the two-dimensional array of data which constitutes the image. Thus for example the analysis and description of shape from edge segments or the use of texture measures to describe region interiors can involve substantial parallelism. At this level, use of pattern matching techniques such as dynamic programming or syntactic parsing may be appropriate, and these can have a parallel implementation. The data set at the information level tends to be a set of features rather than pixels. At the highest levels of processing the emphasis is much more on control and search strategies, and less on raw parallelism. Also the data set should be still further reduced, assuming successful use of processing at the data and information levels.

In looking at the implementation of a general image processing system, one is thus led to the concept of the processing cone. At the lowest levels of this cone are highly parallel processes, relatively inflexible in nature, which operate on large data sets, namely the pixels themselves. Higher up the parallelism and data quantity are reduced, and instead flexibility and control strategies become important.

This concept of a processing hierarchy can also be introduced from the system point of view. In order to illustrate this, we consider a general purpose image understanding system, which must be capable of constructing a description of objects in its environment, and their relations to each other. The input to the system consists of raw data from some imaging device. To pass from this lowest level representation to its final output, the system must construct a hierarchy of intermediate representations at different levels of abstraction. This hierarchy is illustrated in Fig. 2.1.

It may be necessary first to apply a process of restoration and/or enhancement in order to compensate for image degradation caused by noise and by blurring resulting from the image formation process. For many applications this may be the only processing employed. For our image understanding system the next level of representations which may be needed include edge maps and various 'intrinsic images' formed by mapping the value of some physically significant parameter such as surface orientation, range or velocity over the image array. These intrinsic images are an important stage in the process of scene understanding, since they display structural information about the scene which is only contained implicitly in the original data. It is then possible to create representations at the next level of abstraction, namely segmentations of the image into labelled parts likely to correspond to significant entities in the scene. Processing at these levels of abstraction corresponds closely to the information level introduced previously. Indeed much of the processing will consist of feature extraction and manipulation. Some of the intrinsic and segmented images could be used as the basis of an analysis/synthesis coding scheme.

At a still higher level, representations may be constructed which provide quantitative information about two- and three-dimensional shape. For example, three-dimensional objects may be described in terms of generalised cylinders or, using techniques of constructive solid geometry, as combinations of primitive solids. While representations lower down the hierarchy are viewer centred, those at this level and above tend to be based on object centred models. At the top of the hierarchy of representations, knowledge about the scene will be stored in the form of symbolic relational descriptions of image and object structures. Processing at these abstraction levels will involve knowledge based system techniques.

The algorithms by means of which one representation is transformed into another will, in general, vary considerably in structure according to the level

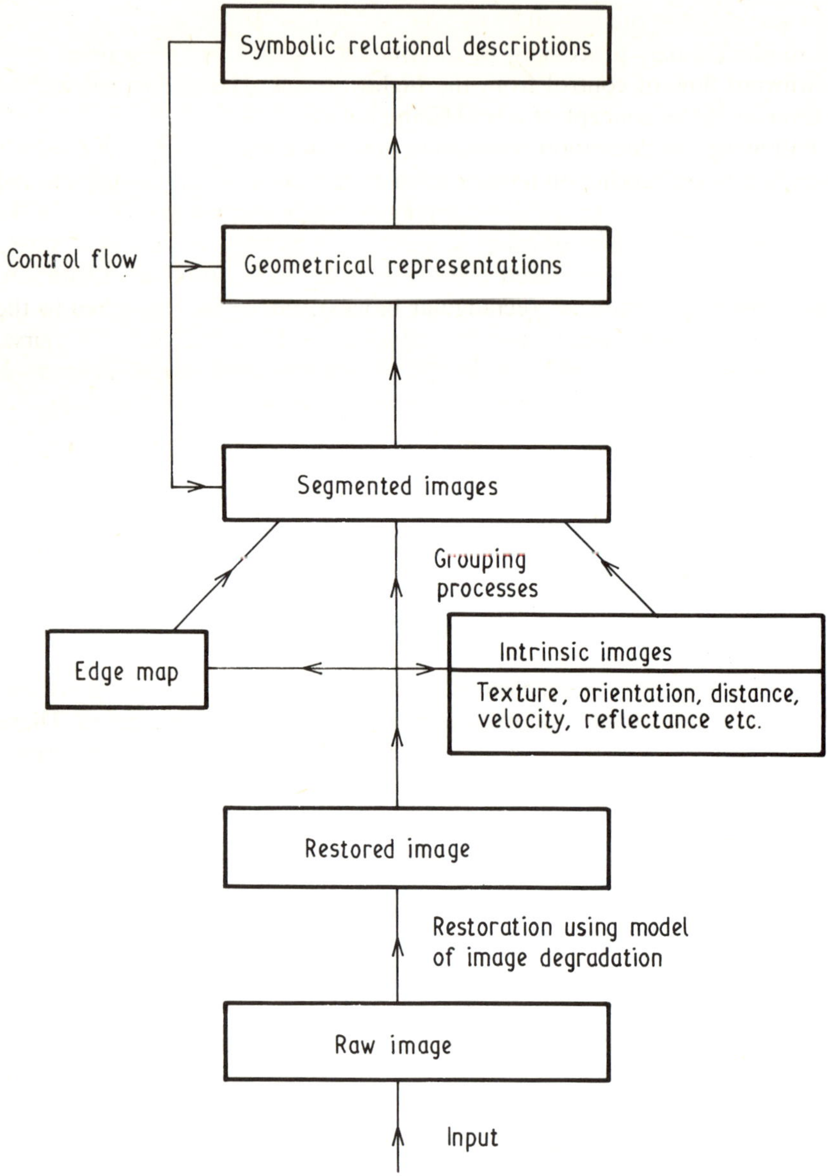

Fig. 2.1 Schematic illustration of the hierarchy of possible processing levels in a general purpose image understanding system.

at which they operate. Transformations between low level representations are often domain independent processes and can usually be implemented as parallel local computations. Those operating at higher levels are likely to require *a priori* knowledge about the 'world model', and to involve a greater degree of serial processing. Information will generally be transferred upwards through the hierarchy of representations, but this passage will not be unidirectional; processing at intermediate stages will be guided by a downward flow of control from the highest levels. Here again we see the relevance of the concept of a processing cone.

Following this discussion concerning processing hierarchies, it is possible, even before embarking on a more detailed treatment of the algorithms and techniques, to make a number of comments and observations. First, we see the inherent inefficiency of using 'waveform processing' techniques to derive codes. Thus, block structured transform codes have an inbuilt limitation. The geometry is artificial (rectangular blocks), rather than matched to the structures within the image, and the same is true of the transform. Of course, it is theoretically possible to define a non-block-structured generalised transform that is related to the information content of the image. Indeed, any coding process may be defined as a transformation. Equally, approaches to image segmentation that are based on analysing and identifying similarities between neighbouring rectangular blocks of picture elements have the same inbuilt inefficiency due to a mismatch with the information content.

Another observation concerns the need for research into new architectures to proceed in conjunction with the development of new algorithms, since each affects the other. This is particularly important for the lower levels of processing, where the parallelism is greatest. Possible VLSI implementation, and the effect this has on algorithm structure, also need to be studied. These matters will be taken up in subsequent chapters, where the interplay between VLSI implementation and algorithmic techniques will be examined in much more detail.

A further observation concerns the prevalence of hierarchies in other types of signal processing activity. To illustrate this, we shall consider a typical speech recognition system. Most current systems take the incoming sampled speech waveform, divide it into a number of frames (of duration 10–20 ms) and extract a number of features for each frame, these being typically related to the spectrum and to the fundamental frequency. For any given test utterance the set of features thus generated is then compared with each of a set of reference templates corresponding to the words in the system vocabulary, using a combination of a distance measure and a pattern matching algorithm. The closest match is deemed to be the recognised word. Most systems can deal with connected sets of words, and rudimentary syntax may be included. Obviously, such a system only makes substantive use of the lower levels of the processing hierarchy, and its performance is accordingly limited, typically to speaker dependence and to vocabularies of the order of 100–200 words. Much work is now being undertaken to learn how to incorporate into such systems higher level knowledge and techniques, based for example on phonetics, linguistics and semantics, thereby obtaining dramatically improved recognition performance. The parallel with image processing is clear.

Of perhaps much greater significance is the parallel that may be drawn with the human visual system. Thus the same sort of feature extraction processes that we have referred to as part of the low and intermediate levels of the processing hierarchy, for example the detection of edge contrast and orientation, are thought to be central to the early stages of human visual processing. Indeed, the influential work of Marr and Hildreth (1980) on edge detection was strongly motivated by studies of biological visual processing, and formed the basis for a model of physiological mechanisms. New insights into the nature of biological visual systems can provide extremely valuable guidance for the development of techniques for image processing and computer vision. However, the analogy should not be taken too far, since machines should complement, rather than substitute for, human capabilities.

2.3 ALGORITHMS AND ARCHITECTURES AS TOOLS

In this section we shall consider a number of fundamental techniques which are in common use in many diverse application areas. The techniques are generally well known, and the discussion presented here is not intended to be either exhaustive or in any way tutorial. The aim is rather to present these basic image processing tools in the context of the hierarchy of processing levels introduced in Section 2.2, and to highlight cases in which algorithms map naturally onto parallel processing architectures.

Algorithms at the lowest levels of the processing hierarchy operate on the image pixels themselves, transforming the original intensity array into another array of grey levels. Often the aim is to enhance or restore a degraded image, either for cosmetic purposes or as a preliminary to feature extraction and recognition procedures. At this stage the processing is data driven, involving no higher level knowledge, and the same operation is performed at each pixel. In Section 2.3.1 we shall outline a number of common early processing techniques which can be broadly classed under the heading of filtering. They are all local, and amenable to parallel implementation on SIMD architectures.

At the next stage of the hierarchy come algorithms which attempt to extract information from the image. Identification of object boundaries is a crucial stage in the process of passing from the raw data to a final interpretation of the scene, and accordingly a vast degree of effort has been devoted to extracting edges from the image, and linking them to form boundaries. Local edge operators and linking procedures are discussed in Sections 2.3.2 and 2.3.3.

Edge linking is often the first stage in the process of segmenting and labelling the image, as discussed in Section 2.2. An extremely versatile technique, applicable not only to boundary extraction but also to a number of other grouping and parameter estimation problems, is the Hough transform, described in Section 2.3.4.

In Section 2.3.5 we outline an alternative approach to edge detection, developed by Marr and Hildreth (1980) and motivated in part by the findings of psychophysical investigations into human early visual processing. This leads naturally into a discussion in Section 2.3.6 of the continuing quest for optimal edge detectors, and of the implications of this work for the design of fast edge extraction hardware.

At higher levels of the processing hierarchy algorithms become more complex and less general purpose, with a greatly reduced degree of parallelism.

2.3.1 Filtering techniques

The purpose of most filtering operations is to transform the image intensities in such a way as to enhance or suppress certain features. For example, image smoothing procedures are widely applied with the aim of removing high spatial frequencies which are dominated by the effects of noise. Alternatively, it may be required to enhance high frequencies in an image in order to sharpen its edges. There is an extensive literature on the design and implementation of digital filters (see Gonzalez and Wintz, 1977; Andrews and Hunt, 1977; Pratt, 1978), and this topic will not be dealt with here. In Section 2.5.1 we shall merely illustrate the approach by describing two linear restoring filters which may be implemented efficiently in the Fourier domain.

In this section we shall discuss a range of simple filtering operations which are particularly suited to implementation on SIMD architectures of the type described in Section 3.4. These are defined by so-called local neighbourhood operators, which act by transforming each pixel grey level in a manner determined solely by the current values of that pixel and its immediate neighbours. In a typical linear procedure, the new pixel value at a given location in the image array is a weighted sum of the pixel values in a symmetric neighbourhood centred on that location. Let the digitised image be represented as an array $\{X(i,j),\ 1 \leqslant i,\ j \leqslant N\}$. Then such a procedure results in the formation of a new image array X^* where

$$X^*(i,j) = \sum_{k=-n}^{n} \sum_{l=-n}^{n} X(i+k,j+l)w(k,l) \tag{2.1}$$

and $\{w(k,l),\ -n \leqslant k,\ l \leqslant n\}$ is a $(2n+1)\times(2n+1)$ array of weights.

For example, a simple smoothing operation can be carried out by using the 3×3 array of weights:

$$\{w(k,l)\} = \begin{matrix} 1 & 1 & 1 \\ 1 & w & 1 \\ 1 & 1 & 1 \end{matrix}$$

where increasing the parameter w has the effect of reducing the degree of blurring produced.

The enhancement of edges in an image can be achieved by applying an operator which simulates the spatial differentiation of the grey levels. The 3×3 array of weights:

$$\{w(k,l)\} = \begin{matrix} 0 & 1 & 0 \\ 1 & -4 & 1 \\ 0 & 1 & 0 \end{matrix}$$

is a commonly used approximation to the Laplacian, which has the effect of outlining edges in an image.

Increasing the window size results in a greater degree of smoothing, at the cost of added computational complexity. On a SIMD parallel array processor whose processing elements are in one-to-one correspondence with the image pixels, correlations of the form of equation 2.1 can be computed by a sequence of point-wise multiplications, shifts and summations. For example, in the case of a 3×3 mask ($n=1$) the entire image is multiplied by $w(-1,-1)$, shifted diagonally south-east by one pixel, and added to $w(0,0)$ times the original image. Shifted and weighted contributions from the other seven directions are then added in the same way to obtain the final result.

In addition to the above linear techniques, a number of non-linear local neighbourhood operators have also been proposed. Here the new pixel value is again computed from grey levels of pixels in a surrounding window, but the computation involves some non-linear procedure(s). For example the median filter (Pratt, 1978) replaces each pixel value by the median of the pixels within the window. Special purpose systolic architectures have been proposed for implementing this filter (see Section 3.5.1), which has the effect of reducing noise without the loss of edge information.

A class of grey level transformations which can have a considerable effect on the appearance of an image are those which act by modifying the grey level histogram. Of these, the best known is probably histogram equalisation (see Ballard and Brown, 1982; Gonzalez and Wintz, 1977) which produces an image whose grey levels have a uniform density. Efficient histogramming techniques require the use of special hardware counters.

Template matching is a simple filtering technique aimed at detecting the occurrence in an image of a particular feature or local pattern of grey levels whose appearance can be precisely specified. A correlation is performed between the image and a template, or window operator, which contains the pattern to be recognised. Intensity peaks in the resulting transformed array correspond to positions of locally optimum match between template and image, and the presence of the required features is thus highlighted. On a SIMD parallel processor array the correlation may be performed by repeated multiplication, shifting and summation in the manner described previously. In the case of large templates it may be cheaper to perform the correlation in the Fourier domain (see Section 2.4.6), noting that correlation with the window $w(k,l)$ is equivalent to convolution with $w'(k,l)=w(-k,-l)$.

2.3.2 Extraction of local edges

In many computer vision processes, the first level of the processing hierarchy at which information is extracted from the image is that at which edges are detected. The position and shape of object boundaries provide important guiding information for many high level recognition and interpretation tasks, and vision modules such as binocular stereo, optic flow, and shape-from-texture may require an initial edge detection stage. The importance of edge detection will be further discussed and illustrated in Chapter 5. Here we shall briefly summarise some of the techniques which have been developed for the extraction of local edges. An extensive literature exists on this topic, and detailed treatments of the wide range of existing approaches can be found in, for example, Pratt (1978), Rosenfeld and Kak (1982), Andrews and Hunt (1977) and references therein.

The passage from an image intensity array to an object boundary map is usually effected via an intermediate representation in which local edges, or discontinuities in grey level, are made explicit. Since these local edges are functions of the grey levels in only a small surrounding neighbourhood, they may be extracted by local neighbourhood operators of small spatial extent. Local edge operators may be broadly classified (see Ballard and Brown, 1982) according to whether they match the image to a range of edge templates at different orientations, fit some form of parametric edge model to intensities in a local neighbourhood, or carry out a digital approximation to a gradient operator. In all cases the output of the operator is an orientation and a magnitude associated with each putative edge point.

Gradient operators such as those of Roberts and Sobel measure intensity differences along two orthogonal directions, and combine these measurements to form an estimate of edge strength and orientation. In Sobel edge detection, for example, the image is filtered with the 3×3 arrays of weights

$$\begin{matrix} -1 & 0 & 1 \\ -2 & 0 & 2 \\ -1 & 0 & 1 \end{matrix} \quad \text{and} \quad \begin{matrix} 1 & 2 & 1 \\ 0 & 0 & 0 \\ -1 & -2 & -1 \end{matrix}$$

If the resulting responses at a pixel are X and Y respectively, the local edge strength is taken to be $G = (X^2 + Y^2)^{1/2}$, or, if the taking of square roots results in undesirable computational overheads, $G = |X| + |Y|$. An estimate of edge orientation can be computed as arctan (Y/X).

In the template matching approach to edge detection the image is convolved with a set of masks, corresponding to different possible edge orientations. The orientation corresponding to the mask that gives the best match at a point is taken to be the edge orientation at that point, and the magnitude of this match provides a measure of the edge strength. For example, the Kirsch operator uses eight masks corresponding to the eight principal directions of the compass.

The third class of techniques referred to above involves the fitting of an ideal edge model to the image data. This approach was first used by

Hueckel, who used a set of eight basis functions defined on a disk to describe an ideal step edge, expanded the image in terms of the same set, and found the coefficients of the best fitting edge by a minimum mean-square error procedure. In Hueckel's algorithm an edge is deemed to exist only if this computed best edge is a sufficiently good approximation to the actual image data, and if, in addition, the computed edge contrast is greater than some pre-set threshold.

Since both template matching and gradient edge operators involve convolution of the image with a small array of weights, these operators may be efficiently performed on a SIMD processor array as described in Section 2.3.1. The Hueckel algorithm is considerably more complex, and does not map in a straightforward way onto a simple parallel architecture.

2.3.3 Edge linking procedures

The output of any one of the operators discussed in Section 2.3.2 is an array of local edge elements, each in general labelled with an orientation and a magnitude or contrast. These local edges indicate positions where the image intensity is discontinuous, or at any rate undergoes rapid change. Some of them will be due to noise or quantisation effects; for example the application of a local edge operator to a broad intensity gradient results in a number of parallel edge elements being detected, rather than the required single edge. Moreover, some edge elements may reflect fine textural detail contained within regions of the image. In any event, not all of them will lie along the boundaries of significant regions, and a subsequent processing stage in which certain edge elements are linked together to form continuous boundaries one pixel wide, and spurious edge features are removed to produce a clean, segmented image, will usually be required.

Before a linking operation of this type is attempted, it may be helpful to perform a co-operative adjustment of the strengths of local edge elements, taking into account the context which is provided by neighbouring edges. The method of relaxation labelling, discussed in more detail in Section 2.4.2, has been used with some success to this end (see Prager, 1980; Ballard and Brown, 1982). An initial confidence value is computed for each local edge element, based on its normalised contrast. This value is regarded as representing the probability that an edge actually exists at that position. The values are then updated in a parallel, iterative procedure whereby the confidence of a given edge element is increased if its existence is supported by appropriate configurations of nearby edges, and reduced if its neighbours provide conflicting evidence. Details can be found in the above references, and in others cited there.

A feature of relaxation labelling processes is their suitability for implementation on parallel hardware (see Section 2.4.2). By contrast, most approaches developed for edge linking and contour extraction suffer from a high computational complexity. They are often based on combinatorial search algorithms operating over a huge search space, namely the set of all image pixels. As an example, the boundary extraction problem can be formulated as one of searching for minimal cost paths through a graph whose nodes consist of the local edge elements (see Ballard and Brown, 1982, and references cited therein). An evaluation function related to, for example, the overall edge strength or curvature of the boundary is used to guide the search.

A number of problems are associated with such search procedures, one of which is the difficulty of ensuring that the path begins at a node which corresponds to a genuine boundary point of the image. An alternative approach, using the technique of dynamic programming (see Section 2.4.5), can be used to construct paths from multiple starting points, evaluating candidate boundary portions according to an appropriately defined merit function. Detailed accounts of dynamic programming edge linking strategies, and further references to the published literature, can be found in Ballard and Brown (1982). Elliott and Srinivasan (1981) use dynamic programming to estimate short boundary segments, and subsequently apply a graph searching procedure to link these segments together into complete boundaries.

In an attempt to develop a highly parallel implementation of an edge linking strategy based on a graph labelling model, Diamond *et al.* (1983) have reformulated the problem in an integer programming framework, and proposed that the search be carried out in an associated Lagrange dual space. The convex dual function can be optimised by a descent algorithm which is guaranteed to converge, and which can be partially decomposed to facilitate parallel solution.

For applications where domain-specific knowledge is available, the complexity of edge linking procedures may, of course, be greatly reduced. Information about the approximate shape and location of boundaries can be used to prune the search space, and will dictate the optimum choice of technique. In cases where the boundary to be extracted can be described in terms of either a parametric curve or a tabulated shape function, a procedure known as the Hough transform may lead to a practical and robust solution. The Hough transform, which has wide applicability in image processing and computer vision, and which is amenable to parallel implementation, will be discussed in Section 2.3.4.

2.3.4 The Hough transform

The Hough transform was originally developed for the detection of straight lines in noisy images. The technique has since been generalised, however, and now encompasses a range of grouping procedures and parameter estimation strategies which may be applied at various levels in the processing hierarchy.

The basic principle underlying the approach is that instances of some phenomenon in the input data are transformed into a parameter space in which features of interest will form clusters. An accumulator array is set up, with bins corresponding to the different possible combinations of discretised parameter values. For each instance of the phenomenon, a 'vote' is placed in every bin whose corresponding parameter set could have given rise to that instance. The mode of the accumulator histogram then represents the set of parameter values for which there is most evidence in the input data.

Since only the mode of the distribution is used, the procedure works well in the presence of noise, or of partial and conflicting evidence. Its use will be illustrated here in connection with the problem of detecting straight line image segments when the input data consists of a set of candidate edge points produced by some local edge detector. As discussed in Section 2.3.3, the linking of local edge features to form boundaries is a frequently needed stage of early image processing.

Any line through the point (x_1,y_1) will satisfy the equation

$$\rho = x_1 \cos\theta + y_1 \sin\theta \qquad (2.2)$$

for some unique ρ, θ with $0 \leqslant \theta \leqslant \pi$.

This defines a sinusoidal curve in the (ρ, θ) parameter space (see Fig. 2.2). A second point (x_2,y_2) will have a second associated curve in parameter space, and the intersection of these curves will define the parameters (ρ, θ) of the line joining the two points.

In practice it will be possible to estimate lower and upper bounds for values of ρ and θ that may arise. The permissible parameter ranges are divided into discrete intervals

$$\rho_0 < \rho_1 < \ldots < \rho_n,\ \theta_0 < \theta_1 < \theta_2 < \ldots < \theta_m$$

and a corresponding accumulator array $A(\rho_i, \theta_j)$ is set up and initialised to zero. Each edge feature at (x,y) in the image 'votes' for all the lines that could have caused it, by incrementing all the bins of A whose indices (ρ_i, θ_j) satisfy Equation 2.2 within the limits of the discretisation. Often, the local edge operator which produced the feature will also have provided an estimate of its orientation; such information can be used to restrict the range of values which need be considered.

At the end of the voting process, each local peak in the accumulator array will correspond to a group of colinear edge elements in the image. The parameters of the line on which the elements lie will be given by the indices of the peak, and an estimate of the number of points on the line by the sum of accumulator values at, or very close to, the local maximum.

The method can be generalised to cover the detection of any curve having a description of the form

$$f(x,y,a_1,a_2,\ldots a_n) = 0$$

in terms of certain parameters a_i. It is impractical unless n is small, however, since the storage requirements increase exponentially with the number of parameters. Since the computation involved grows linearly with the number

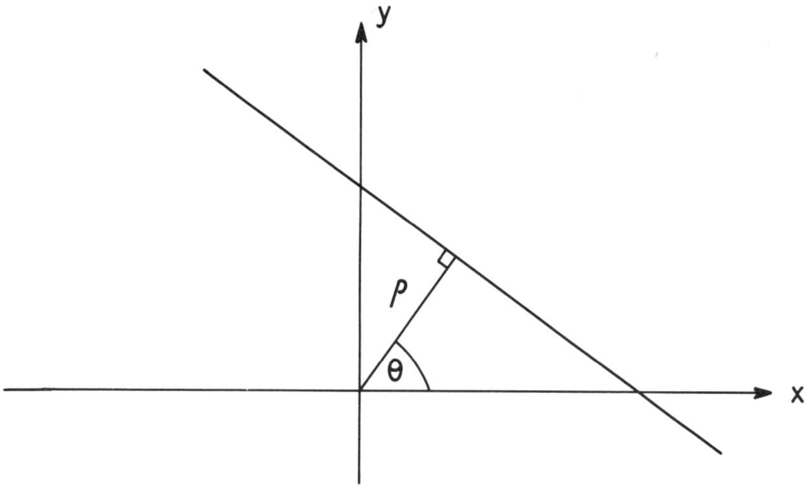

$$\rho = x \cos \theta + y \sin \theta$$

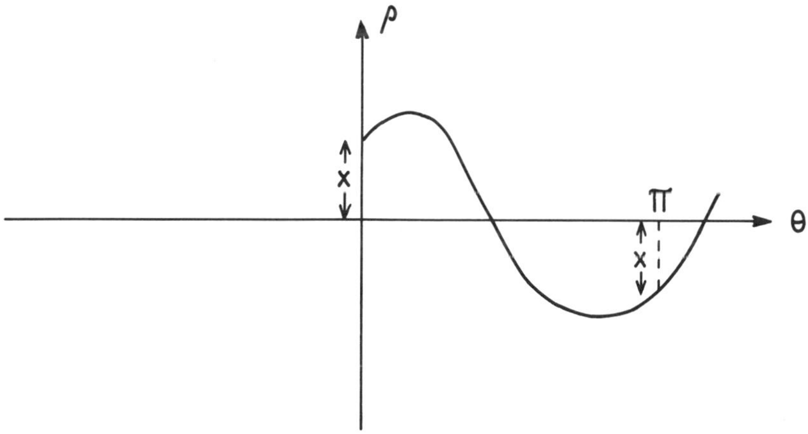

Fig. 2.2 (a) The pair (ρ, θ) defines a straight line in $x - y$ space by Equation 2.2. (b) The pair (x, y) defines a sinusoidal curve in ρ θ space by Equation 2.2.

of edge elements (Duda and Hart, 1972), in cases where this number is large the technique is clearly preferable to an exhaustive search of all possible lines joining pairs of edge elements.

A generalisation of the Hough transform (Ballard and Brown, 1982) can be used to detect objects whose shape function, while having no simple analytic description, can be represented in tabular form. The approach has also found wider application in areas such as the detection of motion parameters (Thompson and Barnard, 1981), and the computation of surface orientation from shading.

A useful feature of the Hough transform is its amenability to parallel implementation. Shapiro (1978) proposes an array of local processors, each associated with a small region of the image and performing the appropriate parameter space mapping on significant features within that region. More ambitiously, it is possible to conceive of a massively parallel computing network in which all the voting processes demanded by the Hough transform are prewired and can proceed simultaneously. The current interest in neural nets, which contain large numbers of simple processing units interconnected in a highly structured manner, has led to the suggestion (e.g. Brown, 1984) that the Hough transform might be implemented using nets of this type. Such an implementation would present a highly structured and complex wiring problem, although the signals passing through the network would be simple excitations between edge features and those units in parameter space with which they were most consistent.

Architectures of this nature are at present the subject of active research (see also Section 2.4.1).

2.3.5 The Marr–Hildreth theory of edge detection

The approach to edge detection proposed by Marr and Hildreth differs in several important ways from the local neighbourhood operators discussed in the preceding section. It makes use of significantly larger convolution masks, and its output consists of a set of connected edge contours, thus rendering unnecessary the subsequent application of edge thinning and linking procedures.

Full details of the underlying theory of the approach, together with discoveries about human early visual processing mechanisms which helped to motivate it, may be found in Marr and Hildreth (1980). The basic idea is that an edge, or sharp intensity change, gives rise to a zero-crossing in a second directional derivative of the image. Moreover it can be shown that, under fairly weak conditions, the computationally expensive procedure of applying a number of differently oriented directional derivatives can be avoided by the use of a single non-directional operator, namely the Laplacian ∇^2. Since differential operators are extremely sensitive to noise, it will be necessary to begin by smoothing the image. Marr and Hildreth propose the use of a Gaussian smoothing filter, since the Gaussian is the unique function having minimal bandwidth $\times$ frequency product. They further point out that intensity changes in natural images occur at many different scales, and that an optimal edge detector should take this into account by looking for edges separately over a range of scales.

The smoothing and Laplacian operations can be combined to form a single convolution filter

$$-\nabla^2 G(r) = \frac{1}{\pi\sigma^2}\left(1 - \frac{r^2}{2\sigma^2}\right)\exp(-r^2/2\sigma^2)$$

where r is radial distance from the origin and σ is the standard deviation of the Gaussian smoothing function. This is a circularly symmetric, 'Mexican hat' operator, which is approximately bandpass. It is shown in cross section in Fig. 2.3, together with its Fourier transform. The zero-crossings of the convolved image are located, and are regarded as describing the edge structure of the original image at the particular scale corresponding to the choice of Gaussian smoothing function.

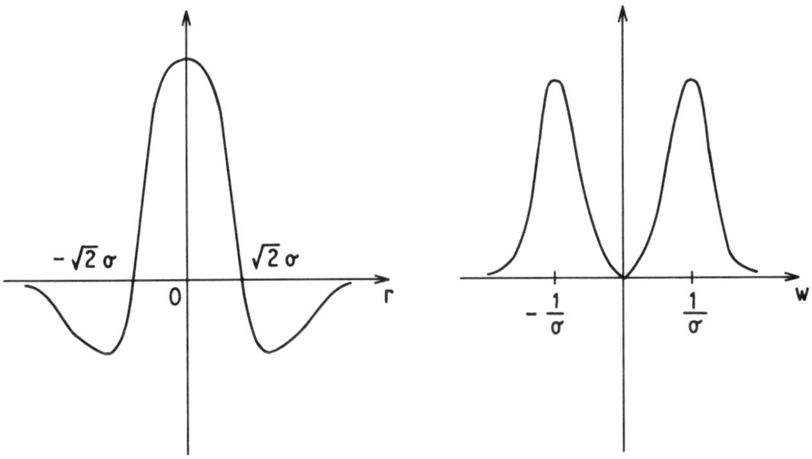

Fig. 2.3 Cross sections through (a) the Marr–Hildreth operator $-\nabla^2 G$ and (b) its Fourier transform.

The Marr–Hildreth operator has a number of attractive properties. As mentioned above it yields connected edges without recourse to edge thinning or linking procedures. Unlike some other edge operators it does not produce spurious edge features when applied to regions of uniformly changing intensity. Being isotropic it responds equally well to edges in all orientations, and moreover the location of isolated zero-crossings can easily be estimated with sub-pixel accuracy by means of, for example, least-squares fitting techniques. The zero-crossing slope, a useful parameter which provides a measure of edge contrast, can similarly be estimated.

The Marr–Hildreth operator will be further discussed in Section 5.2.1.

2.3.6 Towards an optimal edge detection operator

For many processes of computer vision, a preliminary edge extraction stage providing accurate localisation of edge elements is an essential prerequisite. No edge-based stereo algorithm, for example, will yield precise depth information if its input edge data has significant positional errors. A great deal of time and effort has therefore been spent in trying to find good edge detection operators.

Sadly, the above requirement for accurate spatial location of edges is in direct conflict with the other fundamental criterion governing the selection of an edge detector, namely that it should have a good signal-to-noise ratio: all real edges should have a high probability of being detected, while very few non-edge points should be marked as edges. A trade-off, or uncertainty principle, exists between the two criteria.

It has been shown mathematically (Buxton, 1984), that the optimal operator for detecting ideal step edges as zero-crossings is indeed the Marr–Hildreth operator described in the previous section. There are drawbacks to the use of this operator, however: for example a high density of zero-crossing contours is produced in response to an image consisting purely of noise, and in practice some form of thresholding, based on the slope of the zero-crossings, must be employed. Recently Canny (1983) has shown that better performance can be obtained by an alternative approach, in which edges are detected as maxima in the output of an appropriate operator.

Modelling a typical edge as a step intensity change in Gaussian noise, Canny derives an expression consisting of a product of localisation and signal-to-noise terms. He shows that the value of this product does not depend on the amplitude and scale of the applied filter function, and is maximised by a function of unique shape. This function, whose analytic form is derived, can be approximated by the first derivative of a Gaussian.

The optimal two-dimensional mask for an edge of given orientation will thus consist of the first derivative of a one-dimensional Gaussian function in a direction perpendicular to the edge, convolved with a projection function parallel to the edge. A simple procedure for implementing this consists of convolving the image with a symmetric Gaussian, using the gradient of the resulting smoothed intensity image to estimate the edge direction, and non-maximum suppressing the gradient magnitude in this direction.

An adaptive thresholding procedure is applied to the operator output, using threshold values based on local estimates of the image noise. In many images edges will occur over a range of scales, and a coherent way of combining the outputs of operators of different widths is thus required. Canny's approach is to detect the edge at the smallest scale for which the corresponding operator output has an acceptable signal-to-noise ratio. He then suppresses the outputs of larger operators unless they are at a significantly greater level than would be expected to result from the edges already detected in the smaller channels.

The theory and implementation of optimal edge detectors is still an active research topic. Clearly, if the real-time processing rates essential for this vital stage of low level vision are to be realised, algorithms must be implemented in fast hardware. Both the Marr–Hildreth and the Canny operators involve large numbers of convolutions, often with masks of considerable size (typical radii of masks used by the Marr–Hildreth operators are in the range 8–32 pixels). Fortunately, these masks can be expressed as a sum of separable terms and their effect computed efficiently using one-dimensional convolutions. Existing VLSI designs for fast convolvers can be exploited (see Section 3.5.3). An alternative approach, potentially leading to ultra-high speed implementations (Canny, 1983), is to approximate the one-dimensional Gaussian with a pair of recursive filters in an iterative procedure which makes use of the central limit theorem. The complexity of this approach is very low, and independent of the filter size which can be altered by simply adjusting some parameter values.

2.4 GENERIC TECHNIQUES

In this section we shall describe some techniques and approaches which may be applicable to a number of different areas of image processing, as well as, in some cases, to problems involving the processing of more general digital signals.

The underlying theme linking the first three topics to be discussed is the manner in which variational procedures can be used to provide a powerful framework for formulating and solving problems encountered in early visual processing. The reason for this is that many of these problems are essentially underdetermined. There is, for example, in general no unique three-dimensional object yielding a given two-dimensional image. Such an image could have been produced by any one out of a large number of possible combinations of lighting arrangements, object surface orientations and surface reflectivities. Nevertheless, it is apparent that the human visual system is able to extract a considerable amount of information about the shape of an object from its perceived brightness, or shading. This suggests that, in forming an unambiguous percept, various natural constraints are applied which incorporate plausible assumptions about the world. A second example is provided by the problem of extracting optical flow fields from moving image sequences. Optical flow (see Chapter 5) is the distribution of apparent velocities of intensity changes in an image, due, for instance, to relative motion between the viewer and objects in the scene. It contains information about the position and relative velocities of objects, and is known to provide the human visual system with important cues concerning the three-dimensional structure of the environment. It is not, however, possible to calculate the optical flow at a point in the image independently of neighbouring points without imposing additional constraints, since the velocity field at each image point has two components whereas the change in image brightness provides only a single piece of information.

Owing to their underdetermined nature, such problems often fit naturally into an optimisation framework. In order to obtain the desired information, the local point constraints provided by the image measurements must be supplemented by some form of additional global constraint. It may then be possible to formulate a cost or 'energy' function over an appropriate solution space, and to seek the solution which minimises this function subject to the local constraints. There are well established techniques for solving such optimisation problems. Standard gradient descent or hill climbing methods work well for simple convex cost functions, but may fail to find the true optimal solution in the case of a complex cost function possessing a number of local, but non-global minima. In Section 2.4.1 we shall describe a stochastic optimisation technique called simulated annealing, whose effectiveness in solving a range of problems of the latter type has recently been demonstrated.

Section 2.4.2 is concerned with a class of techniques known as *relaxation* labelling, which have been widely applied in a number of areas of image processing. Although the labelling problem was originally formulated in a heuristic manner, more recent work has shown that it can be recast in terms of the constrained optimisation of an appropriate objective function. This optimisation can be performed by a network of processing elements, whose values represent the labels assigned to objects, and whose local interconnections allow constraints to be implemented as excitatory and inhibitory links between the processors. A self-consistent solution is then iteratively approached by the network.

A specific example of a variational approach which has been extensively adopted in image and signal processing is provided by the *maximum entropy method.* This method, outlined in Section 2.4.3, can be applied to image restoration problems by assuming a very general statistical model for the image. Alternative models, incorporating information about the image statistics in a more explicit manner, may well be desirable, and in Section 2.4.4 we describe a class of models known as *Markov random fields* which have this property. In addition, their use leads to the natural definition of an 'energy function' on the configuration space of the image, and hence to optimisation procedures for image restoration in a way that will be illustrated in Section 2.5.

Finally in Sections 2.4.5 and 2.4.6 we shall describe two very powerful techniques, namely dynamic programming procedures for waveform matching, and the use of the discrete Fourier transform to analyse a signal or image into its spectral components. Examples of the application of both techniques will be given in Section 2.5.

2.4.1 Optimisation by simulated annealing

Simulated annealing is a technique for finding minima, or approximate minima, of functions involving large numbers of independent variables. It was first proposed by Kirkpatrick *et al.* (1982) who perceived the analogy between optimisation problems of this kind and statistical mechanical models of complex systems with many degrees of freedom. The phase space of such systems is of enormous dimensionality, and it is thus only possible to predict their behaviour in statistical terms, describing average values and fluctuations about the average of quantities of interest. In thermal equilibrium this average is calculated over a Gibbs ensemble of identical systems, the probability of any configuration with corresponding energy $E(\omega)$ being proportional to the Boltzmann factor $\exp(-E(\omega)/kT)$, where k is the Boltzmann constant.

Metropolis *et al.* (1953) developed a procedure for simulating the thermal motion of a collection of interacting molecules in equilibrium at a given temperature T. At each iteration of their algorithm a small random displacement is made to the system configuration, and the corresponding change in system energy, ΔE, is computed. If the displacement reduces the system energy (i.e. $\Delta E \leqslant 0$), it is accepted. If it increases the energy ($\Delta E > 0$) it is still, however, accepted with probability

$$\exp(-\Delta E/kT)$$

Under this procedure the system evolves into a Boltzmann distribution.

The lowest energy states of a physical system will obviously occur at low temperature. However, to realise these states the temperature must be lowered very gradually, so that the system remains close to thermal equilibrium at all times, as the final state may otherwise be only locally optimal but not a global energy minimum. This procedure is called *annealing*.

In 'simulated' annealing, the energy of the physical system is replaced by the cost function of some optimisation problem, and the temperature becomes a control parameter in the same units as the cost function. The procedure starts off at a high 'temperature' T which is then gradually lowered according to an 'annealing schedule'. At each temperature the Metropolis algorithm is used to sample a number of configurations which is sufficiently large for equilibrium to be attained, a condition which can be tested by monitoring the mean energy $<E(T)>$ or the 'specific heat' $d<E(T)>/dT$. When further reduction in T produces no change in $<E(T)>$ the annealing process is terminated. Some major differences between simulated annealing and conventional gradient descent optimisation algorithms are immediately apparent. The annealing algorithm is prevented from getting 'stuck' at local, non-global minima of the cost function by its ability to make controlled uphill steps. Steepest descent methods allow only downhill steps; this corresponds to rapid cooling or 'quenching' of a physical system which may cause it to end up in a metastable, non-optimal state. Clearly, therefore, annealing would be an inefficient way of attempting to minimise a simple convex cost function. It is important to apply the technique intelligently, with some understanding of the problem at hand, and not to use it in cases where more straightforward approaches are applicable.

A class of problems well suited to the annealing method are those in which 'frustration' is involved. This phenomenon, characteristic of physical systems such as spin glasses, occurs when the system is acted on by mutually incompatible constraints. As a consequence the Hamiltonian (or cost function) has an extremely large number of local minima or 'valleys', separated by energy barriers. For many applications it is required to find one of these which has near-minimal energy, but we are not concerned with precisely which one is selected. In such a case gradient algorithms will evidently fail.

In the simulated annealing procedure, moves causing large changes in the cost function are allowed at high temperatures, while at low temperatures only small changes are likely to be accepted. This suggests that the method may also be usefully applied to problems containing some sort of hierarchical structure, in which choices may be made at a succession of different scales. On the other hand, it will tend to fail in cases where the best solution corresponds to a minimum that is deep, narrow and isolated.

To apply annealing to a particular problem, the following aspects must be considered:

(i) The configuration space of the problem must be defined and parameterised, and a suitable cost function formulated on it. The form of this function, e.g. the possible incorporation in it of mutually conflicting constraints between which trade-offs must be made, will largely determine the appropriateness of the annealing approach.

(ii) A method of generating random 'moves' must be defined. The method chosen can significantly influence the efficiency of the algorithm, but this aspect is very problem-dependent and in need of more theoretical analysis.

(iii) An annealing schedule must be set up. In general terms the system must be warmed until 'melted', and then cooled in slow stages until the diffusion of its components has ceased. Again, little theoretical analysis has been done of the influence on convergence behaviour of different schedules. Kirkpatrick *et al.* (1982) have applied annealing to problems whose exact solution is known to require a computing effort exponentially related to size N, and obtained good results using schedules whose effort scales as N or a small power of N.

There is currently considerable interest in studying the simulated annealing approach to solving optimisation problems, in fields ranging from VLSI design to cognitive science. Some of the most exciting applications may well turn out to lie in the areas of perception and cognition, where annealing is being used to model high level vision and the computational processes of the brain.

The advent of VLSI technology has made practicable the fabrication of massively parallel networks of simple processing elements. A new computing architecture of this type, which implements a parallel form of the Metropolis algorithm to perform a best-fit constraint satisfaction search, is of considerable interest for its resemblance to the structure of neural networks, and for its potential application to tasks of perception and vision. This architecture, which has been called a 'Boltzmann Machine' (Hinton *et al.*, 1984), derives from a system described by Hopfield (1982). Its processors, which are binary, neuron-like units, fall into two categories: the 'visible' units which provide the network's interface with the environment, and the 'hidden' units which can be used to represent hypotheses about the states of the environment. Weak constraints between hypotheses (which incur a cost when violated) are implemented by interconnections between units, and the machine attempts to minimise a global 'energy' function expressing the extent to which combinations of hypotheses violate the constraints and fail to fit the input data.

Use of a stochastic search procedure is equivalent to performing the optimisation in a noisy environment, in which noise levels may be controlled in some way. Hinton *et al.* (1984) suggest a parallel with the stochastic variability exhibited by the firing patterns of neurons in the cerebral cortex. From an architectural point of view, the machine should work well even when implemented using analogue devices with poor signal-to-noise performance.

Long term knowledge is stored as the strengths of connections between units, and the machine has been shown (Hinton and Sejnowski, 1983; Hinton *et al.*, 1984) to be capable of 'learning' the underlying constraints which describe the structure implicit in the input data. To do this, it adjusts its connection strengths to create an internal model which will then generate examples with the same probability as the examples it has been shown. Indeed, if subsequently shown incomplete input data it should be able to generate the full output by first creating an internal model for the partial input and then using this model to generate the remainder. It can thus be used as a content-addressable memory, or, in a perceptual context, as a system exhibiting recognition. The learning procedure uses a domain-independent algorithm and purely local computations, but is at present very slow as it involves a large number of annealing stages.

2.4.2 Local parallel processes and relaxation labelling

When image restoration and analysis problems are reformulated as variational procedures, the result is often a local parallel process, suitable for implementation on a network of processing units with local interconnections. Algorithms which can be implemented by architectures of this type are becoming increasingly significant in computer vision for a number of reasons. First, the requirement for real-time processing often precludes the use of serial architectures. Secondly, with the decreasing cost of logic circuits and advent of VLSI it has become feasible to build large parallel processing networks, particularly when these networks are required to have a uniform structure with only local neighbourhood connections, and the individual processors are of simple type. Thirdly, studies of the structure of biological vision systems have shown the primate visual cortex to contain neural networks which can be modelled as parallel architectures with the above properties of locality, uniformity and simplicity of individual processors. A task which the human vision system performs quickly and effortlessly is thus likely to have a computational solution involving a parallel algorithm with local interconnections.

This local parallel structure is typical of a class of techniques known as *relaxation labelling processes*. Relaxation labelling was originally developed for the purpose of assigning numeric or symbolic labels to objects in the presence of ambiguities. The aim was to use prior knowledge about the problem domain, in the form of the likelihood of co-occurrence of particular label assignments, to resolve ambiguities in a manner consistent with the initial data, and to do so by means of a series of simple, local computations which could be implemented in parallel.

The problem addressed by relaxation labelling can be formulated in general terms as one of labelling the nodes of a graph (S, E), where S is a set of nodes and E a set of arcs connecting pairs of nodes. Each node i has a set Λ_i of candidate labels, and there are constraints on the assignment of labels to groups of nodes which are connected by arcs. These constraints, corresponding to some sort of 'world knowledge', are usually expressed as compatibility coefficients

$$r_{ij}(\lambda,\mu),\ i,j \in S,\ \lambda \in \Lambda_i,\ \mu \in \Lambda_j$$

describing the compatibility of simultaneously assigning labels λ and μ to nodes i and j respectively, although constraints involving groups of three or more nodes may in principle also be incorporated. The aim is to find a labelling of the graph which is consistent with the constraints.

Approaches fall into two categories, according to whether they employ discrete or continuous relaxation. In discrete relaxation, pairs of labels are either compatible or completely incompatible. The relaxation procedure consists of an iterative pruning, in parallel over all nodes, of labels which are incompatible with the entire label set of some neighbouring node. Continuous relaxation processes, on the other hand, permit greater information to be incorporated in the constraints, by defining the compatibility coefficients to be real numbers whose magnitude reflects the strength of the constraint, and whose sign is often used to denote consistency (positive coefficient) or inconsistency (negative coefficient). A set of weights is associated with the labels in each set Λ_i, such that if $p_i(\lambda)$ is the weight corresponding to label λ for node i,

$$p_i(\lambda) \geqslant 0 \text{ for all } i, \tag{2.3}$$

and

$$\sum_{\lambda \in \Lambda_i} P_i(\lambda) = 1 \quad \text{for all } i \in S$$

The weights are often taken to correspond to probabilities. This interpretation is, however, strongly rejected by Hummel and Zucker in a recent theoretical study of relaxation labelling processes (1983) on the grounds that a probabilistic analysis of the relaxation procedure (see Peleg, 1980) requires unrealistic assumptions and fails to provide a sound framework for analysing its performance. It is perhaps safer to regard the weights more generally as representing a measure of confidence in their associated labellings.

In the following discussion we shall treat only the more interesting category of continuous relaxation labelling. This technique was originally proposed by Rosenfeld *et al.* (1976), who used *ad hoc* arguments to define rules for iteratively updating label weights so as to improve the consistency of the labelling. The drawback of such a heuristic approach is that it is difficult to analyse precisely what computation is being performed: convergence of the procedure cannot be guaranteed, and even when it occurs the limit point may not yield the desired result.

A way of overcoming these difficulties is to reformulate the labelling problem in terms of the constrained optimisation of some objective function measuring the feasibility of the labellings. Techniques for solving such optimisation problems are well known, and their convergence is assured. Ullman (1979) takes this approach, and shows that the optimisation can be performed by a Lagrangian gradient process. He describes a method for designing a network of locally interconnected simple processors to carry out the computation. Since each processor computes a partial derivative of the associated Lagrangian, the locality and symmetry of the computation will depend on corresponding properties of the objective function and constraints. Faugeras and Berthod (1981) propose a class of objective functions which measure both consistency and ambiguity, the constraints being given by Equations 2.3 above. Hummel and Zucker (1983) define a consistent labelling in terms of a gradient projection technique. Other authors have constructed similar objective functions, and have drawn comparisons between their own optimisation approaches and the original algorithm of Rosenfeld *et al.* (1976). For example, Lloyd (1983) shows that the Rosenfeld algorithm is a suboptimal way of maximising a special case of the Hummel and Zucker objective function.

Relaxation labelling has been successfully applied to image processing tasks such as segmentation and edge detection. The technique must however be used with caution: when applied to certain image restoration problems it has been observed (Richards *et al.*, 1981) to produce an initial improvement followed by a degradation. Lloyd (1983) discusses the reason for this behaviour. The compatibility coefficients, or consistency requirements built into the cost function, are usually derived from statistical considerations. They thus represent average trends observed in a large ensemble of images, rather than the individual details of a particular image which in fact carry most of the information content of that image. In other words, if the desired labelling does not in fact correspond to a limit point of the relaxation process, then the technique is not an appropriate one to use. An extreme example of this is given by an early linear algorithm proposed by Rosenfeld *et al.* (1976). Under certain conditions this algorithm converges to a limit which is completely independent of the initial labelling. Clearly, this is not always a desirable feature!

Finally, it is of interest to compare and contrast relaxation labelling with the related technique of simulated annealing, described in Section 2.4.1. The approaches are both iterative, and they have similar computing structure. Fundamentally, however, there are significant differences. Simulated annealing models the approach to equilibrium of a complex system, using analogies with statistical mechanics. At each stage of the iteration the system is in a well defined state, and the limiting state is one of near minimal energy. By contrast, the intermediate states, and possibly also the limit state, of a relaxation process will generally only provide a set of label confidences for each node, so that a thresholding procedure is needed to obtain an unambiguous labelling. Moreover, the gradient descent methods used to perform relaxation labelling in its constrained optimisation formulation will ensure that the system converges to that local minimum of the cost function which is most easily accessible from its initial state.

These observations suggest that the technique of annealing should be most effective when applied to problems whose search spaces have many local optima, but where the correct interpretation is much more consistent than any other. Problems of this type are frequently encountered in vision. Relaxation labelling, on the other hand, is better suited to solving local optimisation problems where the influence of the initial conditions must be taken into account.

2.4.3 The maximum entropy method

In its most general form the principle of maximum entropy (ME) is concerned with the problem of estimating a probability distribution $(p_1, p_2, \ldots p_n)$ on a set of n states, when some constraints on the p_i are known, but many candidate solutions exist which are consistent with this information. The procedure adopted is then to choose the distribution which maximises some function S representing configurational entropy, subject to the given constraints. S is usually taken to be the information-theoretic entropy

$$S = -\sum_i p_i \log p_i \tag{2.4}$$

which measures the rate at which information is generated by an information source that produces a variety of symbols i with probability p_i. Alternative definitions of the entropy function have also been proposed, however, and there has been some dispute regarding the most appropriate form for a given problem.

The ME principle has aroused much controversy. It has been justified subjectively by the argument that it selects the simplest possible result consistent with the data, and is 'maximally noncommital with regard to missing information' (Jaynes, 1957). Moreover it yields the probability distribution which can be realised in the largest number of ways, and which (identifying S with Shannon's 'negative information') has in some sense minimal information content. Recent work (Shore and Johnson, 1980; Johnson and Shore, 1983; Tikochinsky *et al.*, 1984) has placed the approach on a more rigorous footing by showing that ME is, in a well defined way, the only consistent method of solving the estimation problem. Without any invocation of information-theoretic concepts, or consideration of experiments involving large numbers of independent trials, it is shown that maximising any other function than S will lead to inconsistency unless that function and S have identical maxima.

The ME principle may be applied to image restoration by assuming a statistical model in which the image f is built up by distributing a fixed number, N, of discrete 'grains' or quanta of intensity over the n pixels of the array, such that the i-th pixel contains f_i grains. Let p_i represent the probability of a grain falling in this pixel. For large N it follows from the law of large numbers that $p_i \sim (f_i/N)$. It is easy to show that maximising Equation 2.4 for these p_i's is equivalent to maximising

$$-\sum_{i=1}^{N} f_i \log f_i$$

The constraints to be imposed are provided by the image data. Because this data is generally noisy, any attempt to fit it exactly would have the undesirable effect of introducing spurious structure into the restored image. Instead, some knowledge of the noise statistics is assumed to be available, and the restoration is constrained to be consistent with the measured data and known noise level (see Andrews and Hunt, 1977; Gull and Daniell, 1978).

The ME solution is guaranteed to be positive: clearly this is an attractive property for an image restoration method. It can handle non-linear image models, although, as we have seen, it incorporates information about the image statistics in only a very general way. Iterative algorithms have been developed for ME image restoration, but the technique is inherently very time consuming on conventional computers. In Section 2.5.2 a connection will be established with an alternative, Bayesian approach to restoration, which models the image as a Markov random field and can be implemented using parallel algorithms.

2.4.4 Markov random field image models

Many techniques for image processing and analysis require the assumption of some form of image model. To be useful, such a model must incorporate reasonably realistic assumptions about the image and noise statistics. A class of models which have been widely studied are those of so-called Markov type, whose statistics are described by representing each pixel grey level as a function of those of its neighbours, together with additive noise. Since in the majority of images pixel intensity values are extremely neighbour-dependent, and indeed concepts of neighbour similarity are fundamental to grouping and segmentation procedures, such spatial interaction models are intuitively appealing. A variety of recursive restoration algorithms, in which the processing at each pixel involves only pixels lying in some neighbourhood, have been developed.

We shall describe a particular class of Markov models known as *Markov random fields* (MRFs). These are a natural generalisation of one-dimensional Markov processes, in which the time index is replaced by a two- or higher-dimensional space index, and their study forms a new branch of probability theory with many interesting applications. The original concept of a MRF resulted from attempts to formulate the Ising model of ferromagnetism in a probabilistic setting, but other physical and biological systems have also been fruitfully modelled in this way. As we shall see, the use of MRF image models enables us to exploit analogies with statistical mechanics systems, and to define such concepts as that of an 'energy function' which are valuable for quantifying and studying image properties.

Let G be a graph defined by the pair (S,E) where S is a set of nodes and E a set of arcs connecting pairs of nodes. E defines a neighbourhood system $\{N_s, s \epsilon S\}$ for S: for $s \epsilon S$ N_s, the set of neighbours of s, consists of those nodes of S which are connected to s by arcs in E. A configuration on S is an assignment to every $s \epsilon S$ of a label from some finite set $\Lambda = \{0,1,2, \ldots, L-1\}$. Let $\Omega = \Lambda^S$ be the set of all possible configurations, and for $\omega \epsilon \Omega$ let ω_s denote the value of ω at $s \epsilon S$ while ω_A denotes the restriction of ω to the subset A of S.

The local characteristics of a probability measure P on Ω are the conditional probabilities $P(\omega_s \mid \omega_{S-s})$. It can be shown that, provided

$$P(\omega) > 0 \text{ for all } \omega \epsilon \Omega \tag{2.5}$$

the local characteristics uniquely determine P. P is said to define a Markov random field (MRF) on G if Equation 2.5 holds and also

$$P(\omega_s \mid \omega_{S-s}) = P(\omega_s \mid \omega_{Ns}) \text{ for all } \omega \epsilon \Omega$$

If the nodes of G are taken to be a set of $N \times M$ pixel sites, with arcs connecting all sites within some given Euclidean distance of each other, and Λ is a set of allowable grey levels, then Ω is the family of all possible grey scale images and a MRF defines a stochastic image model. The attraction of using such models is that an extremely useful parallel can be drawn between them and statistical mechanical models of physical systems, where the concept of an energy function leads to the definition of a Gibbs probability measure.

A potential V is a family of functions $\{V_A, A \subset S\}$ on Ω such that $V_A(\omega)$ depends only on ω_A. V defines an energy function U on Ω by

$$U(\omega) = \sum_A V_A(\omega) \tag{2.6}$$

The Gibbs measure induced by this energy function is then given by

$$P(\omega) = (Z)^{-1} \exp[-U(\omega)/T] \tag{2.7}$$

where

$$Z = \sum_{\omega} \exp[-U(\omega)/T]$$

is the so-called partition function, and the parameter T corresponds to the temperature of the physical system.

A (possibly empty) set of nodes $C \subset S$ is a clique if each pair of nodes in C are neighbours. V is a nearest-neighbour potential if $V_A(\omega) = 0$ whenever A is not a clique. The fundamental relation between Gibbs measures and MRFs may now be stated as follows: a probability measure P on Ω is a MRF if, and only if it is a Gibbs measure with respect to some nearest-neighbour potential.

Proofs of this result can be found in, for example, Kinderman and Snell (1980). It provides a means of modelling a digitised image as a state of a lattice-like physical system whose dynamics are defined by short range interactions. The equilibrium states for such a system are known to be given by the Gibbs distributions, and to be precisely those states at which the specific free energy of the system is minimised. Moreover, the system will under quite general conditions evolve with time from an initial non-equilibrium state into one defined by a Gibbs distribution. These properties, well known in statistical mechanics, can be used to develop image processing algorithms which are well suited to parallel implementation on networks of locally interconnected processors.

There remains the problem of selecting an appropriate MRF model for a given situation. Choices to be made include the nature of the image degradation model, the form of neighbourhood structure, and the type of interaction potential V. Existing treatments generally assume that the degradation is given by a blurring transformation with local support, together with additive Gaussian white noise. Simple, local neighbourhood systems and potentials chosen on an *ad hoc* basis have produced good results, although Geman and Geman (1984) also claim to have developed a method for estimating neighbourhood parameters from noisy data. These authors propose a 'hierarchial' model in which the sites of the graph G consist not only of the image pixels, but also of additional geometrical image attributes such as edge elements. They show that a simple local neighbourhood structure on G can then give rise to a marginal distribution on the image intensities with a completely connected graph.

2.4.5 Dynamic programming

A dynamic programming procedure is a multistage decision process which achieves a logarithmic reduction in complexity by solving an R-stage process as *R* single-stage ones. To be amenable to solution by dynamic programming, the decision process must be Markovian: its future behaviour must be governed entirely by its present state and independent of its past history. In addition, the decision stages must be ordered so that, at any given stage, all other stages whose results are needed have already been processed.

Typical of the type of problems to which dynamic programming is commonly applied are those in which it is required to match two waveforms by means of a dynamic warping of the time axis. Such problems occur in, for example, speech recognition, seismic analysis and binocular stereo (see Section 2.5.3). The waveforms are decomposed into discrete strings of features, each feature having an associated list of attributes. It is then required to match the two strings by forming correspondences between their features. For the technique of dynamic programming to be applicable, correspondences which reverse the order of features must be forbidden (see Fig. 2.4).

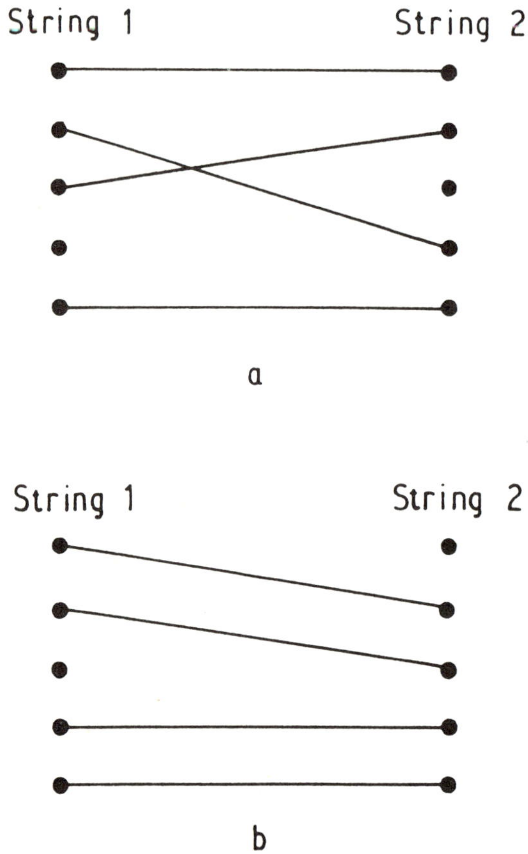

Fig. 2.4 (a) Order reversing correspondence; (b) correspondence with no order reversals.

Consider a two-dimensional Cartesian lattice, whose axes represent the two feature strings to be matched. A match between strings of lengths M and N respectively can be described in terms of a 'warp path' on this lattice:

$$\{(m(k),n(k)),\ 0 \leqslant m(k) \leqslant M,\ 0 \leqslant n(k) \leqslant N,\ 1 \leqslant k \leqslant K\}$$

where for $m,n > 0$ (m,n) denotes that the m-the feature in the first string is matched to the n-th feature in the right string, and, e.g. $(m,0)$ signifies that the m-th feature in the first string is unmatched. This is illustrated in Fig. 2.5.

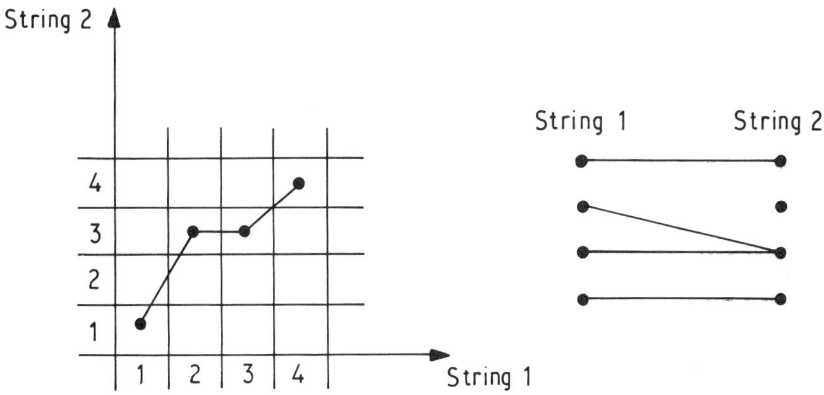

Fig. 2.5 The warp path { (1,1),(0,2),(2,3),(3,3),(4,4) }

A similarity (or dissimilarity) score $s(m(k),n(k))$ is defined for each feature pair $(m(k),n(k))$, and these are combined in some manner to form a total score (or cost function) for the warp path. To evaluate this function for all possible warp paths would in general be computationally very expensive. Often, however, it is possible to derive constraints on the allowed predecessors of a match $(m(k),n(k))$ from the underlying physical problem. For example, a constraint which would ensure that no two consecutive features are left unmatched, and that no feature in one string can be matched with more than one feature in the other string, would be imposed by requiring that the allowed predecessors of (m,n) must be one of the set: $(m-1,n-1)$, $(m-1,0)$ and $(m-2,n-1)$, or $(0,n-1)$ and $(m-1,n-2)$.

The score, $g(m,n)$, of the optimal warp path ending at any candidate match (m,n) can then be calculated recursively, using the fact that this score depends only on $s(m,n)$ and the scores of the optimal warp paths ending at the allowed predecessors of (m,n). This recursive evaluation property relies upon the fact that no order-reversing correspondences are permitted. The value $g(M,N)$, obtained after processing all candidate matches, is a measure of the global similarity between the two feature strings. The optimum warp path can then be found by backtracking from (M,N) along directions of partial score maxima within the lattice.

The power of the dynamic programming technique thus lies in the way in which the difficult problem of finding the best overall correspondence between two strings can be broken up into the simpler one of finding best partial matches of substrings. Processing element and array architectures amenable to VLSI implementation have been developed to perform dynamic programming-based matching algorithms in real-time for connected speech recognition (see Ackland, Weste and Burr, 1981; Wood and Urquhart, 1984).

Ackland *et al.* (1981) have proposed an array of simple processors corresponding to the points of the warp lattice. Each processor carries out two functions: (1) it reads in attribute vectors for the feature pair (m,n) corresponding to its location, and uses them to compute the match score $s(m,n)$, and (2) it uses $s(m,n)$ and input from its neighbours to compute the partial warp path score $g(m,n)$. Attribute vectors pass through the array in orthogonal directions, while partial scores are inserted into the data stream as they are computed.

Wood and Urquhart (1984) describe an alternative processing element architecture, and propose a word level systolic array based on a linear chain of these processing elements to compute the dynamic programming recursion.

2.4.6 The discrete Fourier transform

In general terms, the effect of applying a two-dimensional unitary transform is to decompose an image into a set of orthogonal components, whose relative magnitudes may contain useful information about the spatial variations within the image. For coding applications, components of the image corresponding to transform coefficients with low values may be discarded to achieve bandwidth reduction and image compression without the loss of the most significant information. The transform coefficients may also provide useful image descriptors for pattern recognition tasks.

One of the most widely used image transforms is the Fourier transform, which decomposes the image into a weighted sum of sinusoidal patterns at different spatial frequencies and phases. Low frequency components of the Fourier spectrum carry information about slow variations in image intensity, while high frequency components correspond to rapidly varying intensities such as are found in the edge structure of the image. Another useful property of the Fourier transform is that convolution, a very important image processing operation, is equivalent to pointwise multiplication in the Fourier domain.

Later in Section 2.5 we shall describe the use of the Fourier transform for image restoration. A detailed discussion of its properties, along with those of other unitary transforms, can be found in image processing texts, for example Pratt (1978), Gonzalez and Wintz (1977), and Andrews and Hunt (1977). Here we present only a brief introduction to the discrete Fourier transform, and its use for computing convolutions and correlations on image data. In Chapter 6, efficient algorithms for performing this transform will be described, and their suitability for parallel implementation considered.

The discrete Fourier transform (DFT) of an image array $\{f(k,l),\ 0 \leqslant k \leqslant M-1,\ 0 \leqslant l \leqslant N-1\}$ is defined to be the array $\{F(u,v),\ 0 \leqslant u \leqslant M-1,\ 0 \leqslant v \leqslant N-1\}$ where

$$F(u,v) = (MN)^{-1} \sum_{k=0}^{M-1} \sum_{l=0}^{N-1} f(k,l)\exp[-2\pi j(uk/M + vl/N)] \tag{2.8}$$

$$f(k,l) = \sum_{u=0}^{M-1} \sum_{v=0}^{N-1} F(u,v)\exp[2\pi j(uk/M + vl/N)] \tag{2.9}$$

It is clear from the form of equations 2.8 and 2.9 that the two-dimensional DFT and its inverse are separable transformations. They may be performed by taking one-dimensional transforms first along the columns of an array and then along the rows of the result, or *vice versa.*

The relation between continuous and discrete Fourier transforms is discussed in Cooley, Lewis and Welch (1967) and Brigham (1974). It can be shown that, if $f(k,l)$ is obtained by sampling some continuous image function $f(x,y)$ at the point $x{=}x_0 + k\Delta x$, $y{=}y_0 + l\Delta y$, then the values $F(u,v)$ correspond to samples of the continuous Fourier transform of $f(x,y)$ at the points $u/M\Delta x$, $v/N\Delta y$. This correspondence is not, however exact: the sampled spectrum is in fact aliased relative to the Nyquist frequencies $1/(2\Delta x)$ and $1/(2\Delta y)$ in the u and v directions respectively.

It is important to realise that the arrays $F(u,v)$ and $f(k,l)$, as defined by their transforms (equations 2.8 and 2.9), are periodic with period M in the first variable and N in the second. The use of the DFT to process an image array can thus result in 'wraparound errors' unless care is taken to augment the data by adding zeros so that different periods do not overlap. Consider for example the operations of convolution and correlation, frequently encountered in digital image processing and commonly implemented using transform techniques. Let $f(k,l)$ and $g(k,l)$, $0 \leqslant k \leqslant M-1$, $0 \leqslant l \leqslant N-1$ be arrays with discrete Fourier transforms $F(u,v)$ and $G(u,v)$ respectively. The cyclic convolution, $f{*}g$, and the cyclic correlation, $f{\circ}g$ are periodic arrays defined by

$$f{*}g(k,l) = \sum_{m=0}^{M-1} \sum_{n=0}^{N-1} f(k-m,l-n)g(m,n)$$

$$f{\circ}g(k,l) = \sum_{m=0}^{M-1} \sum_{n=0}^{N-1} f(k+m,l+n)g(m,n)$$

whose discrete Fourier transforms are the products $F(u,v)G(u,v)$ and $F(u,v)G^*(u,v)$ respectively, where G^* is the complex conjugate of G.

It is often more efficient to compute convolutions and correlations by transforming the arrays to the frequency domain, multiplying the transforms (or their conjugates) point by point, and then inverse transforming the result. However, if the original arrays were of size $A \times B$ and $C \times D$ respectively, they must be extended to size $M \times N$ by the addition of zeros before transformation where, to avoid wraparound error, the conditions $M \geqslant A+C-1$ and $N \geqslant B+D-1$ must be satisfied.

2.5 APPLICATIONS

The techniques which were introduced in the previous section will now be illustrated in the context of two specific application areas. The first of these concerns the restoration of degraded images, a goal to which extensive effort has been devoted because of its fundamental importance in image processing. In Section 2.5.1 some linear restoration procedures are described, which can in many cases be implemented efficiently using Fourier transform techniques. Linear image restoration is widely used and often produces a significant improvement in image quality, but it has inherent limitations and as a result non-linear restoration techniques are also of considerable interest. In Section 2.5.2 a non-linear, Bayesian approach is outlined which employs a Markov random field image model and a stochastic optimisation approach based on simulated annealing.

Image restoration procedures will of necessity occur at the lowest levels of the processing hierarchy. By contrast, our second application area, namely binocular stereo imaging, involves a rather higher level transformation which operates on extracted information (e.g. edge structure) in a pair of images taken from slightly differing viewpoints, and combines this information in order to obtain an intrinsic image which maps the range of visible objects in the scene. Stereo matching algorithms exploiting the techniques of dynamic programming, relaxation labelling and simulated annealing will be discussed in Section 2.5.3.

2.5.1 Image restoration: linear techniques

Many low level digital image processing systems have as their final objective the restoration of a degraded image, while for more general applications of computer vision, image restoration is an important first stage process. Details of the physical mechanisms of the image formation and recording processes, and a discussion of sources and models of image degradation can be found in Andrews and Hunt (1977), Gonzalez and Wintz (1977), or Pratt (1978). These references also contain a comprehensive overview and classification of image restoration techniques.

Let $\mathbf{g}$ and $\mathbf{f}$ denote the vectors of length n^2 obtained by lexicographically ordering the $n \times n$ degraded and original image arrays, $g(i,j)$ and $f(i,j)$, respectively. A standard degradation model applicable to many space invariant imaging systems is given by the equation

$$\mathbf{g} = \theta\,([\mathbf{H}]\mathbf{f}) + \mathbf{n} \tag{2.10}$$

where **H** is an $n^2 \times n^2$ blurring matrix constructed from the point spread function $h(k,l)$ of the imaging process, **n** is an independent noise field and θ is a (possibly non-linear) transformation acting pointwise on [**H**]**f**. As is well known (e.g. Andrews and Hunt, 1977) the problem of recovering **f** from **g** is extremely ill conditioned, having no unique solution. In keeping with the philosophy of Section 2.4, most approaches to image restoration consequently attempt to find an approximate solution to equation 2.10 which optimises some appropriately formulated criterion, with or without imposed constraints.

We shall illustrate this by describing two categories of restoration procedure. The first assumes the image degradation to be not only space invariant but also linear, leading to a convolution model. A further assumption of periodicity for the input functions results in a great simplification to this model by allowing the matrix **H** to be approximated by a matrix of block circulant form. Restoration can then be performed efficiently in the frequency domain, using a two-dimensional fast Fourier transform. Two approaches are described in this section: a direct inverse filter method, and a Wiener filtering procedure which uses a minimum mean-square error criterion to overcome the ill conditioned nature of the restoration problem.

The second category consists of non-linear approaches to image restoration. In Section 2.5.2, after discussing the need for such procedures, we shall illustrate their use by describing some stochastic restoration techniques which model an image as a Markov random field.

By restricting θ in equation 2.10 to be the identity transformation we obtain the linear image formation model

$$\mathbf{g} = [\mathbf{H}]\mathbf{f} + \mathbf{n} \tag{2.11}$$

Viewed in a deterministic framework, the restoration problem is one of solving a system of linear equations in the presence of noise. Assuming that we have no specific knowledge about the noise **n**, we may reasonably adopt the criterion that an estimate $\hat{\mathbf{f}}$ of the restored object distribution should be sought which is consistent with the norm of **n** being as small as possible. We thus have an unconstrained optimisation problem with cost function

$$(\mathbf{g} - [\mathbf{H}]\hat{\mathbf{f}})^t\,(\mathbf{g} - [\mathbf{H}]\hat{\mathbf{f}}) = \mathbf{n}^t\mathbf{n} \tag{2.12}$$

to be minimised with respect to $\hat{\mathbf{f}}$, where the superscript t denotes the matrix transpose. Differentiating in equation 2.12 with respect to $\hat{\mathbf{f}}$ and setting the result to zero yields

$$\hat{\mathbf{f}} = [\mathbf{H}^t\mathbf{H}]^{-1}\,\mathbf{H}^t\mathbf{g}$$

or, assuming that **H** is square and invertible

$$\hat{\mathbf{f}} = \mathbf{H}^{-1}\mathbf{g} \tag{2.13}$$

A detailed analysis of the space invariant image formation process (see, for example, Appendix A of Andrews and Hunt, 1977) shows that, if restoration errors associated with deconvolving a segment cut from a larger image can be localised to a small number of border elements and safely neglected, then the matrix **H** in equation 2.11 is of a form known as block Toeplitz.

A Toeplitz matrix is a square matrix whose elements T_{ij} satisfy

$$T_{ij} = T_{kl} \text{ if } i-j = k-l$$

A block Toeplitz matrix is a square matrix of partitions arranged in Toeplitz form, such that each partition is a Toeplitz matrix. The significance of Toeplitz matrices in digital signal processing is that they may be approximated closely by so-called circulant matrices, which have the very useful property of being diagonalised by the unitary discrete Fourier transform matrix. An $N \times N$ circulant matrix **C** has the form

$$\mathbf{C} = \begin{pmatrix} c(0)\; c(N-1)\; c(N-2) \ldots c(1) \\ c(1)\; c(0) \qquad c(N-1) \ldots c(2) \\ \cdot \\ \cdot \\ c(N-1) \quad \cdot\; \cdot \qquad \ldots c(0) \end{pmatrix}$$

Its rows are formed by cyclic shifts of the sequence $c(0), c(1),\ldots, c(N-1)$. Let $\boldsymbol{J}$ be the $N \times N$ unitary matrix with ik-th element

$$J_{ik} = \exp[2\pi jki/N]$$

where $j^2 = -1$. Then $\boldsymbol{J}^{-1}\,\mathbf{C}\,\boldsymbol{J}$ is a diagonal matrix **D** whose elements D_{kk} are the eigenvalues $\lambda\,(k)$ of **C**.

The block Toeplitz matrix **H** may be approximated by a block circulant matrix $\mathbf{H}_B$, which is a matrix of partitions arranged in circulant form, each partition being a circulant matrix. $\mathbf{H}_B$ is constructed by adding elements to **H** to produce a cyclic structure in the rows. This approximation is close in the sense that $\mathbf{H}_B$ tends to **H** in the Euclidean norm as the size of the image array **g** tends to infinity. It follows from the structure of $\mathbf{H}_B$ and properties of the DFT (see Section 2.4.6) that equation 2.13 may be computed in the Fourier transform domain as

$$F(u,v) = G(u,v)/H(u,v) \tag{2.14}$$

where F, G and H are the two-dimensional DFTs of the arrays f, g and h (after these have been suitably extended to some size $M \times N$ by the addition of zeros to suppress wraparound error).

This inverse filter method does not help to overcome the ill conditioned nature of the restoration problem. It is clear from equation 2.14 that difficulties will be encountered in the restoration process if $H(u,v)$ vanishes or becomes very small in any region of interest in the uv-plane. Moreover, there is likely to be excessive amplification of noise at high frequencies where $H(u,v)$ tends to zero. These problems are overcome, at the expense of some rather unrealistic assumptions about the statistics of the original image, by the stochastic restoration procedure of minimum mean-square-error (MMSE) estimation.

In this, stochastic approach to restoration **n** and **f** are modelled as independent, weakly stationary random processes with known means and covariances. The mean of **n** is usually assumed to be zero. The linear image formation model, equation 2.11, is again adopted, and the criterion to be optimised is taken as the closeness of the restored image $\hat{\mathbf{f}}$ to the original **f**, in the sense of the expectation over the ensemble of possible images. Thus we seek $\hat{\mathbf{f}}$ minimising

$$\mathbf{E}\,[\mathbf{f} - \hat{\mathbf{f}})^t\,(\mathbf{f} - \hat{\mathbf{f}})] = \mathbf{E}\,[\mathrm{Trace}\,(\mathbf{f} - \hat{\mathbf{f}})\,(\mathbf{f} - \hat{\mathbf{f}})^t] \tag{2.15}$$

The solution of this unconstrained minimisation problem is given (Papoulis, 1965) by the conditional expectation

$$\hat{\mathbf{f}} = \mathbf{E}\,(\mathbf{f} \mid \mathbf{g}) \tag{2.16}$$

but this is impractical to calculate since it depends on knowledge of the joint statistics of **f** and **g**. Instead, therefore, we constrain the problem by requiring that $\hat{\mathbf{f}}$ be a linear function of **g**. Substituting $\hat{\mathbf{f}} = \mathbf{Lg}$ into equation 2.15, where **L** is a linear operator, differentiating with respect to **L** and setting the result to zero yields

$$\mathbf{L} = \mathbf{R}_f\,\mathbf{H}^t\,(\mathbf{H}\,\mathbf{R}_f\,\mathbf{H}^t + \mathbf{R}_n)^{-1} \tag{2.17}$$

where

$$\mathbf{R}_f = \mathbf{E}\,[\mathbf{f}\mathbf{f}^t],\ \text{and}\ \mathbf{R}_n = \mathbf{E}\,[\mathbf{n}\mathbf{n}^t]$$

are the autocorrelation matrices of the image and noise processes respectively.

The following observations may be made about the form of **L** in equation 2.17. As before, **H** is a block Toeplitz matrix. Since the noise process is weakly stationary with zero mean, it follows (Andrews and Hunt, 1977, p. 135) that $\mathbf{R}_n$ is block Toeplitz. Similarly, weak stationarity of **f** implies that, by subtracting the constant bias term **H[Ef]** from **g** we may also take $\mathbf{R}_f$ to be of this form. By making the block circulant approximations to **H**, $\mathbf{R}_n$, $\mathbf{R}_f$ as outlined earlier, the large matrix computation $\hat{\mathbf{f}} = \mathbf{L}g$ may be reduced to a much simpler pointwise multiplication in the Fourier domain, where the MMSE or Wiener filter takes the form

$$F(u,v) = \frac{H^*(u,v)G(u,v)}{|H(u,v)|^2 + [P_n\,(u,v)/P_f\,(u,v)]}$$

where P_f and P_n are the signal and noise power spectra respectively, and * represents the complex conjugate.

This filter may also be derived (Andrews and Hunt, 1977, p. 150) from the solution to the problem of maximising the signal-to-noise ratio of $\hat{\mathbf{f}}$, in the case when the noise **n** has known norm.

2.5.2 Non-linear image restoration

Although it avoids the problem of ill conditioning, and clearly reduces to the inverse filter (equation 2.14) as the noise tends to zero, the minimum mean-square-error (MMSE) filter does not perform well in cases of low signal-to-noise ratio. This is due in part to the often highly unrealistic assumption of weak stationarity for **f**, and to the linearity of the image formation model, which is also unrealistic in many practical cases. Indeed, non-linearities may be introduced not only by the imaging process but also by the way in which the restoration problem is formulated; the corresponding optimisation problem may not have a linear solution, and even linear optimisation problems are made non-linear by the addition of constraints.

Non-linear restoration techniques are therefore of importance, and a number of these have been developed. They are in general considerably more costly in computation than their linear counterparts, and the development of fast algorithms for implementation on parallel hardware is thus crucial to their effectiveness. Motivated by this consideration, we shall describe a Bayesian approach to restoration which models the image as a Markov random field (MRF), and which can be implemented by means of highly parallel algorithms. We shall also indicate a connection between this approach and that of maximum entropy image restoration outlined in Section 2.4.3.

Bayesian analysis is a stochastic restoration technique in which **f**, **g** and **n** are described by random fields and an estimate of the restored image is derived from consideration of the conditional densities $p(\mathbf{f} \mid \mathbf{g})$ and $p(\mathbf{g} \mid \mathbf{f})$. The approach allows explicit handling of sensor non-linearities, and does not require a weak stationarity assumption for the image.

In maximum *a posteriori* (MAP) estimation, the restored image estimate $\hat{\mathbf{f}}$ is taken to be that **f** which maximises the *a posteriori* density function $p(\mathbf{f} \mid \mathbf{g})$. By Bayes' theorem,

$$p(\mathbf{f} \mid \mathbf{g}) = p(\mathbf{g} \mid \mathbf{f})p(\mathbf{f})/p(\mathbf{g})$$

Since $p(\mathbf{g})$ does not depend on **f**, and for given **f** the variation in **g** is just that due to the noise **n**, it follows that the MAP estimate depends only on the statistics of **f** and **n**. By contrast, the MMSE estimate is the mean of the *a posteriori* density (see equation 2.16) and thus requires knowledge of the statistics of **g**. The effectiveness of MAP restoration on non-linear problems has been demonstrated (Hunt, 1977), with algorithms such as steepest descent being used to perform the maximisation.

Due to the severe computational problems presented by MAP estimation, some simplifying assumptions about the image and noise statistics must be made. Hunt (1977) uses multivariate Gaussian models, a choice which is probably more realistic for the noise than for the image in general. Geman and Geman (1984) model the image as a MRF (see Section 2.4.4) on a graph G whose nodes S consist of the image pixels and also the family of line sites (possible locations of edge elements, lying in between adjacent pixels). The label set for the pixels consists of the allowable grey levels, while that for the line sites in some set of line descriptors (e.g. orientation). A configuration on S thus has the form $\omega = (\mathbf{f},\mathbf{l})$ where $\mathbf{f}$ is a grey scale image and $\mathbf{l}$ is a coded line process. An image degradation model of the form of equation 2.11 is assumed, where the noise $\mathbf{n}$ consists of independent one-dimensional Gaussian variables with mean μ and standard deviation σ, and $\mathbf{H}$ corresponds to a local blurring operation.

It follows from results described in Section 2.4.4 that the joint pixel and line process can be defined by specifying a neighbourhood system $\{Ns\}$ for S and a nearest neighbour potential V on the space of configurations. The probability of a configuration ω is then given by

$$P(\omega) = (Z)^{-1} \exp[-U(\omega)/T]$$

where U and Z are defined as in equations 2.6 and 2.7. Let μ be the vector which has all its components equal to μ. Geman and Geman (1984) show that the *a posteriori* distribution $P(\omega|\mathbf{g})$ of the original image given the data $\mathbf{g}$ is again Gibbsian, with energy function

$$U^P(\mathbf{f},\mathbf{l}) = U(\mathbf{f},\mathbf{l}) + \| \mu - \{g - ([\mathbf{H}]f)\} \|^2/2\sigma^2 \qquad (2.18)$$

and a system of neighbourhoods which are unchanged for the line sites, and for the pixel sites are slightly enlarged to take into account the additional interaction due to the blurring function.

The motivation leading to the choice of this hierarchical model, and the selection of the neighbourhood system $\{Ns\}$ and potential V, were discussed in Section 2.4.4. If a simple, local neighbourhood system is adopted, a MAP restoration algorithm can be designed which is implementable on a parallel network of processing elements with local interconnections. Geman and Geman (1984) suggest that, by augmenting the image process with additional geometrical attributes, it is possible to model images with long range correlations while retaining the computational advantages associated with local neighbourhood systems.

It is clear from the form of the Gibbsian distribution that the most probable states under the *a posteriori* distribution are those at which the energy function U^P assumes its minimum value. However, application of standard techniques to the minimisation of equation 2.18 are computationally impractical. Geman and Geman (1984) use instead the stochastic procedure of simulated annealing, which was described in Section 2.4.1. In a serial version of their technique the nodes of the graph G are visited in some fixed order, and each move consists of replacing the label at a node with one chosen from the conditional Gibbsian distribution at that node, given the observed states of neighbouring nodes. The calculation involved is generally very simple, due to the simplicity of the Gibbsian local characteristics. By placing a processor at each node s of the graph, and connecting it to the processors in the (local) neighbourhood Ns, a highly parallel implementation may be envisaged. Intermediate degrees of parallelism could be realised by assigning collections of nodes to individual processors, each of which would update its own nodes in serial fashion.

We shall now discuss the connection between the above restoration approach, based on MRF image modelling, and techniques which make use of the maximum entropy (ME) principle. As discussed in Section 2.4.3, ME restoration procedures can only take into account information about the image statistics in a very general way (e.g. as in Andrews and Hunt (1977) by imposing the constraint that the noise **n** has known norm). This is in direct contrast to MAP restoration, which takes explicit account of the statistics of the image to be restored. However Trussell (1980) has shown that ME restoration using the constraint on the noise norm described above is equivalent to MAP restoration in the special case when **n** is Gaussian, stationary and uncorrelated with zero mean, and the prior distribution of **f** is

$$(\text{constant}) \times \exp(-\beta \sum f_i \log f_i)$$

The connection with MRF models is more direct, since the Gibbs measures on a finite lattice are known to be precisely those which minimise the free energy, a condition which is equivalent to maximising the entropy for a given average energy. This observation provides further justification for the use of MRF models, since ME restoration is notoriously computation intensive, whereas highly parallel Bayesian restoration algorithms have recently been developed which exploit the analogy between images and statistical mechanics systems (Geman and Geman, 1984).

2.5.3 Stereo imaging

In binocular stereo imaging, three-dimensional structure is recovered from a pair of images of a scene taken from slightly different viewpoints. The basic principles of stereopsis are discussed in Section 5.3.1, where also one particular stereo algorithm is described.

Central to stereo imaging is the need to match corresponding primitives in a stereo image pair, as a prerequisite to the measurement of disparities and subsequent computation of depth/range data. In this section we shall outline two alternative approaches to solving the correspondence problem, which illustrate the application of generic techniques discussed in Section 2.4, namely dynamic programming, relaxation labelling and optimisation by simulated annealing.

We assume that knowledge of the viewing parameters enables us to simplify the image-to-image correspondence problem into a set of one-dimensional matching problems involving primitives in corresponding rows of a registered stereo image pair. An additional consistency constraint, however, is provided by the mutual dependency between rows which is due to figural continuity (see Section 5.3.1).

A number of edge-based stereo algorithms have been developed which start by matching each pair of corresponding rows independently, using a dynamic programming technique (Baker, 1982; Arnold, 1983; Lloyd, 1985). The rows may thus be processed in parallel, an attractive feature from the point of view of possible high speed implementations. We shall briefly describe the method of Lloyd (1985).

In order to simplify the handling of occlusions, the candidates for matching are taken to be the regions between edges, rather than the edges themselves. For each region four possibilities are considered: it may match a region or group of adjacent regions in the other image, be occluded, merge with one or more neighbouring regions to match a region in the other image, or be unmatched because that part of the object is not in the field of view of the other camera. For each possible match a similarity score is calculated based on the lengths of the regions and properties of their bounding edges.

Using the terminology of Section 2.4.5 a 'warp path' between two rows, one from each image, containing M and N regions respectively, is defined to be an ordered set of region matches

$$\{ (m(k),n(k)),\ 0 \leqslant m(k) \leqslant M,\ 0 \leqslant n(k) \leqslant N,\ 1 \leqslant k \leqslant K \}$$
$$\text{with } m(k-1) \leqslant m(k) \text{ for all } k$$

where (m,n) denotes the match between region m in the left image and region n in the right image for $m,n > 0$, and, e.g., $(m,0)$ signifies that interval m in the left image is occluded. The score for the warp path is taken to be the sum of the scores of the region matches it contains.

The imposition of an ordering constraint (see Section 5.3.1), namely that there cannot be reversals of region order from one image to the other, now enables dynamic programming to be used as described in Section 2.4.5 in order to find the best set of correspondences between regions in the two rows. Some additional constraints are also imposed on the possible predecessors of any matching region pair, so as to limit the search space to physically reasonable match sequences.

This procedure, with matching carried out independently for each line, works well on good quality images. In the presence of noise, however, each image may contain a number of weak edges which have no counterpart in its stereo partner. A coarse-to-fine matching strategy (Lloyd, 1984), which starts by matching high contrast edges, and uses these matches to constrain subsequent matching stages at which successively weaker edges are considered, has proved to be effective in this case.

A very powerful constraint on stereo matching, which has not been exploited at all in the row-by-row matching described above, is that of continuity between rows. This figural continuity constraint can be incorporated by adding a second phase to the stereo algorithm, in the following way. The dynamic programming algorithm can be modified (Arnold, 1983) so that it produces, not only the best warp path between each pair of rows, but also all paths whose score is at least a given proportion of the optimum value. These suboptimum row matches may nevertheless contribute to the best overall image match in terms of optimising edge continuity. A relaxation labelling procedure (see Section 2.4.2) can now be applied, with the stereo row pairs as nodes and the set of possible warp paths for each pair as its label set. The compatibility of simultaneously assigning two given warp paths to two successive row pairs is taken to be proportional to the number of closed loops which are created by linking (a) matched edge elements in corresponding rows and (b) edge elements in successive rows of each image which can be regarded as forming part of a single connected edge in that image. Good results have been obtained with this approach, which is currently under development (Lloyd, Haddow and Boyce, 1985).

As an alternative to relaxation labelling, it is possible to use dynamic programming again in the second phase of the algorithm. In view of the fast hardware architectures which have been developed to implement dynamic programming matching algorithms (see Section 2.4.5), this approach might well lead to the design of efficient stereo matching hardware.

In Section 2.4 the potential power of variational approaches to provide a unifying principle underlying much of computer vision was emphasised. We shall end this section by describing some exciting current research which supports this thesis by showing that the stereo matching problem can be formulated in an optimisation framework, and solved very effectively by means of the simulated annealing technique. The approach will only be outlined in general terms here; more details can be found in Trivedi (1985a; 1985b).

The choice of cost function is motivated by the observation that nearby points in the 'world' will always give rise to image points which are in some sense close to each other. Moreover it will also be true in general that neighbouring pairs of features in one image of a stereo pair will correspond to features which are neighbouring in the other pair, the latter property holding without any critical dependence on the nature of the neighbourhoods used.

The above discussion suggests that the stereo correspondence problem may be solved by seeking to maximise the number of pairs of neighbouring features in one image which are matched to pairs of features which are neighbours in the other image. Let features i_1 and i_2 in the first image be matched to j_1 and j_2 in the second. Define

$$\begin{aligned} s(i_1, i_2) &= 1, \text{ if } i_1 \text{ and } i_2 \text{ both lie in some neighbourhood} \\ &= 0 \text{ otherwise} \end{aligned}$$

and similarly for $s(j_1, j_2)$. Then we must maximise

$$H = \sum_{i_1} \sum_{i_2} s(i_1, i_2)\, s(j_1, j_2) \tag{2.19}$$

where the summation is over all pairs of features i_1, i_2 in the first image. H will attain its maximum value when features in every local neighbourhood of an image are assigned matches which give each other the greatest possible mutual support. The result should be a stable, global solution to the correspondence problem exhibiting cooperative behaviour.

The form of H is similar to that of (minus one times) the Hamiltonian of a lattice gas with short range interactions. Such systems have been extensively studied in statistical mechanics, and much is known about their properties. For example, their stable equilibrium states are not sensitive to the precise form of the interaction, provided that it is short range. This suggests that $s(m,n)$ could be taken to be any function which falls off rapidly with increasing separation between m and n, and indeed preliminary investigation has shown that consistent, stable solutions are obtained for a broad range of neighbourhood sizes. Moreover, the ordered phase of a lattice gas in three or more dimensions is expected to be stable against thermodynamic fluctuations, suggesting that the cooperative solution to the stereo matching problem obtained by this method will not suffer from microscopic instabilities.

The optimisation procedure is carried out using simulated annealing (see Section 2.4.1), a robust technique which develops the analogy with statistical mechanical system models. With this approach the accuracy of the final solution does not depend on the quality of the initial 'guess', which may be a significant advantage for many practical applications. Preliminary results using the simple merit function (equation 2.19) are encouraging (Trivedi, 1985a). Modifications to equation 2.19 to take explicit account of the imaging geometry, and a detailed discussion of the underlying theory of the approach can be found in Trivedi (1985b). This is one of the first applications of the simulated annealing technique to stereo vision. It is particularly significant in view of the current interest in annealing, and in architectures such as the 'Boltzmann Machine' (see Section 2.4.1) for carrying out stochastic search procedures of this kind.

2.6 DISCUSSION

The study and development of algorithms is clearly at the heart of research and innovative work in image processing. However, the purpose of this book is also to consider implementation techniques, particularly system architectures and VLSI, together with the relationships between the implementations and the algorithms themselves. Hence, rather than providing an exhaustive description of known image processing algorithms and techniques, we have concentrated on discussing classes of algorithm, their implementation, and their use in image processing systems.

The chapter began with a general discussion of types of algorithm. Here, an attempt was made to provide a framework for classifying techniques, and the very important concept of a processing hierarchy was introduced. Parallels were drawn with the human visual system and with other areas of signal processing. In addition, the important difference between treating the image merely as a source of data, and treating it as a source of information concerning objects and their relationships, was stressed.

A number of specific algorithms were examined from the point of view of basic building blocks. Various architectural modules were identified and related to the algorithms. Some generic techniques were then considered in more detail. In particular, the value of variational principles in providing a unifying approach to a number of problems of early visual processing was discussed, and some recent research results were included. Finally, a small number of specific applications were described. The purpose of this was to illustrate how algorithms are chosen and configured, and how this procedure is affected by system constraints. Two more specific applications, computer vision and synthetic aperture radar, are discussed in detail in Chapters 5 and 6 respectively. Prior to this, the realm of VLSI achitectures and its relationship with image processing algorithms have to be further explored.

REFERENCES

Ackland, B.D., Weste, N. and Burr, D.J. (1981) 'An Integrated Multiprocessing Array for Time Warp Pattern Matching', *Proc. Int. Symp. Comput. Arch.*, Minneapolis, MN, 197–215.

Andrews, H.C. and Hunt, B.R. (1977) *Digital Image Restoration,* Prentice-Hall.

Arnold, R.D. (1983) 'Automated Stereo Perception', *Stanford Univ. Comp. Sc. Rep.*, STAN–CS–83–961.

Baker, H.H. (1982) 'Depth from Edge and Intensity Based Stereo', *Stanford Univ. Tech. Rep.*, STAN–CS–82–930.

Ballard, D.H. and Brown, C.M. (1982) *Computer Vision,* Prentice-Hall.

Brigham, E.O. (1974) *The Fast Fourier Transform,* Prentice-Hall.

Brown, C.M. (1984) 'Computer Vision and Natural Constraints', *Science,* **224**, No. 4655, 1299–1305.

Buxton, B. (1984) *Early Image Processing – Structural Techniques Motivated by Human Visual Response',* Notes on lectures given at the University of Surrey, Mar.

Canny, J.F. (1983) 'Finding Edges and Lines in Images', *MIT AI Lab. Tech. Rep.*, TR–720.

Cooley, J.W., Lewis, P.A.W. and Welch P.D. (1967) 'Application of the Fast Fourier Transform to Computation of Fourier Integrals, Fourier Series, and Convolution Integrals', *IEEE Trans. on Audio and Electroacoustics,* **AU–15**, No. 2, 79.

Diamond, M.D., Narasimhamurthi, N. and Ganapathy, S. (1983) 'Optimization Approaches to the Problem of Edge Linking with a Focus on Parallel Processing', *Proc. 8th Int. Jt. Conf. on AI,* Karlsruhe, W. Germany, 1003–1009.

Duda, R.O. and Hart, P.E. (1972) 'Use of the Hough Transformation to Detect Lines and Curves in Pictures', *Comm. ACM,* **15**, 11–15.

Elliott, H. and Srinivasan, L. (1981) 'An Application of Dynamic Programming to Sequential Boundary Estimation', *Comp. Graphics and Im. Proc.*, **17**, 291–314.

Faugeras, O.D. and Berthod, M. (1981) 'Improving Consistency and Reducing Ambiguity in Stochastic Labelling: An Optimisation Approach', *IEEE Trans. Pattern Anal. Mach. Intell.*, PAMI–3, 412–423.

Geman, S. and Geman, D. (1984) 'Stochastic Relaxation, Gibbs Distributions, and the Bayesian Restoration of Images', *IEEE Trans. Pattern Anal. and Mach. Intell.,* PAMI–6, No. 6, 721–741.

Gonzalez, R.C. and Wintz, P. (1977) *Digital Image Processing,* Addison-Wesley.

Gull, S.F. and Daniell, G.J. (1978) 'Image Reconstruction From Incomplete and Noisy Data', Nature, 272, 686–690.

Hinton, G.E. and Sejnowski, T.J. (1983) 'Optimal Perceptual Inference', In: *Proc. of IEEE Comp. Soc. Conf. on Comp. Vis. and Patt. Recog.*, Washington DC.

Hinton, G.E., Sejnowski, T.J. and Ackley, D.H. (1984) 'Boltzmann Machines: Constraint Satisfaction Networks that Learn', *Carnegie-Mellon Univ. Tech. Rep.*, CMU–CS–84–119.

Hopfield, J.J. (1982) 'Neural Networks and Physical Systems with Emergent Collective Computational Abilities', *Proc. of the Nat. Ac. of Sc. USA,* **79**, 2554–2558.

Hummel, R.A. and Zucker, S.W. (1983) 'On the Foundations of Relaxation Labelling Processes', *IEEE Trans. Pattern Anal. and Mach. Intell.*, PAMI–5, 267–287.

Hunt, B.R. (1977) 'Bayesian Methods in Nonlinear Digital Image Restoration', *IEEE Trans. on Computers,* **C–26**, No. 3, 219–229.

Jaynes, E.T. (1957) 'Information Theory and Statistical Mechanics I', *Phys. Rev.*, 106, 620–630.

Johnson, R.W. and Shore, J.E. (1983) 'Comments on and Correction to "Axiomatic Derivation of the Principle of Maximum Entropy and the Principle of Minimum Cross-Entropy" ', *IEEE Trans. on Inf. Th.*, **IT–29**, No. 6, 942–943.

Kinderman, R. and Snell, J.L. (1980) *Markov Random Fields and Their Applications,* Providence: *Am. Math, Soc.*

Kirkpatrick, S., Gelatt, C.D. Jr. and Vecchi, M.P. (1982) 'Optimisation by Simulated Annealing', *Science,* 220, 671–680.

Lloyd, S.A. (1983) 'An Optimization Approach to Relaxation Labelling Algorithms', *Image and Vis. Comp.,* **1**, No. 2, 85–91.

Lloyd, S.A. (1984) 'Waveform Matching Using Multiresolution Dynamic Programming', *Pat. Rec. Letters,* 2, 345–352.

Lloyd, S.A. (1985) 'A Dynamic Programming Algorithm for Binocular Stereo Vision', *GEC J. of Res.*, **3**, No. 1, 18–24.

Lloyd, S.A., Haddow, E.R. and Boyce, J.F. (1985) 'A New Binocular Stereo Algorithm', *Proc. of Fourth Scand. Conf. on Im. Anal.*, Trondheim, Norway.

Marr, D. and Hildreth, E.C. (1980) 'Theory of Edge Detection', *Proc. Roy. Soc. Lond.,* B 207, 187–217.

Metropolis, N., Rosenbluth, A.W., Rosenbluth, M.N., Teller, A.H. and Teller, E. (1953) 'Equations of State Calculations by Fast Computing Machines', *J. Chem. Phys.,* 21, 1087–1091.

Papoulis, A. (1965) *Probability, Random Variables and Stochastic Processes,* McGraw-Hill.

Peleg, S. (1980) 'A New Probabilistic Relaxation Scheme', *IEEE Trans. Pattern Anal. and Mach. Intell.,* PAMI–2, 362–369.

Prager, J.M. (1980) 'Extracting and Labelling Boundary Segments in Natural Scenes', *IEEE Trans.*, **PAMI–2**, No. 1, 16–27.

Pratt, W.K. (1978) *Digital Image Processing,* Wiley-Interscience.

Richards, J., Landgrebe, D. and Swain, P. (1981) 'On the Accuracy of Pixel Relaxation Labelling', *IEEE Trans. Syst. Man Cybernet.*, SMC–11, 303–309.

Rosenfeld, A., Hummel, R. and Zucker, S. (1976) 'Scene Labelling by Relaxation Operations', *IEEE Trans. Syst. Man Cybernet.*, SMC–6, 420–433.

Rosenfeld A. and Kak, A.C. (1982) *Digital Picture Processing,* Vols. 1 and 2, 2nd ed., New York, Academic Press.

Shapiro, S.D. (1978) 'Feature Space Transforms for Curve Detection', *Pat. Rec.,* 10, 129–143.

Shore, J.E. and Johnson, R.W. (1980) 'Axiomatic Derivation of the Principle of Maximum Entropy and the Principle of Minimum Cross-Entropy', *IEEE Trans. on Inf. Th.,* **IT–26**, No. 1, 26–37.

Thompson, W.B. and Barnard, S.T. (1981) 'Lower-Level Estimation and Interpretation of Visual Motion', *IEEE Comp.,* **14**, No. 8, 20–28.

Tikochinsky, Y., Tishby, N.Z. and Levine R.D. (1984) 'Consistent Inference of Probabilities for Reproducible Experiments', *Phys. Rev. Letters,* 52, 1357–1360.

Trivedi, H.P. (1985a) 'Application of Simulated Annealing to the Stereo Correspondence Problem', to appear in the *Proc. of COGNITIVA 85,* Paris.

Trivedi, H.P. (1985b), 'A Computational Theory of Stereo Vision', to appear in the *Proc. of IEEE Comp. Soc. Conf. on Comp. Vis. and Pat. Rec.,* CVPR–85, San Francisco.

Trussell, H.J. (1980) 'The Relationship Between Image Restoration by the Maximum A Posteriori Method and a Maximum Entropy Method', *IEEE Trans. Acoustics, Speech and Sig. Proc.,* **ASSP–28**, No. 1 114–117.

Ullman, S. (1979) 'Relaxation and Constrained Optimization by Local Processes', *Comp. Graphics and Im. Proc.,* 10, 115–125.

Wood, D. and Urquhart, R.B. (1984) 'An Architecture for a Connected Speech Recognition Algorithm', *Proc. 7th Int. Conf. on Pat. Rec.,* Montreal, Canada, 1197–1200.

Chapter 3
VLSI Architecture

I. N. PARKER

3.1 INTRODUCTION

We now come to the second half of the algorithm/architecture duality introduced in Chapter 1: the architecture of a processor for implementing the chosen image processing algorithm(s). In the same way that algorithms can be regarded either in terms of abstract computational models or specific implementations expressed in an appropriate programming language, so a similar dichotomy, between abstract structure and implementation detail, is apparent in the area of processor architectures.

The most common usage of the term 'computer architecture' is in the context of a high level abstraction, as a framework for development, in terms of a system model, rather than a particular instantiation of an implementation design. We will try to make a clear distinction between the type of a system object (the architectural model) and an instantiation of that object (the physical system). To make this distinction more concrete, note the important conceptual difference which exists between a particular architectural style of building design (e.g. Regency) and an actual physical building constructed in that style. The study of computer architectures is concerned with a number of interrelated topics; primarily, topological structure, data and control flows, performance criteria, hierarchical ordering, complexity theory and possible design-to-implementation routes. Some of these topics are explored in more detail in Chapter 4. This is still a young field, only recently released from the constraints of the traditional von Neumann 'world view' by the prodigious advances in VLSI technology and design over the last couple of decades. There is not a lot of formal knowledge as yet, but it is starting to develop as designers come to grips with the design of very large, highly integrated systems.

Section 1.4 has already looked briefly at the impact of VLSI technology on processor design and implementation, and the considerable advantages which can accrue to the system designer. These advantages tend to be subject to certain technological and organisational constraints, such as simple, regular data and control paths, local *vs.* global communications, utilisation of concurrency and making multiple use of data items. In this context the primacy, for low level image processing, of parallel array and systolic processors as important 'canonical' structures is clear. Despite this, it must be recognised that a specific custom solution can still give important advantages when maximum performance is required.

The following sections of this chapter will not only elaborate on some actual architectual examples, as they relate to image processing, but will also look at the influence of technological constraints and emphasise the vital centrality of a sustainable design methodology.

3.2 VLSI DEVICE TECHNOLOGY

It is not proposed to go into any real detail on the general issues of device technology, which are covered well in the literature (Mead and Conway, 1980; Gray, 1981). Most of the examples and assertions in this chapter are relatively 'technology independent'. Indeed it is right and proper that the digital algorithm, architecture and logic designer should be able to detach himself from the device physics and electrical characteristics of the medium, and free his mind to work on applications issues. A computer programmer need not know how virtual memory is implemented in a given computer to be able to use it effectively.

One of the major recent steps along this road in MOS technologies has been the introduction of multi- (overwhelmingly two-) level metallisation, which gives the designer almost complete two-dimensional freedom with almost perfect conductors. This frees him from many of the mainly analogue problems associated with the resistivity, and distributed *R–C* characteristics of polysilicon interconnect, and makes possible the implementation of architectures which are truly symmetrical, at the electrical level, in two spatial dimensions.

Some of the implications of linewidth scaling (*ibid*) are dealt with in Section 3.6, in the context of the architectural and algorithmic techniques discussed in this chapter.

3.3 VLSI DESIGN METHODOLOGY

A structured VLSI design path consists of three major phases: specification, implementation and evaluation.

The aim of the specification phase is the design of a detailed floorplan (down to leaf cells in many cases) and functional simulation of the chip to implement the function or algorithm required by the user or customer. This work can be extremely application specific, and in this case an optimal solution relies heavily upon intimate contact between user and designer.

Global, clock, power, ground and control signals are distributed at this stage, and the inter-module interfaces (imports and exports) are defined and wired, whether or not the modules actually 'abut' in all cases. Using the same methods, high level floorplan blocks are further decomposed to form the full nested physical hierarchy of the emerging chip. Functions which may be represented as floorplan blocks at these levels range from RAMs, ALUs, shifters, registers, etc., to the programmable 'macros' of silicon design, namely PLAs, ROMs, gate matrices and so on, and necessarily include primitive logic functions at the gate level.

Verification of the evolving design is carried out with register transfer or behavioural simulators, and the development of a detailed test strategy for the whole design is inextricably part of this process.

To facilitate the implementation phase, individual intermediate and low level blocks can now be 'coded' via schematic capture, and then simulated. This level of representation also provides a convenient interface to a number of fault simulators and automatic test program generators.

Leaf cells (i.e. those blocks at the furthest extremities of the floorplan 'tree' are then laid out as 'sticks' or mask–level rectangles, and verified/ optimised by means of a circuit simulator (usually some variant of SPICE). They are subsequently constructed into the complete drawing, and globally connected.

The use of any of a number of commercially available design checking software packages can provide a great deal of confidence in design integrity before mask-making costs are incurred. These detect geometrical and electrical rule infringements in the mask data, as well as indicating where the mask level network differs from that previously defined and verified in the logic description.

These tasks are followed by pattern generation, mask making, prototype fabrication, and parametric testing to ensure correct processing, followed by functional testing.

Within the general context of this specification to implementation pathway, it is the concept of 'floorplanning' which is emerging as the technique which best underpins the innovative and productive development of new computer architectures, especially those for high performance image processing.

3.3.1 Floorplanning

Whilst most system and logic designers will readily accept the usefulness of floorplanning as a means of optimising layout in silicon, this activity is often regarded as one performed in a few minutes or hours by the IC designer, when he is already in possession of a logic specification for the desired algorithm or function.

Fortunately, this approach usually works reasonably well in semicustom methodologies, particularly in the generic macrocell design style where chips are formed from a limited library of logic primitives, and floorplanning entails only the *placement* of chosen, predefined macros of known dimensions for minimum wire length and/or routing area.

However, in full custom design, emphasis falls upon the optimisation of speed, area, power dissipation, and general cost of implementation, and the function of most cells remains firmly 'to be decided' until quite late in the design cycle. In this environment, a fully detailed floorplan can take many weeks to develop, and is a natural and essential high–level design representation, providing:

a quick high level check on the overall feasibility, size and performance of a design,
the focus for specification and trade-off in an evolving system,
the hierarchy or tree upon which the fragments of lower–level implementations are to be hung, and
a means by which to plan and monitor the progress of a project.

In addition, consideration of the three-dimensional (*x, y* and *t*) aspects of a VLSI floorplan has historically yielded a large number of highly novel, powerful and effective solutions to traditionally difficult design problems. Perhaps the most famous example of this is the class of designs known as 'systolic arrays' (Foster and Kung, 1980; Corry and Patel, 1983; Wood *et al.* 1983), which exhibit many of the most attractive and desirable features of the VLSI medium (see Section 1.4.2):

They are regular and repetitive in structure, and thus relatively easy to implement.
They exhibit high degrees of parallelism and pipelining, leading to extremely high utilisation of each gate or transistor, and maximal processing throughput per square millimetre, and per chip.
They make extensive use of short range and nearest neighbour interconnection schemes, allowing unusually high clock speeds for relatively complex functions, and giving very dense implementation (in random logic designs, interconnect commonly dominates the chip size).

It is instructive to ponder on the likelihood of the invention of these powerful architectural techniques in the absence of a fundamentally '*x-y-t*' view of silicon design.

The systolic array is not however the only VLSI structure to come from the floorplan view; intelligent, associative and multi-dimensional memories, processor arrays in *N* dimensions, data-path structures, and even the ubiquitous PLA, among others, may be regarded as owing many of their origins and their widespread use in VLSI to their ease of layout, which in turn comes from a well structured floorplan.

3.3.2 Developing a floorplan

It is not enough merely to sketch a few boxes on the back of an envelope and call the result a floorplan; data and control flows must be defined, along with the paths taken by global power, ground, clock and test signals, and other (design specific) global parameters.

Note that the high–level floorplan which we are so far considering is immediately focusing attention on the most critical areas of the design in terms of feasibility; these global wires often dominate the final size of a chip, and given that some will undoubtedly lie on the critical timing path, its speed of operation.

Some software links

The common use of the word 'global' to describe the above signals, itself suggests that there may be more than a passing similarity in these VLSI design phases with the high–level design of software systems. Indeed, many of the concepts, abstractions, and design management techniques map exceedingly well from one discipline to the other.

In PASCAL (Jensen and Wirth, 1975) for example, a program is constructed as a hierarchy of blocks, each of which contains both data items and executable statements, the main part of each program block being used to 'join' the data structures and algorithms visible from inside that block.

PASCAL's recent descendant, MODULA-2 (Wirth, 1982) goes somewhat further in supporting the separate compilation of modules which may each contain their own initialisation code. This language makes a clear distinction between the definition of a module (the way it 'sees' and 'is seen' in the outside world throughout the IMPORT and EXPORT of VARIABLES, PROCEDURES and FUNCTIONS), and its implementation, which is essentially private.

Floorplan decomposition

IC design starts with the definition of a 'box' representing the whole chip or cell to be designed, whose contents (further boxes) may then be added and wired. This continues down the tree, so that a floorplan (or 'cell') is generally a hierarchy of floorplans.

This structure is familiar to most, being similar to those of 'drawing files' in first-generation IC artwork tools. The only major difference is that in this case the drawing is created from the top down, rather than from fully implemented leaf cells in bottom-up fashion.

Figure 3.1 shows the way in which floorplan definitions (by analogy with MODULA-2) are decomposed, by defining and then using instances of lower–level cells, until the leaf cell level is reached.

In parallel with this, functional details may be added as 'attributes' to existing cell definitions. These exist at the same hierarchical levels as their owners, and are shown as circles in Fig. 3.1. They represent the flesh on the bare skeleton of the floorplan.

It is vital that metaphors, dimensions and abstractions do not become mixed at this stage (a common failing in CAD systems). For example:

> Attempting hierarchically to decompose a NAND gate symbol (Boolean and temporal) into its equivalent transistor level schematic (electrical) is dimensionally incorrect; they are in fact alternative representations of the same thing.

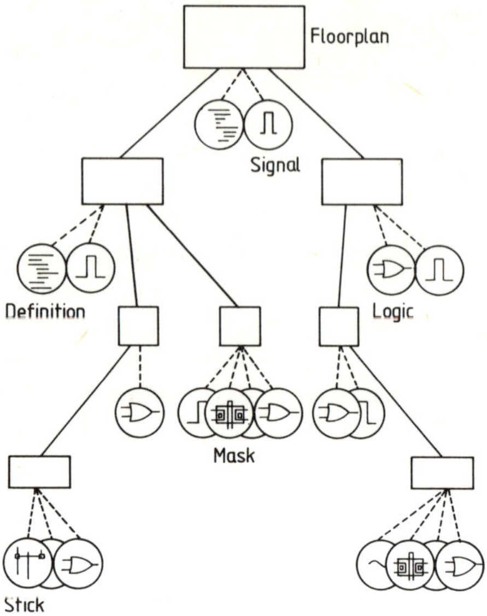

Fig. 3.1 Floorplan decomposition and the addition of implementation details.

Attaching geometrical information (such as width : length ratios) to a STICK diagram (topological and electrical) is similarly misleading; such parameters should be added as electrical gains, resistances, capacitances, etc.

Even the ubiquitous SPICE package falls into these traps in its signal description format. It forces rise and fall times (usually properties of a node) to be defined and performed sequentially in time with pulse width, which is clearly a property of the signal (i.e. the incoming data). Most designers see the mental arithmetic they have to perform here as merely a quirk; in fact it is an unfortunate dimensional inconsistency.

Inspection of Fig. 3.1 will also show that the floorplan acts as the 'trunk routes' by which one can arrive at any desired piece (or combination of pieces) of design data.

Floorplan anatomy

Floorplans at any level, whether they represent a complete chip or just a few gates, must have the same basic representations in a CAD system, so that they themselves may be used as components or macros in a higher level design. They consist of only five fundamental components, shown graphically in Fig. 3.2:

A 'bounding box' (preferably a rectangle) defining the outer dimensions of the cell under consideration.

'Ports' exist on the edges of the boxes, and represent the positions of

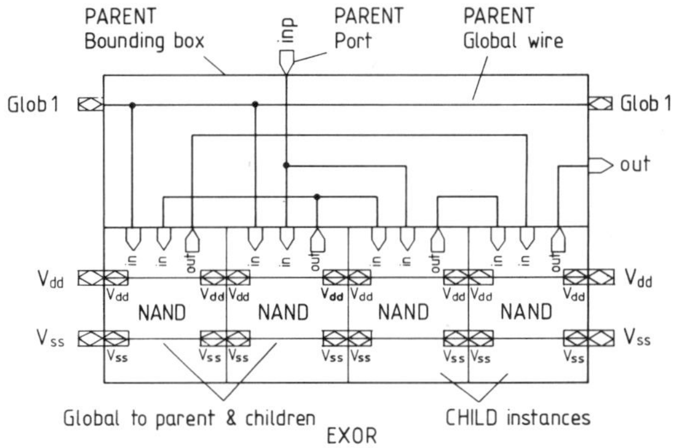

Fig. 3.2 Components of a fully detailed floorplan.

inputs, outputs, bidirectional lines and other signals flowing across cell boundaries.
'Wires' exist to connect to the other wires (via 'solder blobs') and ports which are visible from within the current cell implementation.
'Text' is used to annotate the floorplan, naming cells, ports and wires.
'Instances' of other floorplan definitions are accessible from any point in the hierarchy consistent with normal 'scope' rules. Standard or library functions are global to the whole design (as in PASCAL).

Ports on a component box which abut those on an adjacent box are connected by an 'implied' wire, as are those in a child cell which overlay ports in the parent. Wires which both enter and exit a given cell are considered global in the context of that cell, and become part of its definition.

Checking the design for geometrical, electrical and network errors becomes relatively straightforward in this hierarchical representation. Separation of definition from implementation serves to isolate a cell from its environment for these purposes, making truly hierarchical checking fully possible.

3.3.3 Functional implications

Early emphasis on the physical characteristics of a design through floorplanning often highlights unsuitable features in the target specification. Further, it is at this stage that they must be caught if efficient corrective action is to be taken. Some examples of this lie in the relatively familiar area of general data-path structures, comprising registers, latches, multiplexers, arithmetic and logic units, shifters, etc.

Data-paths are normally arranged on the surface of the semiconductor as a collection of single-bit 'slices' through the logic, where all cells in a slice

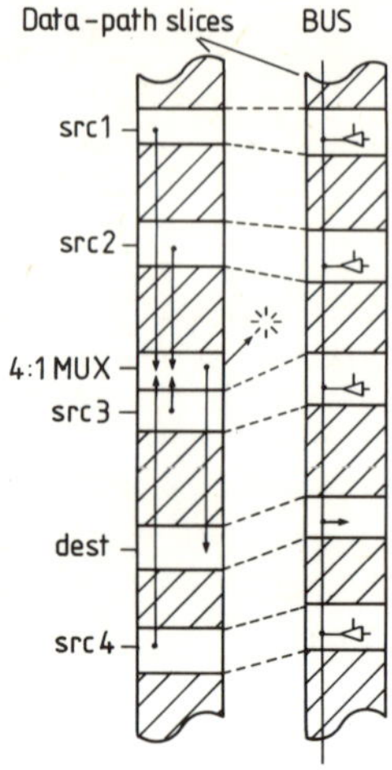

Fig. 3.3 Data-path improvement through inspection of the floorplan.

(e.g. single-bit latches, ALUs) possess a common horizontal pitch, and connect to their horizontal and vertical neighbours by abutment. Figure 3.3 shows a method of converting multiplexers in the logic of a slice into three-state buses in order to reduce the numbers of wires required to pass through many of the data-path cells.

Advantages of this technique are as follows:

it saves space,
it increases the designer's freedom to place cells near other related functions,
it allows the slice to be re-organised easily to contain new functions,
it gives a constant pitch and number of 'through wires' for cell layout, and
it allows the repeated use of a single, simple three-state driver cell.

In a typical data-path, all of these advantages can be achieved without sacrificing the basic functionality required by the specification (it may actually be increased since new architectural opportunities now exist!), and with propagation delays which may improve due to reductions in wire lengths and parasitics.

In other cases, however, conversion from logic to floorplan can be much more painful. It is easy enough to find random logic examples which make life virtually impossible for the IC designer. It is more instructive though to look at a system which is easy in custom VLSI, but which is not viable in other design media.

Figure 3.4 shows the floorplan of a cache memory for use in a microprocessor design. This cache stores recently used memory addresses and their associated data values for faster subsequent access than can be obtained from main memory. An address generated by the microprocessor is matched with recently used addresses in an associative portion of the

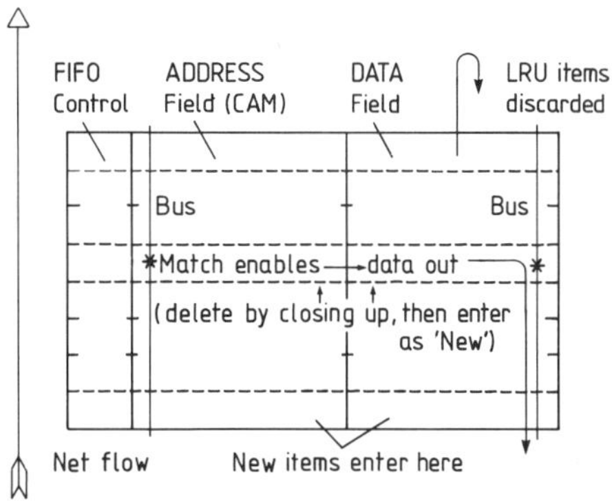

Fig. 3.4 An LRU associative cache memory.

memory (the address field) which, should the address be found, then generates an enable signal for the corresponding word in the standard data memory (the data field) alongside.

A least recently used (LRU) algorithm is superimposed on this structure by adding a third type of cell which implements a first-in, first-out (FIFO) function on the address and data values such that they move from bottom to top in the diagram. New pairs of values enter at the bottom end, with matches being deleted from the middle of the memory as they occur, prior to reinsertion as new values. The whole circuit therefore implements both the associative and the LRU algorithms, and the data structure upon which they operate, directly on silicon, with address and data items being maintained continuously in physical LRU order.

Again, as in the standard data-path, cells are connected by abutment, saving on routing time, complexity, circuit delays, and silicon area. This cache is clearly not a TTL implementable structure; the component parts are just not available. Neither would it be efficient to attempt to implement it in gate arrays, standard cells or macrocells. The power and novelty of this VLSI structure can be directly attributed to a designer visualising control and data flow in three dimensions.

3.3.4 Conclusions

The floorplan is the most powerful and essential high level design representation in modern top-down custom VLSI engineering. It represents to the designer the three real dimensions *x-y-t* in which he is most interested from the outset, and acts as a focus and organisational tool for developing the increasing amounts of detail which must be added to a design as work proceeds. In addition it simplifies the task of design checking greatly by formalising interfaces between blocks in many of the dimensions of design.

Paying heed to the physical characteristics of the design medium at an early stage can, and often does, lead to uniquely powerful design solutions which will otherwise be overlooked. It is an interesting property of floorplanning as a whole that designs tend to 'crystallise' rather than to emerge through sheer force of will.

A floorplan is not a drawing on the back of an envelope, it is the 'scaffolding' by which we can reach higher design efficiency and improved exploitation of VLSI technology.

3.4 A PROGRAMMABLE IMAGE PROCESSOR ARCHITECTURE

A large variety of tasks in the fields of image processing and numerical computation demand very high rates of instruction throughput. As discussed in Chapter 1, figures of 10^8 to 10^{10} instructions per second can be regarded as representing typical computational throughputs for tasks in low level image processing. Such high instruction rates cannot be supported by a conventional serial (von Neumann) computer architecture. A close analysis of many image processing algorithms reveals that the same sequence of instructions is normally repeated, in an essentially separable manner, on every pixel item, pixel by pixel, over the entire image. This inherent parallelism can be exploited to achieve very high computational throughputs. A number of non-von Neumann architectures which seem to offer a suitable solution are under investigation (see Section 1.4.2). One such architecture – the Single Instruction Multiple Data stream (SIMD) type – has shown great promise in the past, but has needed the emergence of VLSI for it to be truly feasible as a cheap and compact system. Such a VLSI system is the GEC Rectangular Image and Data (GRID) processor system.

3.4.1 A parallel processor array implementation

The architecture of the GRID processor system is that of the well known parallel processor array, shown schematically in Fig. 3.5.

It comprises a rectangular array of processing elements (PEs) under common control, all performing the same operations simultaneously on different data items stored in each PE's memory. The interconnections of each processor to its nearest neighbour enable each PE to access the data from its local neighbourhood of processors. A microprogrammed array controller provides for the sequencing and broadcast of both input/output data and instructions to the processor array. The overall system is composed of two distinct processors, a microcomputer and a GRID processor chip based processor array. The microcomputer uses the array as a co-processor by executing a high–level language and passing control to the GRID array whenever parallel constructs are encountered.

The major concentration of hardware in the system lies in the processor

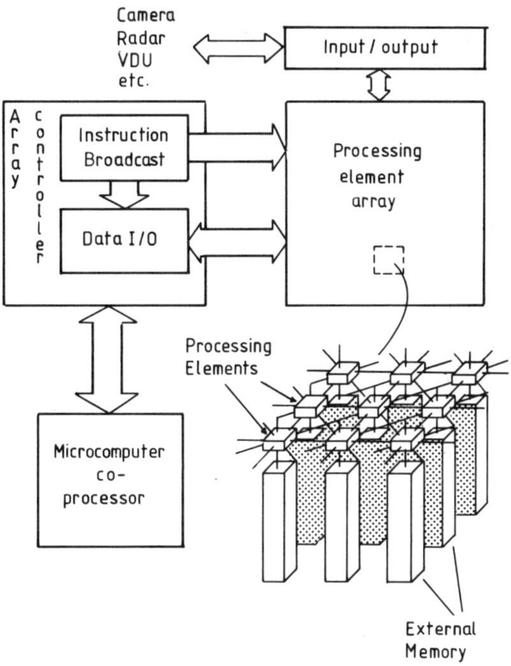

Fig. 3.5 A parallel processor-array architecture.

array. A custom designed VLSI GRID processor chip renders a cheap and compact implementation of this array. Fabricated in a two-level metal 2.5μm bulk CMOS technology, it contains over 50 000 transistors and currently supports a maximum clock frequency of 10 MHz. Arranged as a rectangularly interconnected array of 64 bit-serial processing elements, the GRID64 processor chip includes 64 registers per PE, together with the necessary circuitry for interconnecting many identical chips to construct compact and high performance array processors.

3.4.2 The processing element

The PE design, as shown in Fig. 3.6, has been optimised to implement image processing and numerical computation tasks efficiently. The bit-serial architecture of the PE allows flexible data formats. Variable wordlength data can be processed without any loss in efficiency. The two data-buses of the PE – the 'source' bus and the 'destination' bus – run along the length of the PE. Strung between these two buses are the extended register sets A and B, the ALU, the nearest neighbour switching (NNS) network and the special purpose one-bit registers. The extended A, B registers, implemented as a 64 words × 1 bit dual-port RAM (DPRAM), allow the storage of

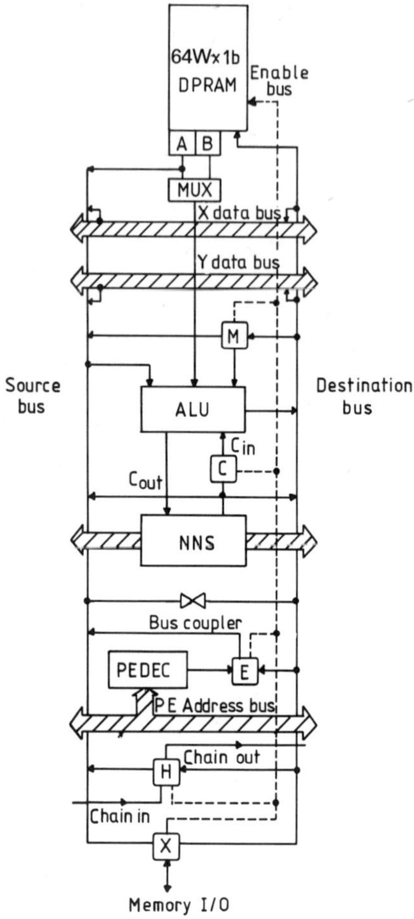

Fig. 3.6 A GRID processing element schematic.

intermediate results during grey level arithmetic operations without recourse to the slower external RAM. The NNS network allows each PE to access the contents of its 3 × 3 neighbourhood of PEs. Through 'corner turning' within the network, diagonal neighbours can be accessed on the square interconnection grid. This has the effect of providing local interconnection to eight neighbours without requiring any extra pins.

The multiplier (M) register holds the multiplier bit during bit-serial multiplication. It can be gated either with the A or B input to the ALU thus permitting the gated full-add (the basic step of bit-serial multiplication) to be performed in one PE cycle.

The enable (E) register supports conditional instructions at the PE level. When set, all write instructions to the associated PE are enabled. This register is set either as a result of a computation or via the PE addressing network. In the latter case, selected chips, rows, columns or individual PEs can be enabled whilst the remainder of the array is inhibited.

Lastly, all data transfers between the PE and its off-chip RAM are latched in the external store register (X). The provision of this port enables external memory expansion to suit particular applications.

3.4.3 The system architecture

The Data-Bus Architecture

The most direct path for data transfer between the GRID processor chip and the array controller is via the 'X-' and 'Y-buses'. They lie along the two axes of the array, with one line to each row (X) and each column (Y) of PEs. Rows or columns are selected by the PE addressing circuitry for enabling onto the X- and Y-buses respectively. The buses can be used to broadcast data from the array controller to every column/row on the chip by enabling the whole chip instead of a single column/row. This ability is useful in matrix manipulations for matrix transpositions.

Histogramming

Histogramming is a frequently used image processing operator for obtaining grey–level statistics or for counting the number of edge pixels for perimeter estimation. The H registers of all the 64 PEs are connected as a linear chain which ends in an on-chip counter. By clocking the data down the chain, a histogram count is obtained of all the bits set in the chain. The contents of the counters are then accessed, chip by chip over the array by the array controller, using the X- and Y-buses. Thus the histogram count over the entire image is accumulated at the controller.

Memory organisation

Typically in image processing applications, the input image will be larger than the PE array. The GRID processor system supports two methods for mapping the image onto the processor memory:

Windowed mapping scheme. In this method, the array is effectively stepped over the image, and windowed regions (the size of the array) of the image are stored in the external memory. Although simpler to implement, this method has a major disadvantage. During a nearest neighbour transfer operation, some data inevitably 'falls off' the edge of the array.

Pyramidal mapping scheme. This scheme overcomes the above problem by storing a $p \times p$ windowed region of the image in the external memory of each PE. The size of this window is a function of the ratio of the input image size and the array size. Moreover, in a pyramidal mapping scheme each PE can access any of the pixels within a $(2p+1) \times (2p+1)$ neighbourhood of PEs as shown in Fig. 3.7. As a consequence the size of the processor array is proportional to the degree of parallelism necessary

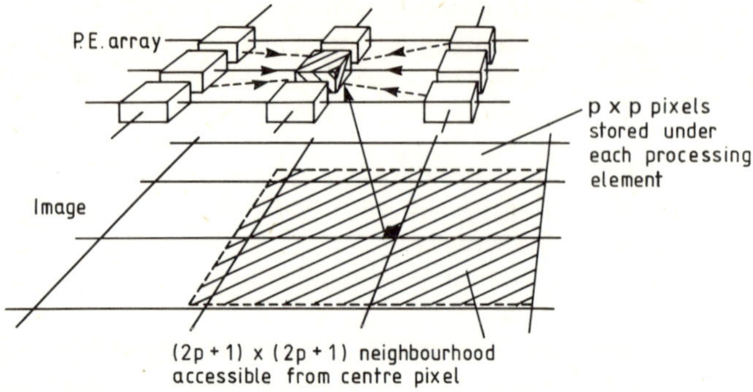

Fig. 3.7 Addressing a $(2p + 1) \times (2p + 1)$ neighbourhood.

to achieve the required computational throughput rather than the size of the image to be processed.

3.4.4 The chip design

Much of the chip design effort was concentrated on the individual PE, to achieve an optimum trade-off between silicon area and efficiency. The 'long and thin' shape of the PE, as evidenced by Fig. 3.6, was influenced by the pitch of the DPRAM and the need to route the wide control bus efficiently. The orthogonal data and control buses were implemented in a double layer metal process to minimise diffusion delays. Once designed, the PE was stepped across the chip, thus completing a large proportion of the layout task. The peripheral circuitry includes the instruction field demultiplexing and decode, and the histogram counter. Much of this was realised by automatically generated PLAs.

3.4.5 The input/output buffer

A 64 × 64 PE GRID array has an overall capacity of more than 40 000 million binary operations per second at a clock rate of only 10 MHz. However, most of this power is worthless if the array cannot be provided with data at sufficient speed to keep it busy (there is an obvious limit to the number of useful operations one can perform on any single image at pixel level only!).

Furthermore, whilst it may initially appear attractive to employ the processor array alone to input and reformat video data, solutions of this type commonly demand close to 100% of the processor time during a video field input, leaving very little time within the field for computation, and therefore making real-time processing impossible.

To attack the problems of providing high bandwidth input/output (I/O) of video data with minimal processor overhead, the GRID system employs a novel, two-dimensional video line buffer, consisting of eight (or optionally more) identical custom designed VLSI chips. These perform the following functions:

bandwidth magnification through the buffering of a complete raster line and subsequent output to the processor over a wider bus (this is pipelining in the macro sense),
reformatting of the image from pixel-parallel to the bit-serial representation required by the processing elements, and
programmable mapping of an incoming raster line into pyramid fragments supporting a wide variety of image and pixel size combinations.

The GRID I/O buffer allows video fields to be input, processed, and output, in three consecutive field periods. These operations are carried out in overlapped or pipelined fashion, such that as the first processed field in a sequence is being output, the second is receiving processing, and the third is being input to the processor array, giving true real-time throughput, with a latency of two video fields. The I/O subarchitecture and timing diagrams are shown in Fig. 3.8. In situations where data rates must be matched, a first-in, first-out (FIFO) buffer is used external to the I/O subsystem proper.

The GRID I/O buffer is a two-dimensional, cascadable, sequential access memory, based on an eight-transistor, differential, two-port static CMOS memory cell. The two ports of the cell are arranged orthogonally and are bidirectional, such that it is possible to read and write data in either the X or the Y directions when the cell is enabled by the associated word line. Figure 3.9 shows a schematic of the memory cell. An array of these memory cells in two dimensions is addressed by programmable shift registers along its top and left sides, holding a single binary 1 which activates each memory word enable line (X or Y) in turn as it is clocked through the registers.

Partitioning the horizontal shift register into blocks of eight stages (see Fig. 3.10(a)), and programming the lengths of these blocks in a global fashion allows varying amounts of data to be spread evenly throughout the line buffer, thus supporting various pyramidal mapping ratios. The path taken by the 1 access bit when programmed to access two locations in every eight is shown in Fig. 3.10(b) and has led to the coining of the description 'bouncing-bit' for this type of access scheme. A similar technique applied to the vertical register gives support of varying pixel sizes.

Figure 3.11 shows the two phases in a video input operation. First, pixels are written into vertical words in the memory, with the access bit proceeding towards the left in the diagram laying down data as it moves. Subsequently, the vertical shift register accesses one bit at a time from each of eight pixels in the memory, selected by 8:1 multiplexers onto the chip outputs. The latter operation is repeated until all active memory locations have been accessed.

As it happens, the 8:1 multiplexers are also controlled by a programmable shift register, which may be thought of as proceeding into the paper in Fig. 3.11. All three dimensions of register are cascadable to increase the length

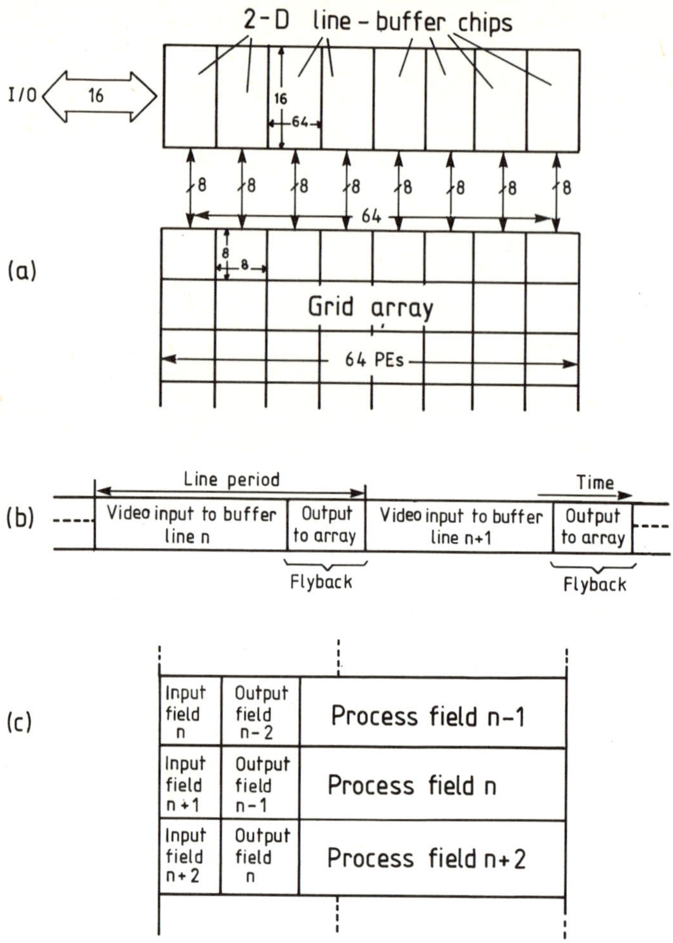

Fig. 3.8 The GRID video I/O organisation. (a) Subarchitecture of the two-dimensional line buffer chips; (b) raster input timing; (c) overlapped field input-process-output operation.

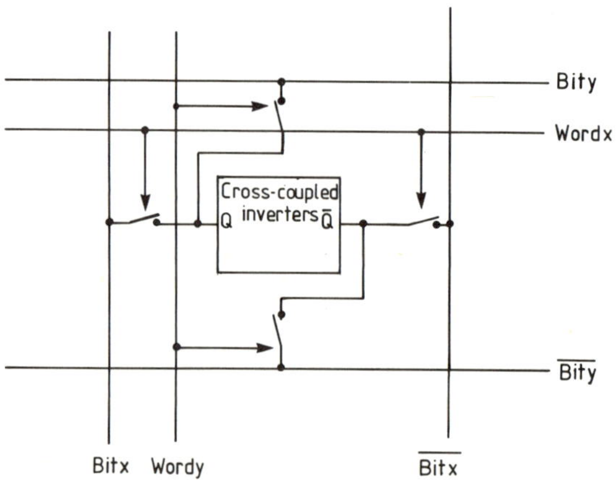

Fig. 3.9 Schematic of a two-dimensional memory cell.

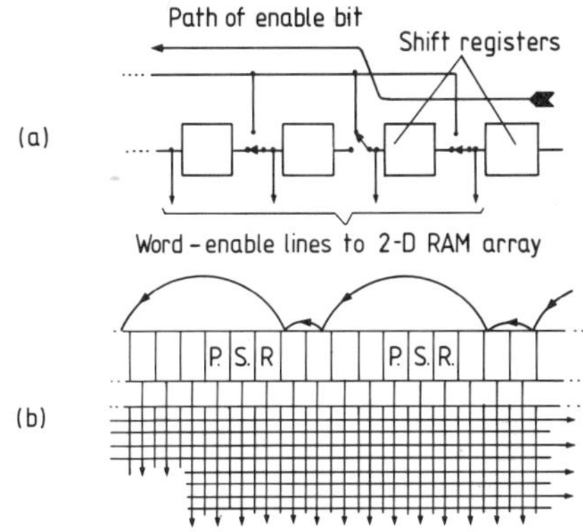

Fig. 3.10 Programmable shift register. (a) A portion of the horizontal shift register; (b) path of a 'bouncing bit' which accesses two locations in every eight.

of a raster line, the width of a pixel, or the number of pixels in each block of memory associated with an I/O line to the processor array. Finite-state machines in each dimension note the passage of the access bit on and off the chip, and control the data output three-state buffers accordingly to allow 'bussing' of the data signals.

The choice of a memory over serial/parallel shift registers was an important design decision. The function of the above circuit is similar in

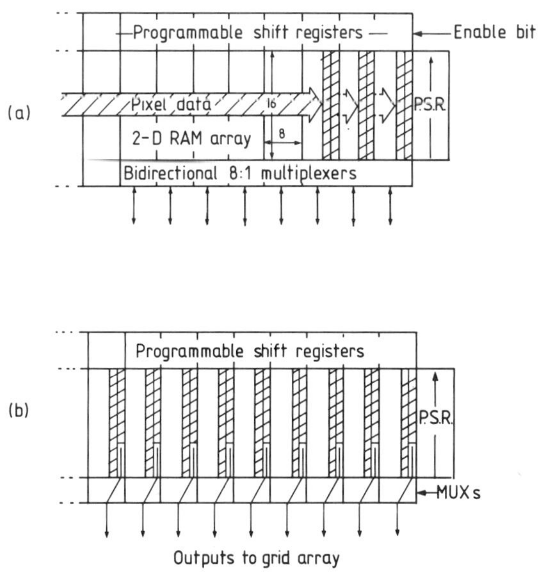

Fig. 3.11 The two-dimensional line buffer chip. (a) Video input phase; (b) GRID output phase.

many ways to that of a serial/parallel shift register, and it may be thought at first sight that the latter form of implementation would perform equally well in this application. However, shift registers possess three major characteristics which make them unsuitable in this case:

They dissipate large amounts of dynamic power since, on average, they will change state on every other clock cycle. By contrast, the memory architecture described here moves each piece of data only twice, inputting it to exactly where it is required for later output.

They consume twice the area of a memory cell, being composed of two transparent latches with switching rather than one as in the case of the memory.

Shift registers require both an input and an output port on each clock cycle for cascading. The memory architecture does not move data around in the same way, and requires only an input or an output port in a given direction at any time, which may be bus connected. This drastically reduces the number of I/O pins on the chip in this implementation.

Given the decision to use a memory architecture, it becomes possible to add the extra mapping and length programming features in a much more straightforward manner than with a shift register, and with very little silicon area overhead.

In conclusion it should be pointed out that the GRID I/O buffer is an example of an intelligent memory chip, where the function is distributed amongst the data storage itself. The passing of 'token-like' access bits between chips has a major advantage in almost completely removing decoding and sequencing logic from around the I/O subsystem, and each chip may be regarded as a fully cascadable subsystem in its own right, or alternatively as a package containing its own share of the algorithms and data structures needed for the whole job.

3.5 SPECIALISED VLSI PROCESSORS

An alternative approach relies on the identification of common operators, which make up more complex algorithms, and their customising as special purpose integrated circuits that operate on an incoming stream of video data.

A considerable amount of experience in low level image processing has already identified many of the algorithms which are 'canonical' at the lower levels of image understanding (see Chapters 1 and 2). Most approaches rely on 'factoring out' the operators necessary to implement these algorithms and integrating these as special purpose/high performance processing elements. A number of potential circuit architectures have been investigated for integrating a variety of kernel functions – convolutions, correlations, spatial filtering – in a form suitable for the support of real-time (TV video-rate) computation.

Specialised VLSI image processors such as those described below frequently operate on streams of data, permitting the use of pipelined and systolic architectures. This is in contrast to SIMD parallel processors, such as GRID, which operate on static frames of data. The following subsections describe some examples of special purpose building blocks for image processing.

3.5.1 A two-dimensional median filter

One example of this approach, a two-dimensional median filter, is described in more detail below. This is a specialised hardwired functional unit which employs extensive pipelining and can operate directly on the serial scan of an incoming image. There is therefore no requirement to process a certain number of pixels within a fixed period as with a processor array. Architectures best suited to this function are based on a design approach utilising 'systolic arrays' (see Section 1.4.2). This type of design is particularly well suited to a VLSI implementation, it being characterised by the use of only a few types of simple cells, an orthogonal data flow and control organisation, the use of extensive pipelining and parallelism to achieve a high computational throughput and the multiple use of input data.

The median filter is a tool often used to reduce the effect of discrete impulse noise on an image. In the one-dimensional case the centre pixel in the process window is replaced by the median of the sequence of pixel values within the window. As an example consider the sequence

180 190 200 110 120

The median is found by sorting this sequence and choosing the centre value of the sorted sequence. Hence the centre pixel of the processing window (here of length five), 200, is replaced by the median of the sequence, 180. In the two-dimensional case a number of geometrics are possible, for instance the 5 × 5 square window or the 5 × 5 plus-sign-shaped window depicted in Fig. 3.12. The basic computational element of the median filter is a sorting network. A nine-input sort array which utilises serial arithmetic is depicted in Fig. 3.13. The array is composed of two types of cell, a delay cell and a sort cell. The delay cell is used for delay equalisation. The sort cell consists of a gated magnitude comparator and a crosspoint switch. Two serial input streams enter a sort cell, initially the crosspoint switch is set in a default

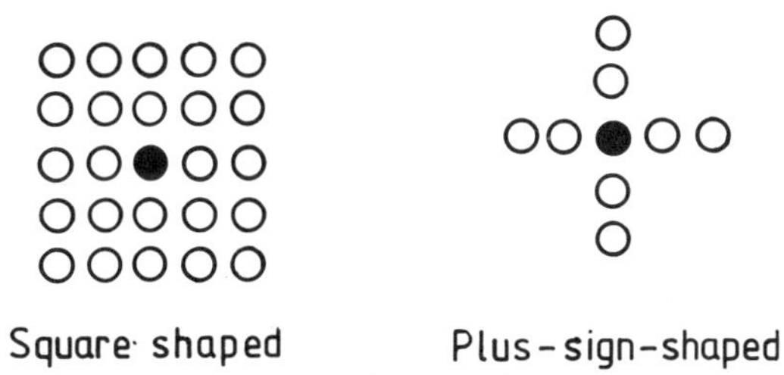

Fig. 3.12 Typical median filter windows.

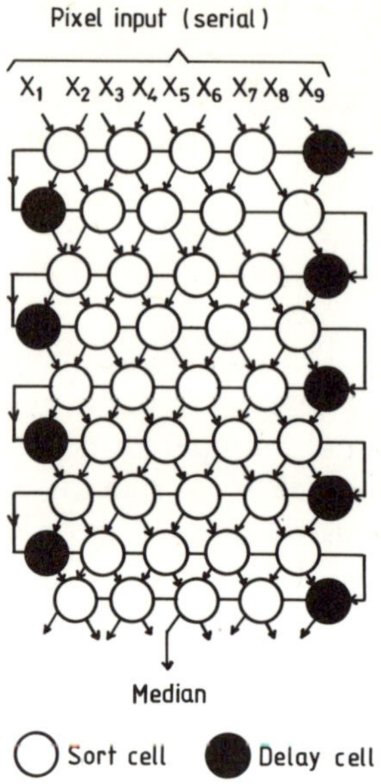

Fig. 3.13 A nine-input sorting network.

direction (i.e. $A_{in} \rightarrow A_{out}$, $B_{in} \rightarrow B_{out}$). Comparison of the two bit streams will result in the switch either remaining in its default position or crossing over to its alternative position (i.e. switching the bit streams $A_{in} \rightarrow B_{out}$, $B_{in} \rightarrow A_{out}$). The sort array implements a conventional 'bubble–sort' algorithm. Each sort cell of the array compares two numbers and routes the larger upward, the smaller downward. A sequence of numbers is grouped in pairs. Successive application of a column of sort cells will cause at each state larger numbers to 'bubble up' and the smaller to 'sink down'. The sorting takes place bit serially and is pipelined at the bit level. The throughput of the median filter is determined by the delay through a sort cell. Once the pipeline is full the pixel delay becomes the cell delay times the pixel word length.

In the case of a 5 × 5 plus-sign-shaped median filter the nine pixels of the plus filter would be sorted with the fifth element of the sorted outputs (i.e. the median of the input sequence) replacing the centre pixel of the plus window. For this case the sorting network requires only thirty sort cells. Increasing the number of inputs to the network by either changing the shape of the filter (5 × 5 square) or increasing the size of the filter kernel (7 × 7 plus) results in a rapid growth of the sorting network (e.g. 94 cells for a 5 × 5 square filter). Alternative network structures exist which have a lower growth factor but these suffer from an inherent irregularity in network wiring. These therefore offer little improvement over the more regular bubble–sort when dealing with the sorting of sequences up to a length of twenty-five pixels.

The technology chosen for implementing the spatial filter chip, capable of video-rate computation, was a 3 μm CMOS silicon-on-sapphire (CMOS/SOS) process. The chip consists of a bubble–sort network capable of sorting twenty-five pixels, together with on-chip circuitry for word length programming pixel storage, and for serial/parallel and parallel/serial conversion.

3.5.2 **One-dimensional rank order filter** (by Dr R. B. Urquhart, GEC Research Laboratories)

Although the systolic median filter described above is an elegant way of sorting small sets of pixels, it is not suitable for sorting data in large windows. The size of the sorting array grows in proportion to the square of the number of elements in the window, thus for a $1 \times k$ linear window, the size of the array is $O(k^2)$. This approach could be classified as being 'window oriented' because to filter the image we require to look at each possible window, sort all the pixels in each window and select the medians; however, this is wasteful because every pixel is sorted separately in every window in which it occurs. An alternative 'pixel oriented' approach is to evaluate the rank of a pixel in every window in which it occurs and to extract the pixel if it has the required rank. This approach also means that we can apply filtering using any rank without incurring extra cost.

Consider a stream of pixels and a one-dimensional window W which is applied to the stream (Fig. 3.14). As the window passes across the pixel sequence a given pixel appears in the k different window positions. Therefore, if the rank of a pixel is evaluated in the first window in which it occurs, its rank in the following $k-1$ windows can be found by modifying its initial rank rather than re-evaluating its rank from scratch. Consider the pixel sequence of Fig. 3.15, where we have a window of length 5 (k=5).

Let us assume that we already know the ranks of the pixels in window W_7. Then the next window, W_8, can be obtained from W_7 by deleting pixel X_3 and inserting pixel X_8. Since pixel X_8 has not previously been examined, its rank must be evaluated from scratch. However, the ranks of pixels $X_4 \cdots X_7$

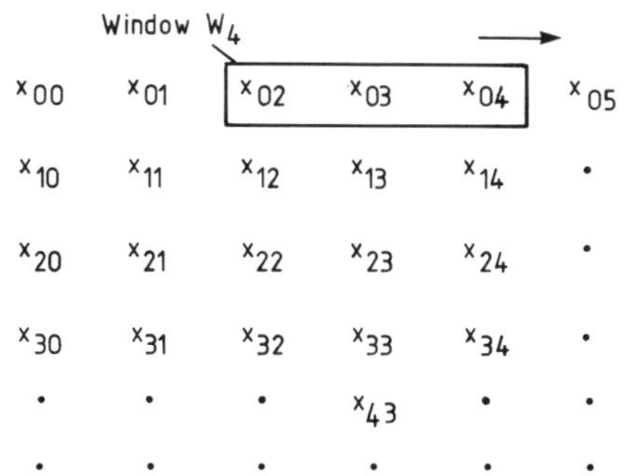

Fig. 3.14 A one-dimensional window applied to a pixel stream.

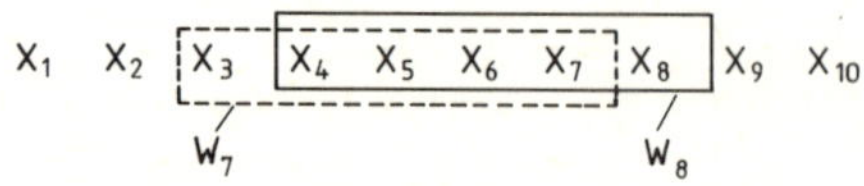

Fig. 3.15 A length-5 one-dimensional pixel window.

have been evaluated in W_7. Therefore, for each of these pixels, the rank r_{i8} in window W_8 can be found from r_{i7} by comparing pixel X_i with X_3 and X_8. Each window requires $O(k)$ comparisons for the pixel oriented approach compared with $O(k^2)$ comparisons for the window oriented approach.

Therefore, rank order filtering using a pixel oriented approach requires the solving of three subproblems: rank evaluation, rank modification and rank extraction. We need to evaluate the rank r_{ii} of pixel X_i in the first window in which it occurs, W_i. Having obtained r_{ii} the rank of pixel X_i in windows W_{i+1} to W_{i+k-1} is obtained by a simple rank modification procedure. Finally the rank extraction problem involves the transferring of pixel X_i to an output stream if it has the required rank for any window W_{i+h}, $h = 0,1,\ldots.,k-1$. Throughout this section we illustrate arrays for $k = 5$.

The rank evaluation problem is defined as follows:

Given a window of length k and an input sequence $(X_1,X_2\ldots..X_n)$ compute the output sequence $(r_1,r_2\ldots..r_{n+1-k})$ defined by

$$r_i = \sum_{h=1}^{k-1} X_i \circledcirc X_{i-h}$$

(where $a \circledcirc b = 1$ if $a > b$, zero otherwise)

Depicted in Fig. 3.16 is one of the many systolic arrays suitable for performing rank evaluation for a window of length 5.

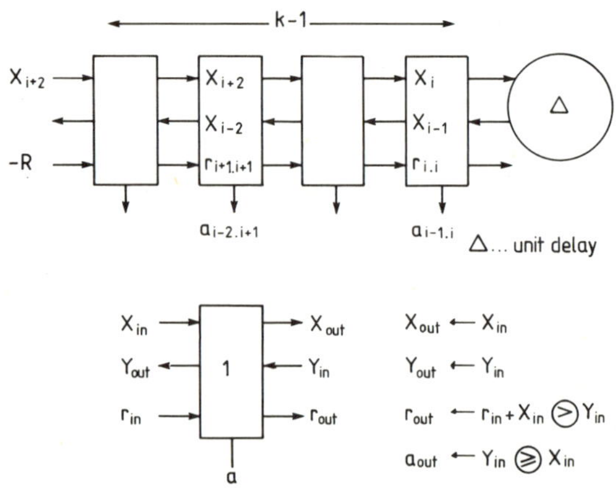

Fig. 3.16 A systolic array for performing rank evaluation for a window of length 5.

The status signal a_{out} is generated for the insert/delete array; it is used to update out the rank held in the insert/delete array.

The insert/delete problem is defined as follows:

> Given a rank r_{ii} of an element x_i within a window W_i compute the rank $r_{i,i+h}$ of the element x_i within windows W_{i+h}, $h = 1$ to $k - 1$. The rank $r_{i,i+h}$ is computed by replacing X_{i-k+h} by X_{i+h} ($h = 1$ to $k - 1$). Updating the rank requires that x_i be compared with x_{i-k+h} and X_{i+h}. The effect on the rank will either increase it by 1, decrease it by 1 or leave it unchanged.

A systolic array suitable for performing this operation is shown in Fig. 3.17.

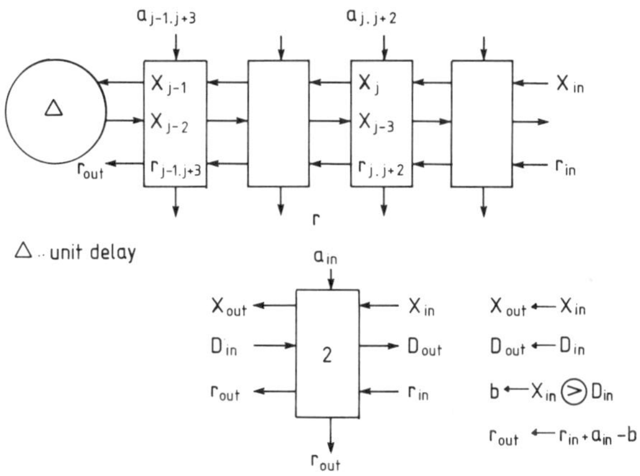

Fig. 3.17 A systolic array suitable for performing insert/delete operations.

To simplify the requirements for the rank extraction array, the rank $r_{i,i+h}$, which the insert/delete array modifies, is first subtracted from the rank which the rank extraction is to extract. Hence, when an element has the required rank then the modified rank produced by insert/delete array will be zero. This zero condition is detected by the array below.

The rank extraction problem is defined as follows:

> Given that the insert/delete array determines whether an element has the required rank, extract the element if it does. Figure 3.18 depicts a systolic array which performs this function. The result stream flows from left to right through the array. The input data sequence, X, flows from right to left. If X_{in} has the required rank, then X_{in} replaces Y_{in}. Otherwise Y is passed through unchanged.

The three arrays can now be formed into a single rank order filter. This processor is shown in Fig. 3.19.

For image processing applications we still require filtering to take place in two dimensions; however, most of the properties of the true two-dimensional median filter are preserved by using a separated median filter.

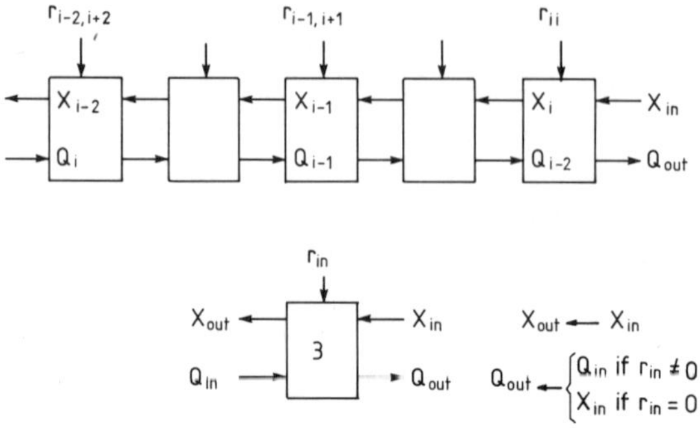

Fig. 3.18 A systolic array which performs element extraction.

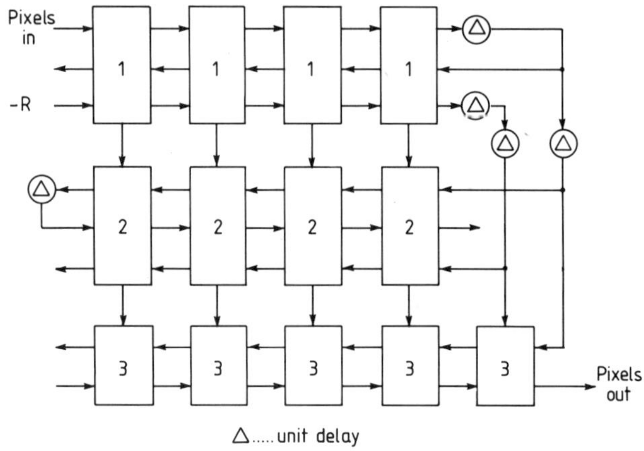

Fig. 3.19 A one-dimensional systolic rank order filter.

A separated median filter filters pixels by one-dimensional median filter along rows, then filters the resulting image along the columns. Thus two-dimensional separated median filtering can be performed by using two one-dimensional systolic rank order filters.

3.5.3 Two-dimensional convolution (by Dr R. B. Urquhart, GEC Research Laboratories)

The two-dimensional convolution of an input $x(m,n)$ and an impulse response $h(p,q)$ is given by

$$y(u,v) = \sum_{i=0}^{M-1} \sum_{j=0}^{N-1} h(i,j)\, x\,(u-i,v-j)$$

A number of systolic solutions are available to implement this operation, but one requires data in a non-raster scan format and another is difficult to cascade. In this case a pipelined convolver architecture with internal data-broadcast is appropriate (Fig. 3.20). Coefficients $h(i,j)$ are stored in registers and input data is broadcast to multiplier accumulators. Partial convolutions pass through the pipeline until all contributions have been added. Figure 3.20 shows a 3×3 convolver which requires 3 pixel data inputs. Given an $N \times N$ image and a $k \times k$ window, a two-dimensional convolver requires the use of $k-1$ line delay buffers (Fig. 3.21), each buffer containing N pixels. This means that only $N(k-1)$ pixels of intermediate storage are required compared with N^2 for a frame store.

3.5.4 **Two-dimensional edge detection** (by Dr R. B. Urquhart, GEC Research Laboratories)

An important application of convolution is edge detection. Low level edge detection can be performed by convolving a window with a template. The pixel can then be assigned to a particular edge direction by finding the maximum convolution among the templates. For example, we could use the following masks to detect vertical, horizontal and diagonal edges:

$$\begin{matrix} 1 & a & 1 \\ 0 & 0 & 0 \\ -1 & -a & -1 \end{matrix} \qquad \begin{matrix} 1 & 0 & -1 \\ a & 0 & -a \\ 1 & 0 & -1 \end{matrix} \qquad \begin{matrix} a & 1 & 0 \\ 1 & 0 & -1 \\ 0 & -1 & -a \end{matrix} \qquad \begin{matrix} 0 & -1 & -a \\ 1 & 0 & 1 \\ a & 1 & 0 \end{matrix}$$

where a typical value of a might be 2.

This means that a general purpose two-dimensional convolver would provide much greater precision in the convolution weights than would be required for two-dimensional FIR filtering. We could therefore consider a special purpose VLSI structure for edge detection (Fig. 3.22). This consists of four two-dimensional convolvers (corresponding to four convolution masks) which are pipelined together. The result of each convolution is compared with the input from the left, and the maximum value and tag passed on to the next stage in the pipeline. At the end of the pipeline the maximum edge strength and an associated tag indicate likely edge direction. Line delay considerations for the edge detector are identical to those for the two-dimensional convolver.

3.6 FUTURE DIRECTIONS

One particular area of current research in VLSI architectures and technologies appears to offer very attractive benefits for future exploitation in low level image processing and handling, and that is wafer scale integration (WSI). This sub-technology promises at least one, and eventually perhaps two to three orders-of-magnitude increase in function on a single substrate over what might be called 'orthodox' VLSI, with single devices

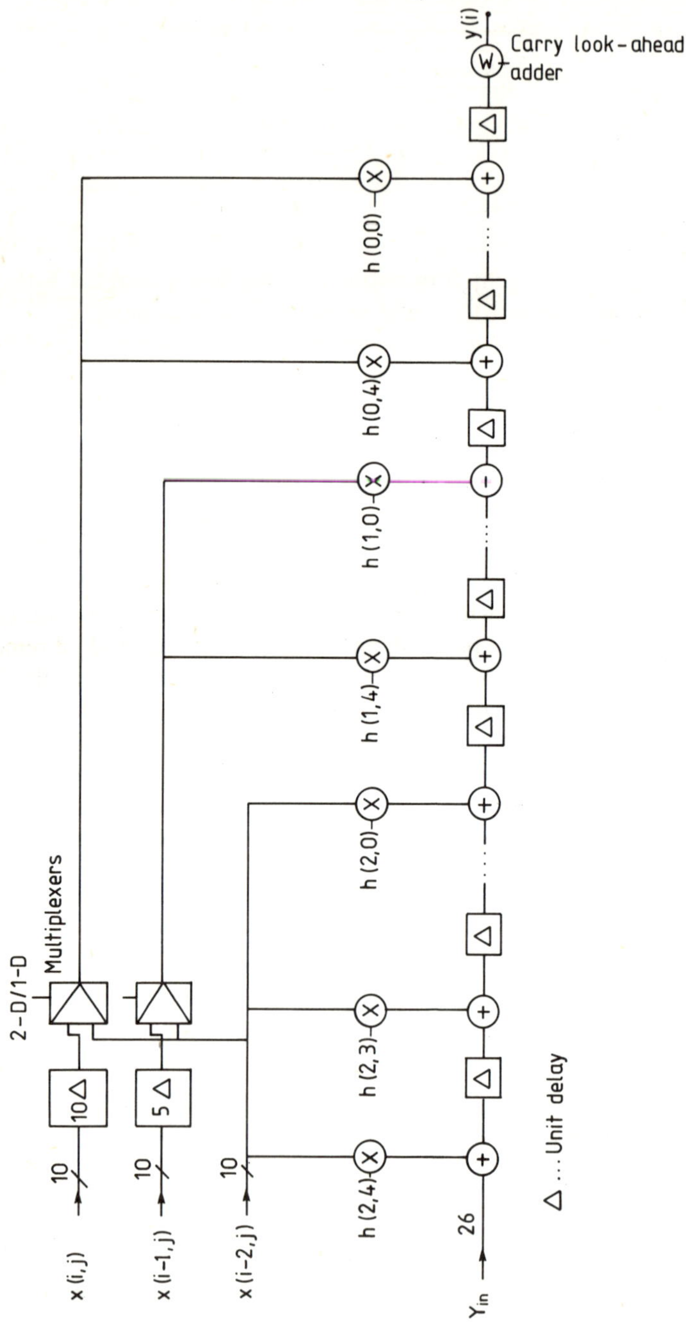

Fig. 3.20 Two-dimensional convolver architecture.

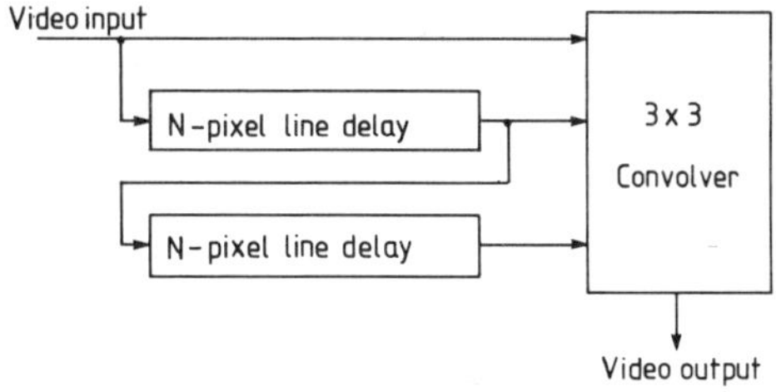

Fig. 3.21 3 × 3 convolver with line delays.

holding, say, an intelligent framestore or even when coupled with CCD technology a smart sensor to mount in the back of a camera.

However, a number of major questions remain to be answered before WSI can become a technical and economic reality in generalised system and logic design for image processing:

Can fault tolerant, redundant and/or self-healing techniques which currently exist for relatively simple structures such as memories and systolic arrays be extended to cover some of the more powerful, intelligent and hence more random architectures?

How does one distribute 'high quality' power supplies, at several amps or tens of amps, on a chip which measures some centimetres on a side (this is widely acknowledged to be much more difficult than the task of removing the heat produced by the operating device – a challenge in itself!)?

Similarly, one must broadcast global control signals, clocks, etc., or suffer the penalties in efficiency of self-timed logic.

At wafer scale dimensions, inductance starts to play a major role in circuit performance. Noise margins are effectively reduced by ringing, and it seems likely that some penalty will have to be paid in interconnect density in order to control its resistivity.

How does one 'test' (and presumably reconfigure) a wafer scale circuit, and treat the results with confidence? What are the reliability issues?

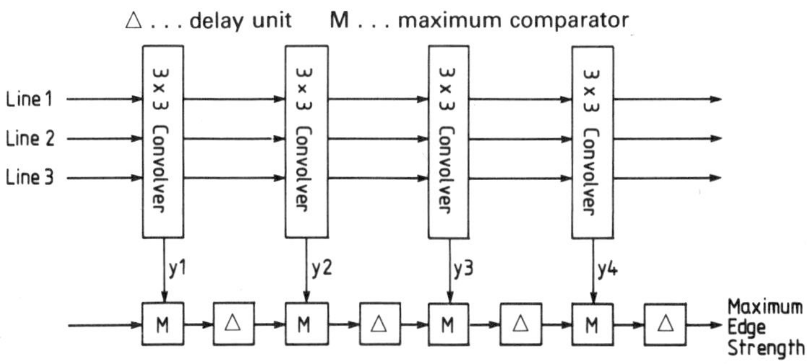

Fig. 3.22 Two-dimensional edge detector.

Few of these questions have anything more than very case-specific answers at present; there is certainly no identifiable design methodology for controlling the problems and their inherent complexity.

Another issue which will impact the future is the success (or not) of attempts to produce three-dimensional integration (initially two layers of transistors, which seem only to offer density rather than architectural advantages).

As far as orthodox VLSI is concerned in image processing, we are sure to see numbers of designs emerging in the class of dedicated image processors, particularly making use of systolic arrays, with their inherent advantages of throughput and ease of design. The basic pixel rate of the input data is likely to remain relatively constant, as dimensions scale, in comparison with the improving speed capabilities of MOS VLSI logic. The main direction of exploitation will therefore be in increased functional density, and improvement in architectures/algorithms for implementation. Due to the vagaries of real, as opposed to 'classical' scaling theory (e.g. power supply voltage remaining constant through several generations of process) the internal power dissipation of a typical chip will be likely to remain fairly constant or increase slightly. Architectural and logical ingenuity will be vital to control device pin-out, and the increasing contribution of this portion of the design to a chips power budget.

This is also true in the programmable device sphere (e.g. the GRID processor array design described in Section 3.4). Connections to external image memory in this design, as described, grow as the square of the number of processing elements p on the side of the sub-array. Nearest neighbour connections grow linearly with p. These are very practical and related limitations on the increase in the number of PEs per chip, which have solution only in algorithm and system–level architectural trade-offs.

Most of this discussion is of course centring on the requirements of low level (i.e. pixel) image processing, filtering, edge finding, etc. Still more work remains to be done at the higher levels of what might be called image understanding before we can suggest that our systems can actually 'see' rather than merely 'look'. Interpretation of low–level objects (edges, blobs, etc.) into scene descriptions is yet another potentially rich area of the application of VLSI technology, with associative processors, content-addressable and other 'smart' memories appearing high on the list of future requirements.

REFERENCES

Corry, A.G. and Patel, K. (1983) 'Architecture of a CMOS Correlator', *IEEE Int. Symp. on Circ. and Sys.*, May.

Foster, M.J. and Kung, H.T. (1980) 'The Design of Special-purpose VLSI Chips', *Computer,* 26–40.

Gray, J.P. (ed.) (1981) *VLSI 81: Very Large Scale Integration,* Academic Press.

Jensen, K. and Wirth, N. (1975) *PASCAL User Manual and Report*, Springer-Verlag.

Mead, C.A. and Conway, L.A. (1980) *Introduction to VLSI Systems*, Addison-Wesley.

Wirth, N. (1982) *Programming in MODULA–2*, Springer-Verlag.

Wood, D., Evans, R.A. and Wood, K.W. (1983) 'An 8-bit Serial Convolver Chip Based on a Bit-level Systolic Array', *Custom Integrated Circuits Conference*, Rochester, NY.

Chapter 4

From Algorithms to Architectures

R. J. OFFEN and P. N. SCHARBACH

4.1 INTRODUCTION

The preceding chapters have been very much concerned with discussing the component parts of two domains, one relating to the algorithmic aspects of image processing and the other to the architectural aspects of possible machine implementations. In this chapter we attempt to draw together, admittedly somewhat incompletely, these two domains by expanding on some of the ideas previously introduced in Sections 1.5 and 1.6. Our eventual goal is to see the development of a formally sustainable, implementable connection between these two domains.

Efficient implementation of complex high performance systems embodying considerable inherent parallelism, as is so clearly appropriate for image processing, will impose some fundamental high level constraints on the system designer. The data structures need to be matched, as far as possible, to the implemented processor–memory–communications architectural topology. High level languages are needed for specification, description and coding. Some effective means of 'concurrency extraction' need to be available, preferably as a tool within the design environment. The designer needs to be encouraged, and constrained, to 'think parallel' from the outset, and the tools he has at his disposal should reflect this. The facility to perform various types of optimisation needs to be available so that complex trade-offs can be constructively explored.

This chapter is concerned with the intellectual underpinnings which can make this design task efficient and straightforward. Many of the concepts discussed in earlier chapters will be drawn together, perhaps in a somewhat abstract way, with the aim of discussing how the difficult route between algorithm and architecture can in principle be negotiated constructively and productively. The abstract nature of some of the discussion reflects the fact that there is currently very little practical experience of such a design methodology.

Figure 4.1 illustrates the structure and the conceptual interrelationships which can be regarded as underpinning the general aims and context of this chapter. Both algorithms and architectures can be represented at varying levels of abstraction, giving rise to the concept of a layered system of representations **R** which are connected by transformations **T**. The algorithmic and architectural domains are linked by the mappings **M**. We are, in effect,

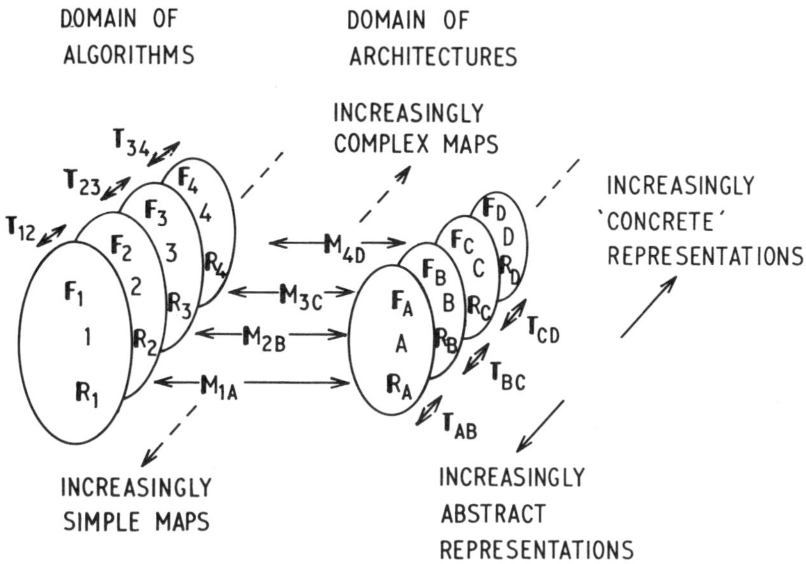

Fig. 4.1 A highly schematic view of the algorithm-architecture mapping problem. **R** is an appropriate representation, **F** a set of formal rules for manipulating the representations, **T** an intra-domain transformation and **M** an inter-domain mapping. In the context of this figure the simplicity or complexity of the mappings refers to their intrinsic detailing. Note the parallels between the figure and Fig. 1.15.

concerned with finding the 'best', in some sense, **R**, **T** and **M** for a given, hopefully large, class of design problems.

It should be emphasised that in the context of this chapter the notions of both algorithm and architecture are as high level abstractions, possibly with a 'deep' common representation. Unfortunately, the term architecture is commonly used to describe a number of different concepts. It can describe an abstract framework for system development, usually in terms of a system model, it can describe a particular form of machine implementation or it can even refer to the actual hardware design itself. We shall attempt to avoid confusion by making a definite distinction between the type of a system object (architectural model) and an instance of that object (physical system). The former will be what we usually mean by 'architecture' as the term is used in this chapter. To make this distinction more concrete the reader should note the impossibility of walking into a modern architecture and the ease of walking into a modern building, constructed according to the architectural 'rules' of the modern model.

Before describing the actual structures and processes inherent in the task of mapping image processing algorithms onto implementation architectures, and the concomitant support tools, the following four sections will discuss some of the basic underlying themes which will need to be taken into account. These include formal methods, complexity theory, representation and models, along with the notion of the silicon compiler as a particular paradigm for what we are more generally seeking.

A formal framework is mandatory if a manageable and sustainable design methodology is to be developed. Theoretical tools, such as complexity theory, are required if appropriate measures of performance and efficiency are to be incorporated into the design path, both at the algorithmic and the architectural levels. The whole question of representation is both basic and difficult. The ease or difficulty of achieving efficient useable mappings will be directly affected by the choice of representation. Silicon compilers, at the present time, represent a first, often faltering, attempt to achieve some measure of automatic control of a fairly restricted VLSI design route, and as such must be regarded as relevant to a more global ambition.

Before proceeding, we should emphasise that in this chapter we do not in any sense claim to present the answers to what is obviously a daunting problem, especially if one tackles the general case. What we hope we do achieve is to show some directions for bridging between the algorithmic and architectural domains, and that such connections, if developed, can provide the bases for advanced and sophisticated design tools. Obviously these tools will be of a wider applicability than just the development of the image processing capability which provides the main focus of interest for this book.

4.2 FORMAL METHODS

This chapter is aimed at exploring the possibility of establishing a systematic approach to VLSI system development involving a verifiable mapping, or set of mappings, from an abstract computational algorithm to an abstract computational structure and hence to an acceptable physical implementation. The criteria for acceptability will of course depend on the circumstances but will generally include such factors as design time, performance, cost, size and reliability. Such an approach will, of necessity, involve a layered, formal, top-down, step-wise design-to-implementation route.

If we wish to design reliably, establish correctness, reason, and use powerful automatic or semi-automatic tools, then the use of formal methods is mandatory. An initial formal specification facilitates making sensible, justifiable design decisions and provides mechanisms for evaluating the consequent designs with respect to the intended system behaviour. It makes possible the verification of the mapping of a behavioural specification (what is to be computed) onto a computational structure (how to compute within the constraints of a realisable technology). An appropriate formal description of all the design phases, with appropriate transformations, means that optimisation measures with respect to various implementations can be readily applied in the full knowledge of their complete and demonstrable correctness and validity.

Formal methods, synonymous with formal techniques, are a means of giving a usable syntactical and semantic coherence to specification, description, transformation and implementation, via a mathematical system providing precise rules for the manipulation of system elements. Switching theory, for example, rests on the formal system of Boolean algebra. An enchanting and very readable introduction to formal systems can be found in Hofstadter (1980). The use of formal techniques brings with it an assured precision and the ability formally to prove correctness of transformations; for example, a computer program can be checked against its specification to see if it correctly implements it. Figure 4.2 shows some of the concepts involved in the application of formal techniques to system design and implementation.

A significant advantage of a formal notation technique is that it enables one to study algorithms as mathematical objects. A formal description of the algorithm enables one unambiguously to define, and then go on to prove theorems about, classes of algorithms. In other words, developing an algorithm and devising a proof of its correctness should be inseparable and

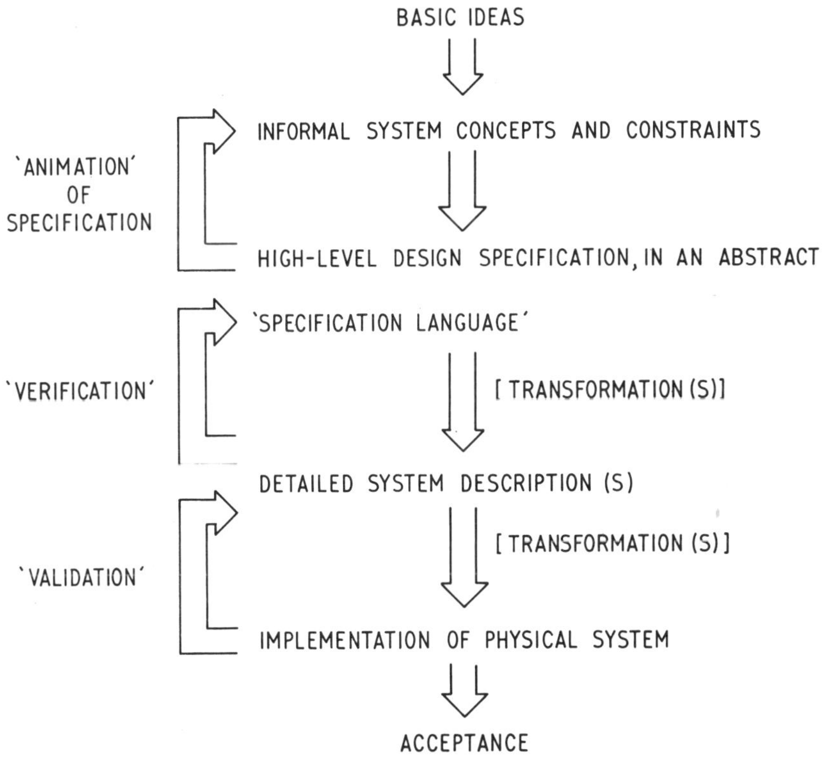

Fig. 4.2 Schematic diagram illustrating the use of formal descriptions of hardware and software, the transformations between them, and the provision of means for studying and optimising these transformations. In practice much of this functionality will be provided by the verification and validation tools intrinsic to a design environment, such as theorem provers, test generators, performance analysis measures and code generators.

complementary components of the same process. This ability to study algorithms as objects within a formal framework is fundamental to being able to automate the design/transformation process. Further, formal reasoning, as opposed to simulation, will have increasingly to be used for proving and testing large systems, be they VLSI circuits or large software packages. Simulation can never show proper functioning under all conditions. The problem generally becomes computationally explosive. The testing of the SIFT computer, described later, is a case in point.

Formal mathematical techniques for the unambiguous representation, specification, analysis and verification of complex system designs have been applied in such areas as software engineering, communication protocol and operating system specification and VLSI design verification. Experience has shown that specifications in natural language tend to be imprecise, inconsistent, poorly structured and incomplete. Formal methods, on the other hand, are mathematically rigorous, consistent, amenable to automation and highly versatile. It follows then that an obvious and fundamental requirement for a set of formal design tools is the proper choice of a formal specification and description language. Three significantly different approaches can be identified.

The *axiomatic* approach causes the meanings of constructs in a language to be explicated solely in terms of what assertions can be made about the effect of executing programs. There is no explicit notion of a machine.

A second approach is that known as *denotational* semantics, where the meanings of language constructs are explicated in terms of various mathematical objects, and in particular functions. There is no explicit concept of a machine or computation sequence. This approach is difficult to extend in order to deal satisfactorily with languages embodying concurrent constructs.

The third and most practical approach is to develop an *operational* definition of a language. This, as a rule, consists of a formal system which models the interpretation of programs written in the language. With this approach, the meanings of language constructs are explicated in terms of an abstract machine with discrete states and fairly explicit computational sequences. This approach embraces the tractable concepts of net theory, extended finite-state machines (finite-state machines, where there exists the possibility of a number of 'state variables' in addition to the basic machine states) and sequence models for the ordering of interaction primitives – e.g. 'temporal logic'. These are all finding increasing use in the more practical sides of system and protocol specification, description and design.

The general notion of a formal system and its application can best be grasped by considering a simple example known as a production system (Post, 1943), which has more recently provided the basis for an important class of pattern-directed inference system, known as a rule-based system (Waterman and Hayes-Roth, 1978).

Generically, a formal system comprises a set of symbols or 'alphabet', some rules or 'productions' for manipulating these symbols, and a set of initial axioms which provide a basis for establishing concomitant rule-derived 'theorems'. It is important to realise that the use of 'theorem' in this context is quite different from its more familiar mathematical context, where it usually means a statement, often in natural language, which has to be proved rigorously true. In formal systems, theorems are basically strings of symbols derived by applying the appropriate productions to the given set of axioms, which themselves consist only of 'strings' made up from members of the symbol set. Figure 4.3 gives a simple illustrative example of an abstract formal system, with only three symbols, four rules and a single axiom, which forms the basis of Hofstadter's (1980) '*MU* puzzle'.

(i) Alphabet: *M, I, U*

(ii) Rules: (1) If *xI* is a theorem so is *xIU*
(2) If *Mx* is a theorem so is *Mxx*
(3) In any theorem *III* can be replaced by *U*
(4) *UU* can be dropped from any theorem

(*x* is an arbitrary string)

(iii) Axiom: *MI*

Fig. 4.3 A simple formal system, from Hofstadter (1980). Readers may like to attempt to answer the question, 'is *MU* a theorem of the system?'

The imperative for some form of formal mathematical proof that guarantees that a system meets the desired functional requirements raises two major issues. First, how does one define criteria sufficient to ensure the correct functioning of the system, and secondly, how does one prove that these criteria are in fact satisfied by the actual system?

The first issue is crucial if formal techniques are to have any practical significance. Non-specialists must have confidence that the formal specifications stating what is meant by the correct functioning of a system in fact reflect the intended behaviour. That a formal specification expresses what the system designer intuitively means must, in the end, be determined by inspection. A formal specification must therefore be believable if rigorous mathematical correspondence to the specification is to ensure the desired effect. The larger and more complex the system, the more acute the problem becomes. Specifications reflecting the detailed behaviour of a system allow the most straightforward formal verification effort, but it is difficult to ensure that low level specifications embody what is meant by the proper functioning of a system. Very high level specifications, abstracted from the details of a system, are necessary if we are to state the overall functional properties of a system in a way that can be understood and believed. The problem then becomes one of reconciling the very high level specifications with the detailed operation of the actual system (Melliar-Smith and Schwartz, 1982).

In order to state high level system specifications that can be shown to be consistent with the actual system, one must formulate not just a single specification of the system, but a hierarchy of specifications. This leads naturally to a layered representation, with successive layers representing more detailed specifications which are rigorously provable, as theorems, using the specifications of precursor layers as axioms. The specification for the model is given by a set of axioms, characterising those properties of the model appropriate for that layer of system abstraction. At each layer in the hierarchy, a specification can be seen as an abstraction of the layer below. Correspondence between successive layers is done by expressing each primitive function and predicate of a higher layer in terms of the functions and predicates of a lower layer. With this mapping one must then show that each property derivable from the higher layer specification can be proved from the lower layer specification. The mapping between levels need not be complete; the mapping itself may be given as a set of axioms, saying only enough about the correspondence to derive the necessary axioms of the higher layer as theorems from the axioms of the lower layer. It is required only that the mapping axioms be consistent; in other words that there exists a complete functional mapping between layers which satisfies the mapping axioms. By demonstrating the correspondence between successive layers, one can conclude by induction that any property provable from the highest layer specification is also provable from the lowest layer specification. Thus, any analysis of the system based on a higher layer specification in the hierarchy is valid and could have been performed on the lowest layer system specification (*ibid*).

This general approach is applicable to any structural design task, be it hardware or software or, as is most often the case, a complex composite of the two.

One of the most elegant illustrative examples (Melliar-Smith and Schwartz, 1982) of a significant practical application of formal specification and proof is the rigorous demonstration that an advanced avionics flight-control computer (SIFT, for Software Implemented Fault Tolerance) met its design requirements. In outline, the methodology employs a hierarchy of design specifications, expressed in the mathematical domain of first-order predicate calculus. The most abstract of these, from which almost all implementation related details have been removed, represents the requirements on the system for reliability and proposed functionality. Successive specifications in the hierarchy add design and implementation detail until the Pascal programs implementing the SIFT executive are reached. At each stage a formal mathematical proof is constructed to show that these more detailed specifications exhibit precisely the properties of the more abstract specifications above them in the hierarchy. The formal proof, that a SIFT system in a safe state operates correctly despite the presence of arbitrary faults, has been completed all the way from the most abstract specification to the Pascal program. This proof has been performed using STP, a semi-automatic specification and verification system designed and implemented at SRI International, Menlo Park, California.

The SIFT computer system was designed to meet the requirements for future passenger aircraft control. The reliability requirements for such a flight control computer typically allow a probability of life–threatening failure of no greater than one part in a billion during a ten hour flight. This is equivalent to a mean time between failures of about one million years of operation, assuming maintenance after each ten hour flight. This extreme reliability requirement on SIFT imposes some very severe problems in demonstrating the achievement of that level of reliability. At the required reliability rate, observation of a large number of systems over an acceptable test period would be quite ineffective. Further, a SIFT system must be able to recover successfully from several million faults for every allowable system failure, and must therefore be able to recover from quite unusual and probably unforeseen malfunctions. Thus, validation by fault injection, while necessary, is unlikely to establish convincingly that SIFT meets its reliability requirements.

The justification that SIFT meets its reliability requirement was based on an extrapolation from fault rates that are easier to establish, such as those for an individual processor. For SIFT, this extrapolation took the form of a discrete Markov analysis, with the numbers of working and faulty processors defining the states and the fault and reconfiguration rates defining the transitions.

The validation of SIFT consisted of two parts. The first of these was a demonstration that so long as a predicate *system safe* is true, the system performs the desired flight control function, even though one or more processors may be faulty. This is a correctness property for the function performed by the system. The second was a demonstration that the Markov analysis computes an upper bound on the probability that *system safe* becomes false.

The layered hierarchy used to represent the interdependent set of system specifications is shown in schematic form in Fig. 4.4. Each level $\mathbf{L}(i)$ in the hierarchy specifies an abstract view of the system in terms of a set of primitive logical predicates $\mathbf{P}(i)$ and functions $\mathbf{F}(i)$ of system variables.

Within the hierarchy, the lowest level specification of the system is the actual SIFT system executed by the hardware, while the highest level specification reflects the intended overall function performed by the fault tolerant system. The higher layer specifications represent, in effect, system requirements, stating properties to be possessed without defining method of attainment. As one moves down the hierarchy, each lower layer specification successively introduces additional mechanism in the design specification to achieve the fault tolerance and expresses a more detailed and operational view of system transformation. Between successive specification layers one can perform incremental design verification, proving that the more detailed design specification at the lower layer supports the abstracted view at the

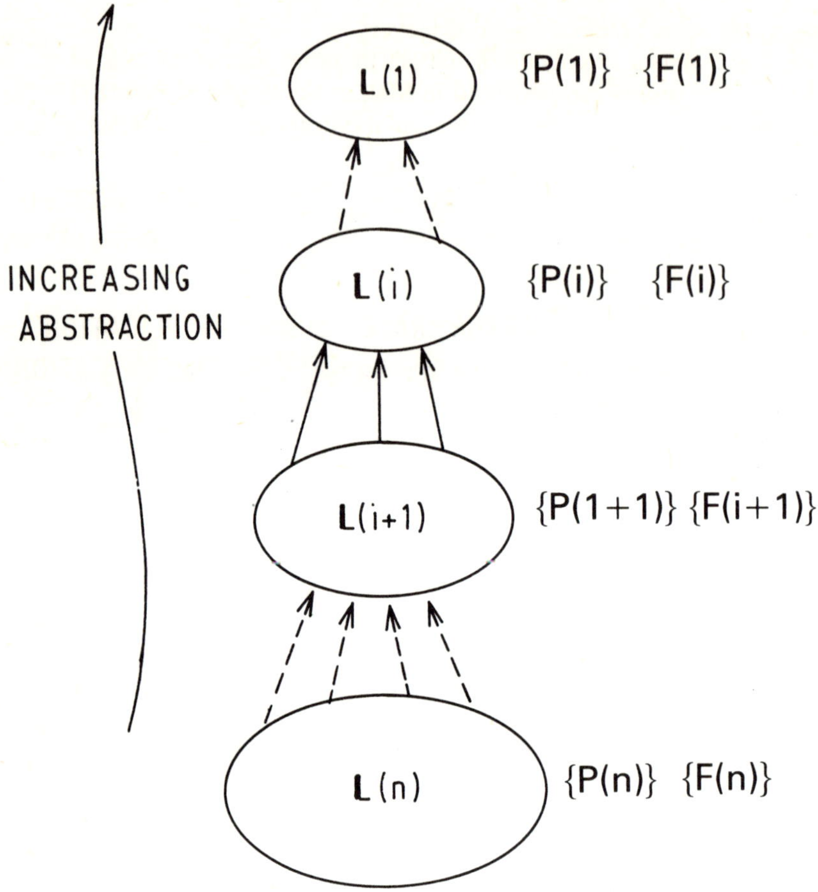

Fig. 4.4 A layered hierarchy of system specifications. See the text for an explanation of the symbols. Adapted from Melliar-Smith and Schwartz (1982) – copyright © 1982 IEEE.

higher layer. By gradually introducing the algorithms used to achieve fault tolerance, it was possible to verify each aspect of the design at the highest level of abstraction containing the necessary concepts.

Figure 4.5 shows an outline of the various specifications and analyses that are used in the justification of the reliability of SIFT. On the right of the figure is a hierarchy of specifications of the correct functional behaviour of SIFT, while on the left is a set of analyses that yield the probability of that correct behaviour. The models at the bottom of the figure describe the hardware and software of SIFT, upon which the more abstract analysis is based.

To cut a long story short, this set of specifications and analyses was used to demonstrate formally that the SIFT distributed system did indeed satisfy its basic high level specifications. In fact the processes of formal specification and verification resulted in the discovery of four design errors that would have been almost impossible to detect by other means. Details of the full analysis and proofs can be found in Melliar-Smith and Schwartz (1982).

It should be apparent by now that, in its various guises, a formal approach

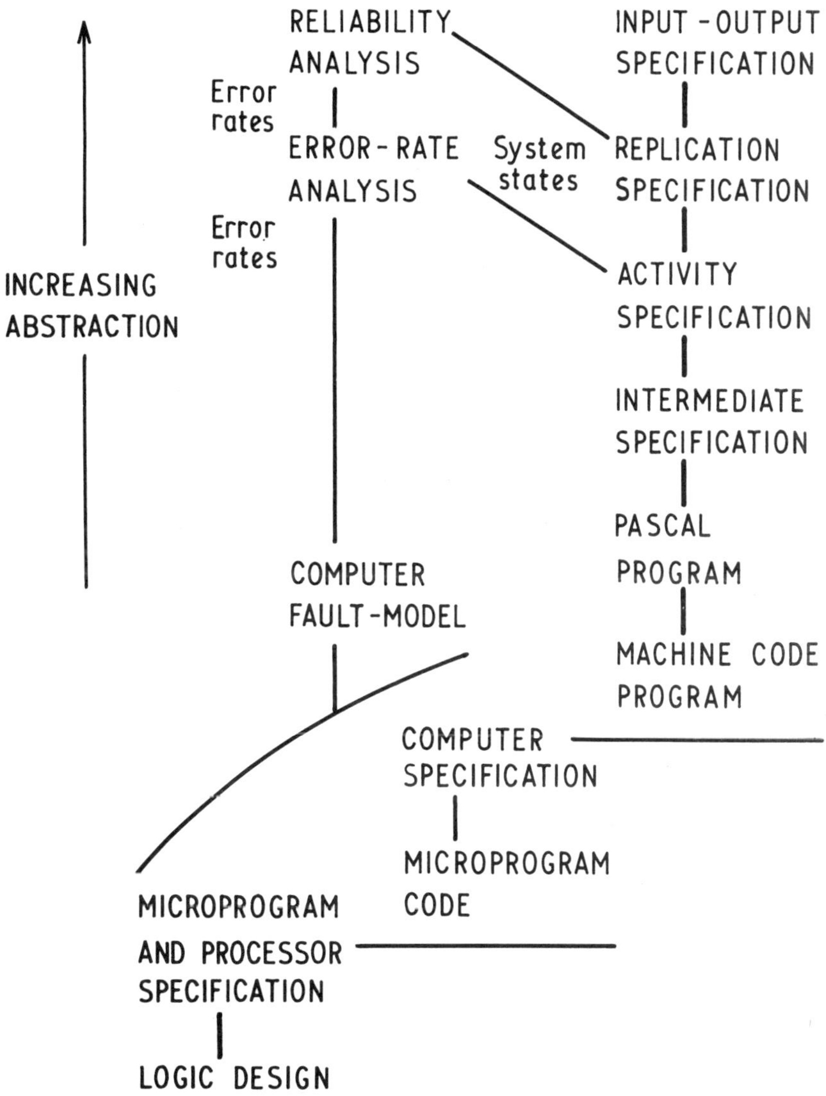

Fig. 4.5 The hierarchy of specifications and analyses used to establish the reliability of the SIFT computer system. The input/output specification is the most abstract functional description of the system. Lower layers make the system implementation details increasingly explicit, until at the bottom of the figure an actual system instantiation is represented. Adapted from Melliar-Smith and Schwartz (1982) – copyright © 1982 IEEE.

to system design and development has very considerable potential. Later in this chapter we will explore some of this potential in the context of the algorithm-architecture mapping which lies at the heart of efficient implementation of image processing algorithms.

4.3 COMPLEXITY THEORY

Some of the basic concepts underlying the notions of formal system design and evaluation, and the consequent mapping of computational algorithms onto VLSI structures, have natural intellectual precursors in classical switching and complexity theories.

Classical switching theory provides formal techniques for minimising the number of gates needed to implement a given logic function, while classical complexity theory (Aho *et al.*, 1983; Tarjan, 1978) attempts to relate the overlying algorithmic structure and performance to implementation related abstractions such as space and time. Complexity theory, which is more germane to our general topic, attempts to provide systematic information about algorithmic complexity so that, for example, an informed choice can be made between two algorithms for solving the same problem. To be useful, this information should be largely machine and programming language independent and involve measures which are amenable to theoretical study. Complexity measures can be static or dynamic. A typical static measure is program length, in the sense that program length in some sense measures the simplicity and elegance of an algorithm. Typical dynamic measures are running time and storage requirements. Most efficiency studies emphasise running time, with worst case, upper-bound, or asymptotic measures being commonly used.

Complexity measures are often associated with the size of the problem, which is a measure of the quantity of input data, for example, the largest dimension of a matrix or the number of edges of a graph. If an algorithm processes inputs of size n in a time An^2, for some unit-dependent constant A, then the time complexity $T(n)$ is stated to be of order n^2, represented as $O(n^2)$. More precisely, a function $g(n)$ is said to be $0(f(n))$ if there exists a constant A, such that for all but some finite, possibly empty, set of non-negative values of n, $g(n) < Af(n)$. A well known example (Rabiner and Gold, 1975) is the $O(n \log n)$ time complexity of the fast Fourier transform versus $O(n^2)$ for the discrete Fourier transform, where n is the number of points in the transformed sequence.

Any convincing computational measure must be based on a computer model. Much work on computational complexity has been based on Turing machines and on various models of stored program sequential random access register machines. The spirit of this work has been extended to cover abstract parallel machines, especially SIMD. These classical theories readily yield direct syntheses of a given logic design and basic complexity criteria but do not permit optimisation with respect to area, power, delay, etc. – fundamental parameters of any VLSI implementation.

For VLSI based systems, topological and performance constraints will increasingly dominate. Already the absolute number of logic gates in an implementation has become a second-order consideration. A traditional minimum-gate-count implementation may well require more silicon area than an alternative design with more but optimally, in a VLSI context, connected transistors. Algorithmic structure, as opposed to size, becomes much more pertinent once the above constraints come to be regarded as important design parameters. There is a definite need for an extended formal theory for VLSI system design, embracing and properly incorporating the above parameters while at the same time exploiting concurrency and constraining complexity. One must be motivated to look for simple and regular control and data flows. Communications costs, increasingly using more space, energy and time than processing structures, are a fundamental constraint. These communications costs are likely to become the dominant factor in highly parallel machines. It is important to realise that there is a definite change in emphasis as one moves from classical to VLSI-centred complexity theory. Classical theory is concerned primarily with problem size, whereas the VLSI view is very much communications centred, with the communications overheads inherent in a problem being made quite explicit in the mapping into a silicon implementation. Meaningful measures are then much more concerned with space (topology and silicon area) and time (latency and coherence in memory and communications structures). Concurrent systems spread computations out in space and in time. For these systems, then, the classical approach is inadequate, based as it is on a sequential model of computation. The key to VLSI systems is communication, and the time delay involved in communication across a chip cannot be ignored. A model for VLSI computation is needed to extend complexity theory to highly integrated systems.

In an early work Mead and Rem (1979) showed how an analysis of communication paths could be used to derive a lower bound on chip area and cycle time. Cells are nested hierarchically within successive levels of modules driving short output wires. The trade-off in the resulting improvement in driving time is against the increased area occupied by the wires. Minimisation of an area × time complexity measure gives the optimal number of cells to be grouped together in the recursive structure.

Optimisation of the layout of a network on a chip has been studied by Leighton and applied to a number of networks suitable for parallel computation such as the shuffle-exchange network (Leighton, 1983). Techniques are given for the proof of lower bounds on layout area, crossing numbers (the number of wire crossing points in the layout) and maximum edge length (the length of the longest wire in the layout) in terms of the number of network nodes. Area efficient algorithms for graph layout are given by Leiserson (1983). However, research remains to be done on the effects of relaxing the assumption that graph nodes occupy constant area.

Other circuit performance measures may be considered in addition to those of the area A and the time T required to perform an operation. For example, a parameter P can be introduced to characterise the maximum throughput of an integrated circuit. P, the 'period' of the circuit, is given by the minimum time elapsed between consecutive inputs, and measures the degree of pipelining. Since $T \geqslant P$, stronger lower bounds may generally be obtained in terms of the period than in time. Area, period and time trade-offs for pipeline circuits are given in Baudet (1981).

The VLSI model commonly assumed in layout optimisation is the 'grid model' (Thompson, 1979; Ullman, 1984). A chip is considered to consist of a rectangular, multilayered grid of orthogonal lines along which horizontal and vertical lengths of wire may be routed. Point processors may be sited at line intersections. Discrete time steps, typified by the phases of a two-phase clock, are usually assumed. It is assumed that in one time unit, signals have all the time they need to propagate down wires, as well as to switch active devices. In reality, this is an increasingly unrealistic assumption*.

Within the context of this grid model the most common method of proving computational lower bounds on the area or time, or some combination of these, is by a 'fooling argument'. This involves showing that, in order to solve a particular problem, at least some minimum amount of 'information' must be transferred from one part of the chip to another, or else one half of the chip can be 'fooled' into thinking that the input to the other half was one thing when in fact it was another. Essentially, this is a *reductio ad absurdum* type of argument. Three basic lower bound arguments exist. As illustrated in Fig. 4.6, the history of a piece of information can be represented by considering portions of a parallelepiped traced out in this space-time volume.

First, we can argue about the volume of the solid. At most one bit can be input at any point in the solid, so the number of input bits cannot exceed the volume. The volume is the product of the area and time, so this method gives us a lower bound on the product AT, where T is the time taken by the chip to perform a particular operation, and A is the area. The second possibility is to draw a boundary horizontally, as shown in Fig. 4.6, and argue that a certain amount of information must flow upward across that boundary, if the chip is to solve a particular problem. Since only one bit can flow through a unit area of the horizontal slice, we can sometimes obtain

* Long range interconnection delay times are going to assume increasing significance as device technologies scale down towards the sub-micron regime. For a scale factor of α (Mead and Conway, 1980) these delays scale as α^2, with the concomitant gate-relative delays scaling as α^3. In addition, the phenomenon of 'metal migration' may well impose some fundamental constraints on the distribution of clock, power, control and data signals in highly parallel structures, such as systolic and SIMD arrays. The upshot of these constraints may well have to be the imposition of strong short range ordering in VLSI structures, with special provision for inherently long range connectivity, such as clock, power and bus structures. This may well involve the use of two or more levels of design rules, with the coarsest relating to the longest range connectivity.

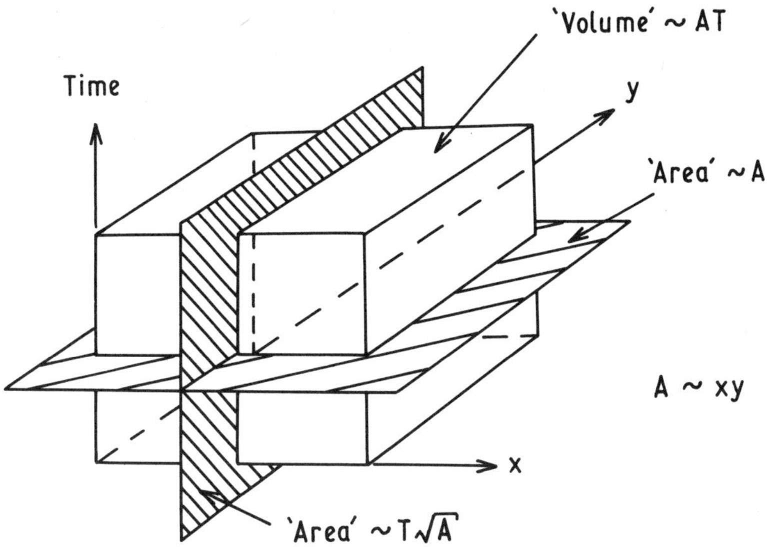

Fig. 4.6 The space-time 'complexity' parallelepiped. Adapted from *Computational Aspects of VLSI,* by Jeffrey D. Ullman, with permission of the publisher, Computer Science Press Inc., 1803 Research Blvd, Rockville, Maryland, 20850, USA.

lower bounds on the area A in this manner. The third method is to make a slice vertically, cutting across the shorter dimension of area; the distance in that dimension is at most $A^{1/2}$. If we can claim that a certain amount of information must flow across that boundary, then we can put a lower bound on the area of that slice, which is at most $TA^{1/2}$. If we square all quantities, we get the lower bound in its more familiar form, as a bound on AT^2. For a detailed discussion on the derivation of lower bounds based on the grid model, with examples, see Ullman (1984).

A more realistic computational model is advocated by Chazelle and Monier (1981). Here, propagation times are required to increase at least linearly with distance. Time performance, then, is strongly dependent on circuit geometry. Lower bounds on complexity measures T (time of computation), P (minimum time between consecutive inputs) and chip area A may be found in Chazelle and Monier (1981). Unfortunately many circuits shown to be optimal within the Thompson model turn out to be less efficient in time complexity when, more realistically, linear propagation time is taken into account.

This is still an embryonic, but obviously fundamental and very important field. Complexity theory is starting to emerge as a tool of use to practising circuit designers. Other areas where complexity theory is starting to produce useful results is in the description, derivation and transformation of systolic algorithms (Ullman, 1984; Leiserson, 1983) and in the analysis of processor organisations where the interconnect area dominates the wire area (Ullman, 1984; Leighton 1983). From the point of view of the system designer the present measures are rather weak in the sense that they do not readily facilitate the solving of design problems involving architecture-cost-performance-size trade-offs. It is still early days for complexity theory and much is still to come. The efficacy of silicon compilers, *inter alia,* may in the long run depend critically on the appropriate measures being developed by complexity theorists.

4.4 SILICON COMPILERS

The silicon compilation of VLSI circuits is one of a number of attempts to come to grips with the rapidly increasing technology–led complexity inherent in high level design specification, logic design, simulation, layout, routing, testing and verification. In essence, a silicon compiler is an optimising transformation program which produces a viable integrated circuit design from an intelligible high level description of the device. Some form of silicon compilation may well be the only way to realise, effectively and economically, the very considerable potential of VLSI for the implementation of computational structures which exhibit high levels of parallelism. The temporal structuring inherent in highly parallel systems can probably only be realistically handled by formal and automated techniques such as would be embodied in a silicon compiler. As this complexity increases, the intuitions of even very capable designers become increasingly fragile.

Silicon compilation is, in broad terms, a design automation methodology which attempts to abstract the process of realising a VLSI circuit towards the architectural level and away from the more traditional classical circuit design considerations. This approach will assume an increasingly more central role as integrated circuit design cycle times and costs become steadily more significant. Currently the major cost, in time and money, is in the translation from an informal functional specification to a two-dimensional layout. It is a highly interactive and labour intensive activity. Current techniques will probably fail completely when three-dimensional VLSI structures become available. Increasingly in the future integrated circuits will be designed by engineers and systems designers who know little about the subtleties and idiosyncrasies of complex technologies. The aim of silicon compilation is a quick, safe, easy route to solutions in silicon for the systems designer. A structured system design is very similar to the structured design of large software packages. The design tends to be partitioned into small manageable pieces, such as modules or cells, which are analogous to procedures in programming languages. This means that VLSI design and software development have much in common, both in terms of philosophy and methodology. Just as software compilers synthesise complex programs from input written in a high level language, a silicon compiler synthesises various abstract design levels, such as layout topology or 'floorplan', functional description and simulation model, from appropriately formalised high level descriptions. These descriptions are of canonical structures (PLA, RAM, data path, etc.) and functions (data path width, how the data path will multiply, etc.) known to the compiler. The structures themselves are synthesised from 'if-then' rules which are embodied in the compiler. These rules, representing an attempt to encapsulate the expert designer's art, when envoked by the compiler yield a layout, functional simulation model, a timing model for the structure and eventually a timing model for the chip. If the intermediate design representations in such a process resemble the hierarchical expansions that human designers typically employ then the approach can be described as 'successive decomposition'. This can be viewed as a collection of distinct tools that operate in turn on the design. On the other hand, if the design system consists of a number of layout generators for a given set of entities then the approach can be described as 'incremental refinement'. Existing silicon compilers use typically several thousand rules that will yield several hundred of the most useful high level VLSI structures grouped in several tens of structure classes. A specific compiler can usually cope with a range of similar fabrication processes with no other user input than a process 'declaration'.

Silicon compilation inherently involves the use of an integrated database capable of automatically supporting functional simulation at any level of the design hierarchy. It also supports the provision of performance reports and critical path analysis of the completed layout.

Two key concepts of silicon compiler methodology are the high level system specification language and the target layout specification language. The high level source language should facilitate ease of system and behavioural specification and the straightforward embodiment of notions such as concurrency, timing, data types, algorithmic structure, hardware/software trade-offs (for example a RISC architecture *vs.* extensive micro- and nano-coding), testing and test generation. The target language should embody a two-dimensional representation which facilitates efficient use of silicon area, maximises speed and minimises power dissipation. In other words, it must be naturally orientated towards performance and efficiency. Pictorial layouts should, apart from their artistic impact, become a thing of the past. Once a layout is produced it should be provably correct without any need to check masks.

This functional separability of the specification and layout phases leads naturally to the concept of an abstract space in which formally to describe the mapping(s) between source and target representations. Such a formally defined space, which may be layered, is essential if automatic transformation and verification is to be naturally incorporated within silicon compilers. These points are illustrated in Fig. 4.7. The structured, layered nature of this space ensures that meaningful simulations can be performed at the high level functional block stages of the formally refined design specification. This is really the whole point behind silicon compilers; they push design upwards towards the architectural and functional levels.

This partitioning of the abstract transformation space into functionally distinct levels with explicit physical counterparts has previously been

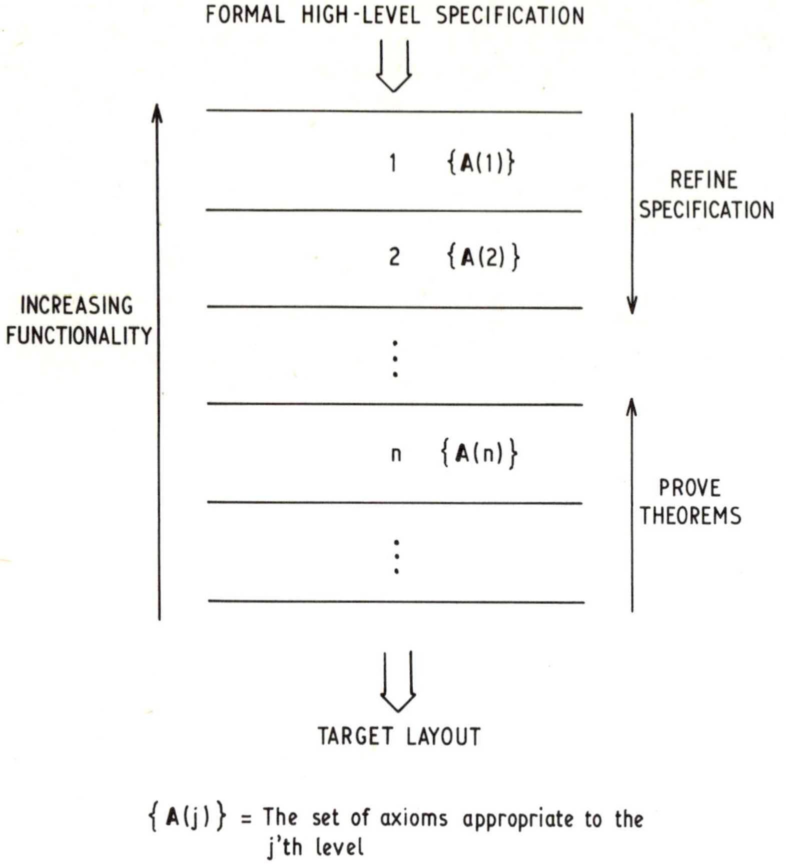

Fig. 4.7 A schematic representation of a layered abstract transformation space linking an input specification and an output VSLI circuit layout. Compare this figure with Fig. 4.1, which in fact contains this figure as a subset.

proposed by Stefik (1981). He points out that the normally accepted model of silicon compilers does not embody such a strongly layered structure, a comment which is still true. In the context of the metaphor 'design is search', such a layering is probably mandatory if automating the searching of design spaces is to become a practical reality. In AI, many problem solving systems are based on the formulation of problem solving as search. In this formulation, a description of a desired solution is called a goal, and the set of possible steps leading from initial conditions to possible solutions is the space to be searched. Problem solving is carried out by searching for sequences that lead to solutions that satisfy a goal. At the Xerox Palo Alto Research Center, Stefik and his colleagues have developed a prototype expert design system called PALLADIO, which embodies many of the concepts and techniques that are going to be needed to support silicon compilation.

The key idea is that designers should design not only circuits, but also knowledge. Knowledge in PALLADIO is expressed in a knowledge representation language called LOOPS. Using PALLADIO, a designer interacts with previously designed circuit fragments and rules taken from knowledge bases. While working on an individual design, a designer is able to discover gaps and errors in the knowledge base by applying it to his own design. The designer can create personal versions and modifications to the knowledge bases, which can later be incorporated. Primarily, PALLADIO is intended as a focus for long term experimentation with design methodologies.

Timing, in a sequencing rather than a continuous sense, can be explicitly introduced via the notion of 'temporal logic'. This is a formal methodology in which orderings such as before, after, until, and eventually, are given clear and unambiguous meanings.

The coherent integrated approach described above contrasts strongly with the current informal and loosely coupled combination of top-down and bottom-up design performed with the aid of a variety of CAD/CAE tools. This informal approach is illustrated in Fig. 4.8.

There are many research issues still to be addressed and solved before true silicon compilers become commonplace, probably in the 1990s. AI techniques, as well as formal methods, are probably going to play an important role, certainly in the development of the user interface.

The basic question of silicon efficiency is fundamental, as are speed, power dissipation, concurrency, timing, hardware *vs.* software trade-offs and architectural diversity. These are the criteria by which emerging silicon compilers will be judged, and most existing implementations fail to satisfy.

Various references (Johnson, 1984; Hicks, 1983; Bryant, 1983; Gray, 1981; Ayres, 1983) give some indication of progress to date in the implementation of silicon compilers. It is worth pointing out that most of these reported implementations have been done very much in parallel with a particular specific chip design so that in no sense can those designs be regarded as silicon compilations. No convincing demonstrations of generality have yet appeared. Benchmarks are sorely needed. The reported advantages of contemporary silicon compilers are more a reflection of the intrinsic structured design processes used in developing both chip and compiler than a reflection of actual compiler performance. Existing products in the area of silicon compilation tend to address the task of mask generation. These instruction generators and module linkers encourage designers to experiment with alternative floorplans and architectures, because varied cell layouts can be produced quickly, repetitively and accurately. These tools in some cases are intended to replace conventional IC design; in other cases, they are seen as an additional approach that should be integrated with existing tools.

It is apparent that silicon compilers, as envisaged here, can in principle be abstracted upwards to the algorithmic level. Consequently it is clear that any general discussion of the algorithm to architecture transformation should take cognisance of the similar intellectual exercise inherent in the development of versatile and efficient silicon compilers.

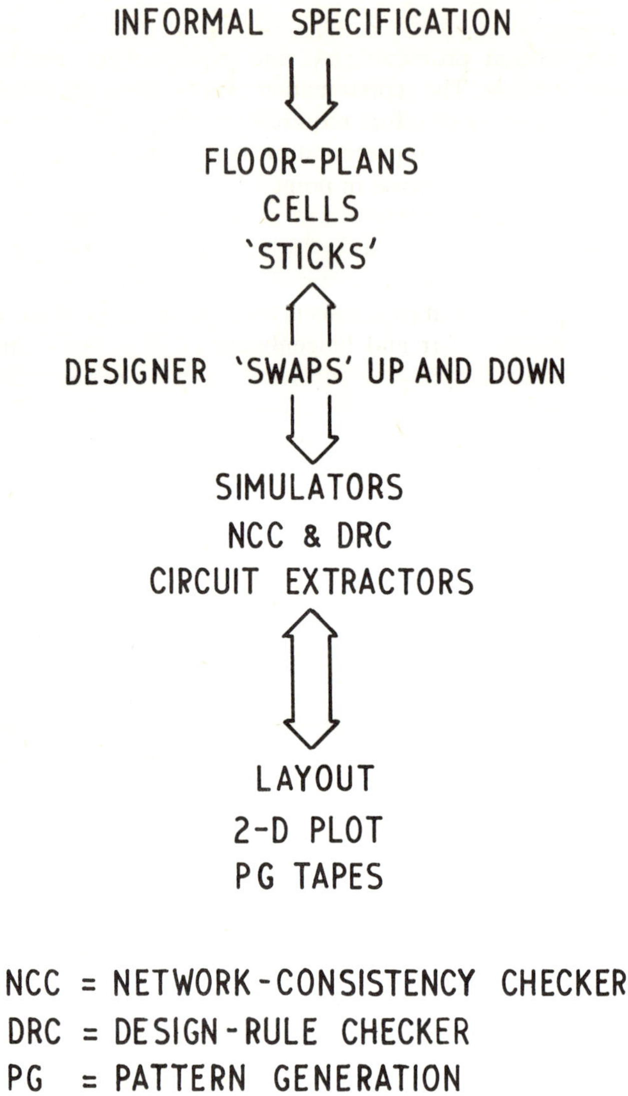

Fig. 4.8 A typical traditional VLSI circuit design environment.

4.5 MODELS AND REPRESENTATION

In this book we are concerned primarily with describing some of the generic structures and algorithms used for representing and transforming image data with a view to discovering straightforward mechanisms for mapping these into VLSI implementations. Efficient computation requires considerably more than just an efficient control strategy. It requires the careful selection of an appropriate representation for problem states, data structures, control flow, and input-output protocols, and the means to use and link these representations properly. The representation of a problem has a considerable influence on the amount of effort required to solve it. The importance to mathematics of the Arabic representation of numbers as opposed to the Roman is an obvious if trite case in point.

The processes necessary properly, in some optimal sense, to represent problems initially and to improve, or transform, given representations are still imperfectly understood. Perhaps the most thorough attention to representation and attendant problems is found in the area of AI, e.g. Marr (1982), Nilsson (1980), Barr and Feigenbaum (1981), where this issue is regarded as having fundamental importance. In a very real sense the recent history of image processing, especially in areas such as image interpretation and understanding, can be looked on as the search for adequate representations for image data and knowledge (Havens and Mackworth, 1983). Expressed loosely, a representation is a formal system (a set of symbols and rules for their manipulation) for making explicit our knowledge about certain entities, coupled with a specification of how the system achieves this. The consequence of using a particular representation to describe a given entity is a description of the entity in that representation. The importance of these concepts to theories of vision and scene understanding are discussed at length in Marr (1982).

As an illustration of how deeply embedded such concepts as representation are, in the general notion of algorithm to VLSI mapping, consider Fig. 4.9 (Parker, 1984) which illustrates two, in some sense equivalent, system

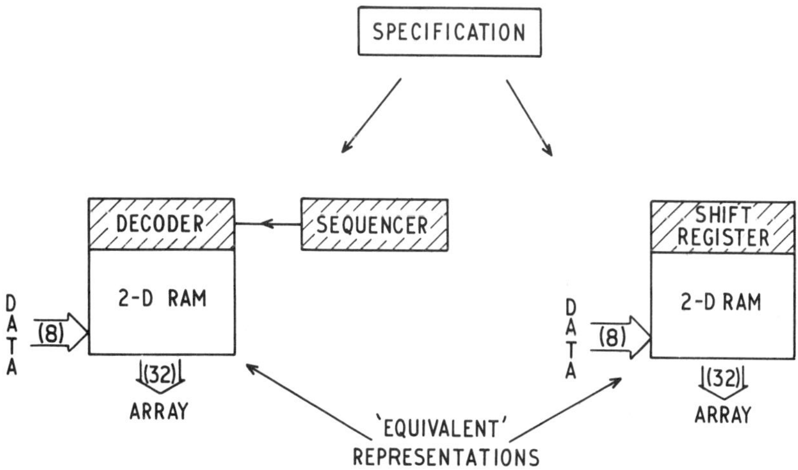

Fig. 4.9 An illustrative example of how a single unique functional specification can yield more than one 'equivalent' system implementation. A decoder-sequencer combination and a shift register are completely equivalent methods for addressing a two-dimensional 'corner-turning' memory used to load a processor array with input data.

implementations arising from an initial unique specification. Are there others? How does one choose between representations? How does the choice of specification language constrain the derived representations? These are cogent and fundamental questions.

The sequencer-decoder combination is obviously directly equivalent to a shift register, but how do we express this equivalence, and what algorithm in an appropriate representation is the common factor? It is these underlying questions and their generalisation which provide the central impetus for our present discussion.

It is clear from this example that an important but elusive aspect of the designer's task is the choice of representations appropriate for particular design or problem spaces. How are such choices to be made and by what criteria are they to be evaluated? At the present time there is no clear answer, though a number of general points can be made. Representation should be as uniform and as consistent as possible. This reduces the complexity of the design or problem spaces and mitigates against unnecessary 'opacity'. In crude terms, a representation can be assessed either in terms of its 'declarative adequacy', the ability of a representation to capture the essential properties of objects and the relationships between them, or its 'procedural adequacy', the capability of a representation to support efficient processes for achieving given goals. Beyond this, there is still much to be done if a general methodology is to emerge.

An issue which is inextricably linked with the problems associated with the choice of an appropriate representation is that of how to model the problem and design spaces. A model can best be regarded as a coherent self-consistent representation of some physical or abstract projection of the actual 'real' universe. The rationale behind seeking models is that they are, at a fundamental level, mechanisms for understanding and tractability, encoding as much, and ideally no more, detail as is relevant for the analysis of the system being modelled. The concepts of model and representation are inseparable in practice. It is not an exaggeration to say that the whole of science, with its powerful inbuilt predictive capabilities, revolves around these twin pillars of model and representation. For example, the wave–particle duality of physics is, in a sense, the consequence of the graininess of different depths of modelling of the physical universe. In this context, the matrix and wave function formulations of quantum mechanics are examples of different representations of the same underlying deep model.

The notion of layered models, which has already been discussed in Sections 1.5, 1.6 and 4.2, is also fundamental to much of science. A layered, or modular, approach provides a comprehensible framework for subdividing a complex problem into individual pieces of manageable size. To date, the best and most widely known example of such an approach relates to the development and definition of international networking standards (Folts and desJardins, 1983).

4.6 ALGORITHMS

As we have seen in Sections 1.5 and 1.6 and in Chapter 2, the first step in attempting to solve a given image processing problem is to formulate properly an underlying computational theory, based on some type of formal model. Once a computational theory has been established, solutions in an algorithmic form amenable to machine evaluation can be sought. In this context an algorithm can be operationally regarded as 'a finite sequence of instructions, each of which has a clear meaning and can be performed with a finite amount of effort in a finite length of time' (Aho *et al.*, 1983). In practice, the algorithms used for image processing are, for the purpose of evaluation, expressed in some form of programming language, which will of course reflect an underlying machine architecture, real or virtual. This corresponds to one possible instantiation of the range of permissible representations of the deep structure intrinsic in the computational theory. For our present purposes the programming metaphor is by far the most convenient because of both its close relationship to machine architectures and its ubiquitous applicability. It is often the case than an initial version of an algorithm is expressed in a few general high level statements which are subsequently refined and expanded into more definite and compact steps corresponding to a formally defined programming language. This process is often called 'stepwise refinement', already discussed in Section 4.2, and it is greatly facilitated if there are proper formal tools to assist the algorithm implementer.

In order to clarify what, in the context of the present discussion, we mean in practice by an algorithm and its execution, the following three sections will briefly outline some of the basic structural and functional characteristics of computational algorithms. An extended and much more detailed discussion can be found in Aho *et al.* (1983).

4.6.1 Data structures

Data is the computational raw material that an algorithm operates on, and in a computational context there are a number of different ways in which the data may be regarded; in terms of data type, data structure and abstract data type. The *data type* of a variable is the set of values that the variable may assume. Typical data types are Boolean, integer, real and string. An *abstract data type* is a mathematical model, together with various operations defined on the model. Typical abstract data types are sets, lists, stacks, queues, trees and graphs. A convenient way of regarding an abstract data type is to compare it with the more familiar notion of a procedure. Procedures extend the notion of a primitive combinatorial operator by virtue of being able to generalise and encapsulate. These two properties apply equally well to abstract data types in that they are generalisations of primitive data types and encapsulate a data type in the sense that the definition of the type and all operations on that type can be localised to one section of a program. In order to represent the mathematical model underlying an abstract data type the notion of *data structure* is invoked. This is a collection of variables, possibly of several different data types, connected in various ways. Typical examples are arrays, records and files.

The actual implementation of an abstract data type is a translation, into statements of a programming language, of the declaration that defines a variable to be of that abstract data type, plus a procedure in that language for each operation of the abstract data type. An implementation uses a data structure to represent the abstract data type, with each data structure being built up from the basic data types of the underlying programming language using the available data structuring facilities.

One of the most useful techniques for representing arbitrary relationships among data objects involves the use of directed and undirected graphs, which are natural models of such relationships. A directed graph **G** consists of a set of vertices (or nodes or points) **V**, and a set of arcs (or directed edges or directed lines) **E**. An arc is an ordered pair of vertices (v,w), often expressed as $v \rightarrow w$, where v is called the 'tail' of the arc and w the 'head'. In many applications it is convenient to attach information to the vertices and arcs of a directed graph so as to convey relationships explicitly. Examples of this can be found later in this chapter. Several data structures can be used to represent a directed graph. The appropriate choice usually depends on the operations which will be applied to vertices and arcs of the graph. A typical and common representation is the 'adjacency matrix', which is an $n \times n$ matrix A of Booleans, where $A(i,j)$ is true if and only if there is an arc from vertex i to vertex j. A 'labelled adjacency matrix' is characterised by having $A(i,j)$ represent the label on the arc going from vertex i to vertex j. An undirected graph is more appropriate for modelling symmetric relationships since it differs from a directed graph only in that the directed arcs are replaced by undirected edges between vertices, such that $(v,w) = (w,v)$. Undirected graphs can be represented in a similar manner to directed graphs.

4.6.2 Data and control flow

The underlying approach to programming computational algorithms in conventional languages is to guide a single locus of control through the maze of assignment, conditional and repetitive statements that constitute the program (Kennaway and Sleep, 1983). At each step the programmer has as a major concern the details of how things are done rather than getting right what is done. The emergence of new machine architectures, especially those exhibiting high degrees of concurrency, is starting to catalyse significant changes in programming languages and style, through the architecturally implicit data and control flows.

The most familiar operational model is that of 'control flow'. In traditional sequential, or von Neumann, control flow there is a single thread of control which is passed from instruction to instruction, with only explicit control transfers permitted via special instructions such as DO WHILE, DO UNTIL, IF and GOTO. There are also parallel forms of control flow, where the implicit sequential flow model is augmented by parallel control operators which allow more than one thread of control to be active at any one time, as well as providing mechanisms for synchronising these threads. Typical instructions are FORK and JOIN.

As a means of circumventing the tyranny of sequential control flow, the data flow model provides one promising implementation architecture suitable for highly parallel machines, as does the reduction model, which is inherently capable of supporting functional and logic programming because of its support for features such as lazy evaluation and higher order function facilities.

In a data flow model, programs are represented as a directed graph called a data flow graph. The graph has several input ports and output ports. Nodes on the graph represent operations. There are two categories of operation: data manipulation and data flow control. The former includes arithmetic operations, list manipulation and memory read and write, while the latter includes switch operation, gate operation and linkage operation. Node operations are enabled when tokens, or operands, arrive at input arcs, and the operation results are set at their output arcs. The result token on the output arc becomes the operand token for its destination nodes. Execution starts when initial tokens are put into the input ports of the data flow graph and is completed when result tokens are output from its output ports. In the course of a computation, a number of tokens flow in the data flow graph.

Control flow and data flow programs are built from fixed size instructions whose arguments are primitive operators and operands. Higher level program structures are built from linear sequences of these primitive instructions. In contrast, reduction programs are built from nested expressions. The underlying reduction model specifies only reduction (or rewrite) rules, which permit the machine to replace elements of an expression which match the left-hand side of an equation with the corresponding right-hand side, and does not explicitly express evaluation control. It is this characteristic of reduction models which gives rise to the possibility of 'referential transparency' through the use of zero assignment, or declarative, languages. A data flow model, on the other hand, can quite explicitly express execution control by utilising an appropriate token driven control mechanism in which tokens are used as data or control signals.

'String reduction' involves the copying of arguments whereas 'graph reduction' involves only copying pointers. 'Eager' and 'lazy' evaluation are evaluation strategy concepts arising in reduction models. If the evaluation of expressions proceeds innermost, it is called an eager evaluation, and if it proceeds outermost, it is called a lazy evaluation. When the eager and lazy evaluation concepts are applied to a parameter mechanism in a function application, it is termed 'call-by-value', 'call-by-name' or 'call-by-need'. In a call-by-value mechanism, each argument of a function is evaluated before the function body is evaluated. In a call-by-name mechanism, a parameter that is called by name is not evaluated until it is encountered in the expression of the function body. In a call-by-need mechanism, a parameter is evaluated only the first time its value is actually needed, even if it is encountered several times in the function body. Thus call-by-need is more efficient than call-by-name. Lazy graph reduction leads naturally to the concept of 'combinators' (Kennaway and Sleep, 1983), which are elementary switches for directing arguments to the graph nodes where they are required to contribute to a function evaluation.

4.6.3 Performance requirements

For solving a particular problem, expressed in terms of a particular computational theory, there is generally no unique computational algorithm. How then do we choose? For present purposes we will assume that the best algorithm is one which runs as rapidly as possible and makes efficient use of the computer's physical resources. The running time of a program depends on a number of factors: the input data, the efficiency of the compiler, the instruction set and the time complexity of the underlying algorithm. As we have mentioned in Section 4.3, it is customary to define a quantity $T(n)$ to represent the running time, in arbitrary units, of a program operating on inputs of size n. Because of the difficulty of quantifying factors such as compiler performance and machine instruction sets it is usual to express $T(n)$ only as proportional to an appropriate function of n and leave any constants of proportionality unspecified. For many programs the running time is a function of the particular input and not merely the input size. In such cases $T(n)$ is defined to be the worst case running time, in that it is taken to be the maximum for all possible inputs of size n. This worst case running time is usually regarded as the principal classical measure of time complexity. It is usual to express the growth rate of algorithms in terms of a 'big-oh' notation. For example, to say that $T(n)$ is $O(n^2)$ means that there are positive constants k and N such as that, for n equal to or greater than N, $T(n) \leqslant kn^2$.

In practice it is assumed that algorithms can be evaluated by comparing their respective running time functions, with no regard to any constants of proportionality. This means that if one is comparing $T(n) = 100n^2$ ms and $T(n) = 5n^3$ ms, where the latter is preferable for $n < 20$, it is usual to consider asymptotic rather than low-n behaviour and hence assert a preference for the former. The justification for this is that as computation becomes cheaper and machines become faster, as is bound to be the case for some time to come, there will be a continuing tendency to want to solve larger and more complex problems. Consequently those algorithms with low asymptotic growth rates will become increasingly more attractive and important.

4.7 ARCHITECTURES

In Section 4.6 we considered some of the distinguishing features of algorithms regarded as a programming language level representation of problem solutions arising from an underlying computational theory. We were concerned primarily with static and dynamic aspects of data organisation, as well as control flow, and performance measures, all without any explicit architectural connotations. In the following three sections we explore essentially the same three areas discussed in Section 4.6, but with the primary emphasis being placed on the computational architecture, regarded as an implementation model, appropriate to a particular class of problems. It will soon become apparent, especially in Section 4.7.2, that the diversity of potential implementation architectures referred to in Section 1.6 is very real indeed.

4.7.1 Abstract architectures and topologies

In keeping with our earlier discussion, we shall view computer architectures somewhat abstractly here, i.e. as a set of computing elements provided with an interconnection network realising the necessary paths for data communication between elements. Efficient concurrent processing relies upon the suitable design of data structures, decomposition of the problem for allocation to processors and choice of interconnection pattern.

As discussed in Chapter 2, low level image processing is largely concerned with the processing of a two-dimensional matrix of pixels, or grey levels. Low level image processing algorithms, e.g. smoothing, typically involve the combination of data localised near individual pixels. The same set of operations is carried out for each pixel. These tasks can readily be performed on SIMD arrays of identical processing elements (PEs) with near neighbour interconnections (see Section 1.4.2). The image, or consecutive parts of it, is distributed over the array and processed in parallel. Parallelism may be increased by scaling up the number of PEs. If there are as many PEs available as pixels, performance may be increased by employing bit-parallel rather than bit-serial operations.

Some problems, typified by the fast Fourier transform, demand access to data which are distributed globally over the processor array. Linking each processor to every other by a fully connected crossbar network rapidly becomes impractical since the cost of providing the necessary data links grows as the square of the number of processors. An appropriate data routing network has then to be devised. The Goodyear STARAN computer, for example, implements a cube network by a multistage set of switching elements connecting processors through memories (Potter, 1978).

Various types of interconnection networks consisting of communication links and switching elements have been proposed; for a review see Siegel (1979). Abstractly, we can represent a network as a graph consisting of nodes (switches) and arcs (communication paths). Links between processors may be static (see Fig. 4.10) or the topology may be dynamic, with links being reconfigured by appropriate setting of the switching elements. The systolic array is a well known example of a two-dimensional static topology.

Dynamic networks may be divided into two classes: single-stage and multistage. The single-stage network consists of N input selectors (demultiplexers) and N output selectors (multiplexers). The desired communication routing is determined by the pattern of connections of input to output selectors. Depending on the network connectivity, data may have to be recirculated a number of times through the single stage to achieve the required final routing. Multistage networks, on the other hand, employ many stages of interconnected switches, in general n stages, each of $N/2$ switch boxes, where $N = 2^n$ is the number of input and output lines.

Our concern here is the logical (rather than the physical) interconnection topology; network control strategy and switching methods, and the communication protocols by which these are implemented by the switching elements, are discussed in Feng (1981). Formally, a general SIMD machine may be represented by the quartet (N,F,I,M) where

(i) N is the number of processing elements in the machine
(ii) F is a set of interconnection functions representing the interconnection network of inter-PE communication links
(iii) I is the set of machine instructions executed by the PEs
(iv) M is the set of masking schemes, partitioning the PEs into enabled and disabled subsets.

Stone (1971) has evaluated and compared a number of SIMD architectures in terms of this model. Techniques for assessing the suitability of particular topologies for given problems will be discussed in Section 4.8.3.

At the higher levels of the image processing hierarchy, i.e. image analysis and understanding, MIMD, or even mixed SIMD/MIMD schemes, may be used to advantage. A number of proposed schemes are reviewed in Reeves (1984). An MIMD architecture usually consists of a number of processors which share memory through a processor-memory module interconnection network (Fig. 4.11).

The majority of processor-memory interconnections discussed or implemented to date have been crossbar or delta networks (Patel, 1981). A delta network is a multistage network in which each stage consists of several crossbar switches. Thus the network topology may be reconfigured dynamically.

The progress of VLSI technology makes it feasible to consider implementing interconnection networks directly onto a chip or module of chips. A problem attracting increasing attention is that of embedding an interconnection topology into a plane in such a way as to optimise the fundamental resource parameters of area (reflecting the cost of processing and communication) and computation time. The VLSI model of complexity discussed in Section 4.3 can be used to study the trade-off between area and

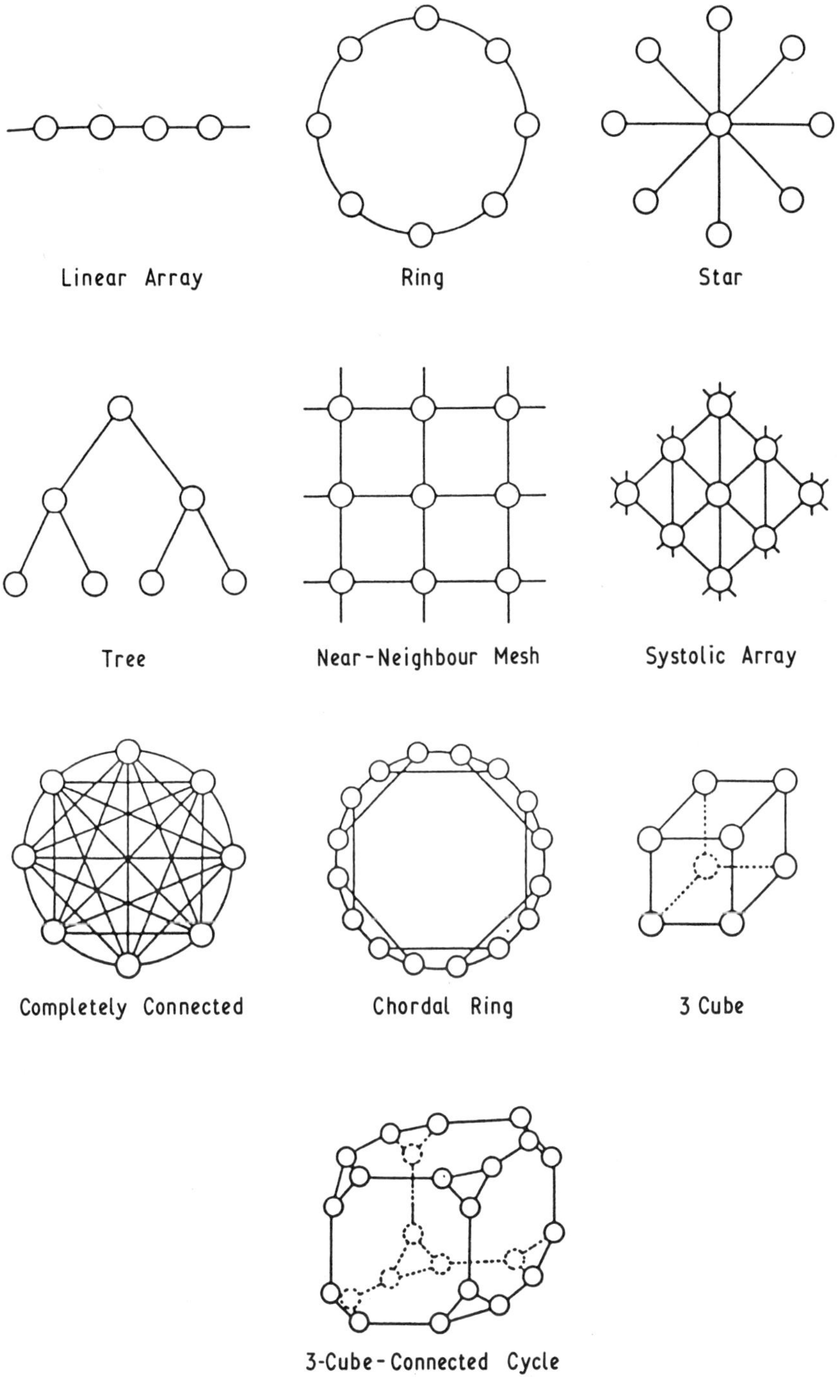

Fig. 4.10 Some static interconnection network topologies. Taken from Feng (1981) – copyright © 1981 IEEE.

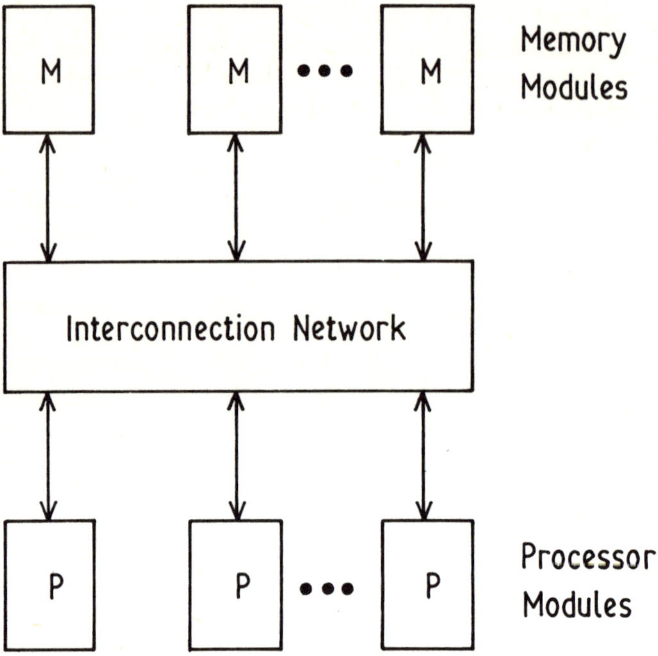

Fig. 4.11 A generic MIMD machine architecture.

time in attempting to implement efficiently algorithmic data flow requirements (Ullman, 1984; Preparata, 1984). The goal may be either a special purpose chip embodying directly the data flow required for a particular algorithm or function, or a simple and versatile interconnection pattern suitable for a range of problems.

4.7.2 Architectures: data and control flow

Computing is moving rapidly from an emphasis on sequential, centralised systems to parallel, decentralised systems. The nature of both the data and the processing of data is changing. Applications increasingly involve significant amounts of non-numeric, or symbolic, data items such as sentences, symbols, speech, graphics and images as the emphasis moves from the scientific towards more general problems and applications. Correspondingly, the programming and underlying computational styles are moving from the imperative (procedural) to the denotational (declarative).

There are a large number of potential contenders for future parallel machine architectures. Lacking any properly consistent taxonomy we shall attempt to group these contenders into broad generic groups based primarily on how the data and control flows are managed. In broad terms there are two possible approaches to implementing a parallel machine. First, packet communication architectures, such as those currently being used for data flow and reduction. Secondly, lattice architectures, where the local memory of each component computer collectively forms the address space of the machine. A packet communication architecture tends to destroy locality in an algorithm whereas a lattice architecture can be designed in such a way as to take advantage of algorithmic locality. It is this latter class of machines which are of greatest utility for the processing of images, especially at the lower, data driven levels of computation, with a consequent minimisation of communication overheads.

In the design of any parallel machine four important architectural mechanisms should be considered: memory, and consequently data, structuring, control strategies, programming for concurrent operation and the interconnection topology. In this section we are primarily concerned with the dynamic aspect of data structuring and control mechanisms.

It should be noted that from a static point of view there are effectively only two basic ways of connecting computing elements together in such a way as to obtain parallel activation, as illustrated in Fig. 4.12. Most systems exploiting some form of parallel hardware use a combination of the two.

The first uses sequentially cascaded components in which the output of one is connected to the input of the next. This structure, often referred to as a 'pipeline', is used whenever a large task can be decomposed into several smaller tasks of roughly equal complexity which can be performed independently in sequence to achieve the overall computational task. This overlapping of the execution of successive tasks in the various stages of the pipeline means that once the pipeline is 'full', completed tasks emerge at a rate governed by longest subtask time, which is usually much shorter than the overall task time. If the subtask times are all equal then the overall task execution time is shortened by a factor equal to the number of stages in the pipeline. This idealised performance is rarely achieved in practice because of program branch instructions resulting in pipeline breaks and consequent 'purging'.

The second basic structure is that of the parallel array. This can be used when it is not possible to decompose a task into sequential subtasks and when there are many tasks, some of which can be activated simultaneously. It may be activated in one of two possible ways. The first is to apply an input to each module of the system in turn, with the output appearing cyclically from each module at the same rate as the input is presented. This scheme is known as the 'cyclically activated' or 'systolic' array. The second method involves supplying all of the parallel modules with input at the same time, and after one complete task execution time all the modules produce output simultaneously. This scheme is known as the 'lockstep' parallel array. In both cases the execution speed-up is proportional to the number of modules in the array.

It would be very convenient to be able to refer to a rigorous established taxonomy of parallel computer systems in which several critical characteristics are used as parameters to divide the possible range of structures into representative classes. Unfortunately there is no such taxonomy available.

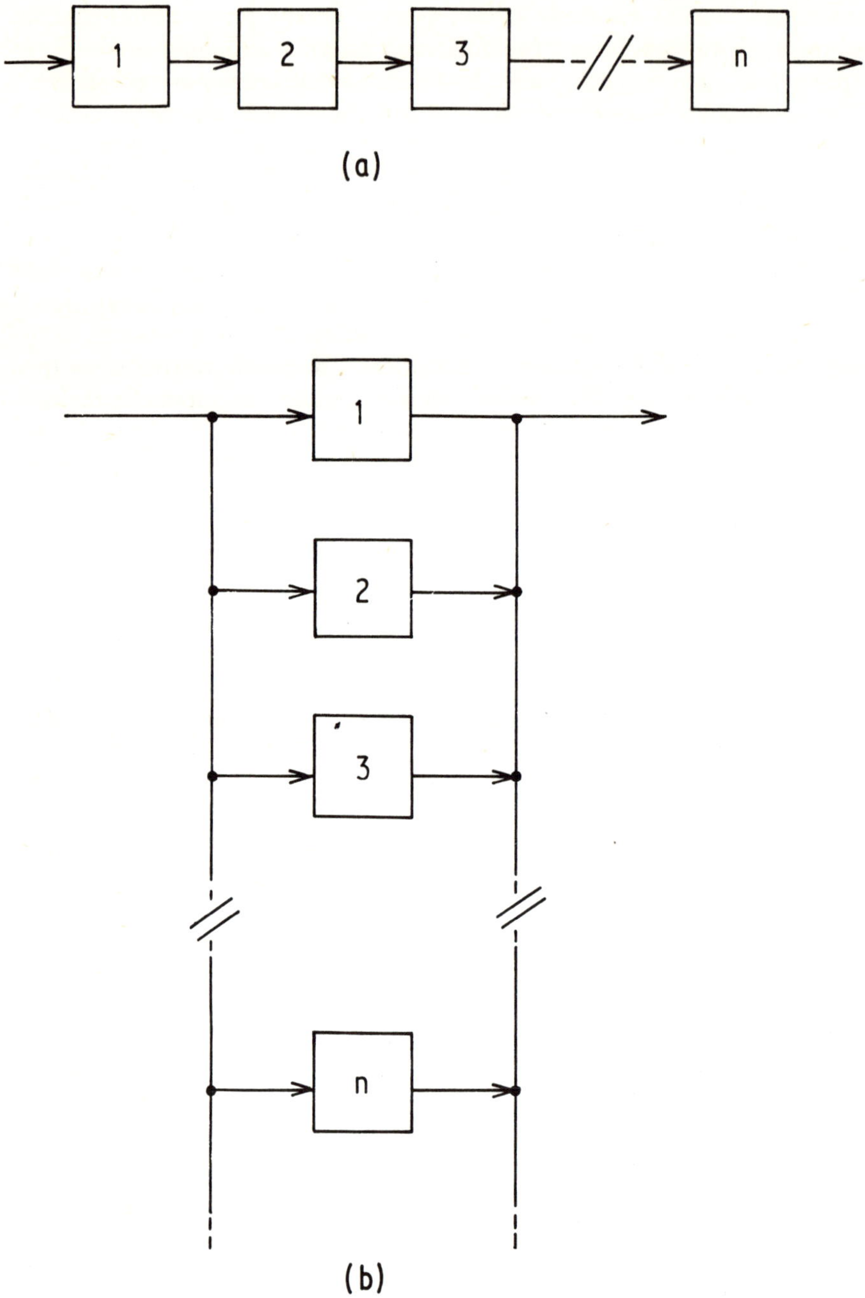

Fig. 4.12 The two effectively orthogonal patterns of connectivity for multiple processor systems which exploit parallelism. (a) Pipeline connectivity, and (b) parallel-array connectivity.

See Chapter 1 of Hockney and Jesshope (1981) for a representative approach to this problem. The following informal grouping presents its representative constituents roughly in order of increasing intrinsic parallelism. (*This discussion of machine types is based on a classification contained in Annex 5 of the UK Alvey Architecture Strategy Group's Interim Report* (*June 17, 1984*).)

Single-instruction multiple-datastream (SIMD) machines

SIMD machines are closely related to von Neumann machines in that they implement a single state-transition per instruction executed. They differ from conventional systems in the amount of storage that can be changed by one instruction. Von Neumann instructions are highly restricted in this respect. Typically they change the program counter, an arithmetic register or a store location. SIMD instructions have the potential to change many registers and store locations in a systematically defined relation to each other. There are two approaches to implementing SIMD instruction sets. The first, vector processing, uses pipelined arithmetic units linked to a cyclically accessed parallel array of appropriately interleaved store modules. The second approach uses a lockstep parallel array of processors which simultaneously access independent stores. They can be further subdivided according to whether the individual processors are linked with or disjoint from the independent store modules.

Shared-memory multiprocessors

The current generation of multiple-instruction multiple-datastream (MIMD) machines, in which many simultaneous and independent state transitions can occur, tend to be systems containing several von Neumann instruction processors sharing a large common memory space which is divided into independent store modules. The usual arrangement for sharing the memory is a hardware crossbar switch by which any processor can be connected to any memory module with equal time delay. Individual processors may contain cache memories, which window portions of the common store, or local memories which are not used for communication. An alternative is to provide processors with local store only linked together by memory mapping hardware which distributes the shared memory address space amongst the constituent processor-memory pairs. In either case the fundamental issues of 'memory latency' and 'cache coherence' have to be addressed and resolved (Arvind and Iannucci, 1983).

Message-passing multiprocessors

The key to connecting large numbers of processors together is thought to lie in the idea of basing process-to-process communication on the passing of messages rather than providing explicitly shared memory. The most obvious way of achieving this is to connect von Neumann computers together via some form of communications network. The topology of the network is of some interest since there is an obvious trade-off between network size and communication time. The richest connection is afforded by a full interconnect, in which all computers are linked directly to one another, with a constant communication time from any one process to another. Examples of low capacity networks are linear links, rings and rectangular arrays in which communication time is linearly bounded. Intermediate schemes, such as hypercube connectivity, can provide interconnection delays that are bounded approximately logarithmically by the number of interconnected computers. Figure 4.10 shows some representative static interconnection topologies.

Data flow computers

A large problem with process-based MIMD systems is the partitioning of applications programs into suitable networks of communicating processes. Responsibility for decomposing the program is usually passed to the programmer who all too often finds the task overpoweringly complex. Data flow attempts to sidestep this problem by lowering the complexity of the communicating processes to the point that the decomposition process can be successfully automated. An appropriate level for this seems to be about the level of the conventional von Neumann machine instruction. Data flow machines incorporate a parallel processing architecture in which each processor acts, according to appropriate instructions, when the data needed becomes available for computation. The basic principle of operation is quite straightforward. The data flow program is effectively a directed graph which shows where the data are to go and what operations are performed on the data when they get there (Gurd and Watson, 1980). A compiler uses the program to distribute groups of instructions to the various processors before the compiled program runs. During calculations, data are packaged into tokens which include labels to identify the process for which the data are intended. Processors continuously receive data and when they have acquired a complete set of data for a given label the processors take the data and execute the appropriate instruction. The results are passed on to other processors, with the token changed so as to reflect the task for which these new results are to be used. This continues until the calculation is complete. By considering each instruction as a process which communicates its results to its successor instructions, data flow researchers have been able to demonstrate extensibly parallel multicomputers for a range of applications with sufficient inherent parallelism, with parallelism extraction performed automatically. The price paid for relieving the programmer of the partitioning duty is the increased communications traffic required by such low level processes. Consequently, data flow multicomputers tend to use higher bandwidth communications networks than von Neumann multicomputers, with parallel data path and relatively rich network topologies. Individual data flow processors still rely on the use of conventional state transition techniques.

An alternative way of viewing data flow is as a method of exploiting irregular software parallelism by providing a two-dimensional programming model. Instructions which must be executed in sequence are written one under the other, as in conventional programs. Instructions which can be executed concurrently are written alongside one another, to indicate their lack of interdependence.

There is certainly some controversy over how well data flow machines will cope with structured data with a high degree of locality, as would be the case with most image processing tasks, certainly at the lower levels of processing (Lerner, 1984). It may well be that in attempting to avoid the von Neumann bottleneck the data flow protagonists have inadvertently introduced new bottlenecks, especially with well-structured problems involving considerable locality.

Declarative machines

Declarative languages have assumed increased importance in recent years. They represent a fundamentally different approach to computer design in that the languages are not based on the operational characteristics of the computers. Instead the languages are effectively designed first, based on mathematics, and the machine design follows. It has been observed that implementing such languages on sequential von Neumann computers involves unexpected difficulty, especially in handling higher order functions, logical unification and large declarative data structures. Further problems are anticipated when implementing these languages on parallel computer systems such as those discussed in the preceding sections.

Reduction techniques have been proposed in order to reduce the difficulties for both serial and parallel realisations. The reduction process involves systematic transformation of a program, represented either as a string or as a graphical data structure, until it is reduced to a minimum form (its 'meaning'). Thus a reduction machine needs a memory, which holds the current representation of the program, together with a number of transforming 'agents' which recognise and process reducible substructures within the graph. This can be implemented either as a shared store multiprocessor or by means of message passing. It should be emphasised that reduction is simply a model of computation particularly suited to the evaluation of higher order functional programs. Although the model is fundamentally different from the von Neumann model, it may still be implemented using von Neumann techniques.

Figure 4.13 is an attempt to group the above machine 'styles' in terms of control and data mechanisms, regarded as orthogonal coordinates. This representation is a convenient one in the sense that it does provide a succinct basis for classification. The 'logic' and 'actor' styles have not been explicitly discussed here, but are included for completeness.

At the present time it seems that the computer systems which will exploit parallelism in their hardware will continue to be based on concurrently activated state transition machines. These machines are characterised by uniquely indentifiable storage locations and combinatorial processing capability interconnected by logical pathways under the control of a storage-

CONTROL MECHANISMS	DATA MECHANISMS	
	SHARED MEMORY	MESSAGE PASSING
CONTROL DRIVEN	Control flow (von Neumann)	Control flow (Communicating processes)
DATA DRIVEN	—	Data flow
DEMAND DRIVEN	Reduction (string)	Reduction (graph)
PATTERN DRIVEN	Logic	Actor (object oriented)

Fig. 4.13 Relationships between data and control mechanisms for various types of machine that will support parallel computation. Adapted from Treleavan and Lima (1984).

state influenced controller operating in finite time steps. These basic and highly ubiquitous building blocks are known variously as a state transition machine, finite state machine, a finite automaton or a sequential state machine. In order for the desired computations to be distributed properly in time, system communication must be provided with an adequate scheme for the control of data flow. The simplest scheme involves the use of two clock signals, each of which alternates between high and low phases in such a way that the phases of the two clock signals do not overlap: clock signal 1 is high (low) when signal 2 is low (high). The shift register is an elementary example of this type of control for the propagation of data.

Systems in which the present state depends not only on the input of the present step but also on the state of the system in the previous step are known as *Moore machines,* a particular class of finite state automata. The relation between the state E and output O of the Moore machine at clock cycle t is given in terms of its state and input I at cycle $t-1$ by

$$E(t) = F(E(t-1), I(t-1))$$
$$O(t) = G(E(t-1), I(t-1))$$

where F and G are transition functions. A programmable logic array (PLA) implementation is shown in Fig. 4.14. For a detailed discussion of PLAs see Mead and Conway (1980). The feedback signals and primary inputs are

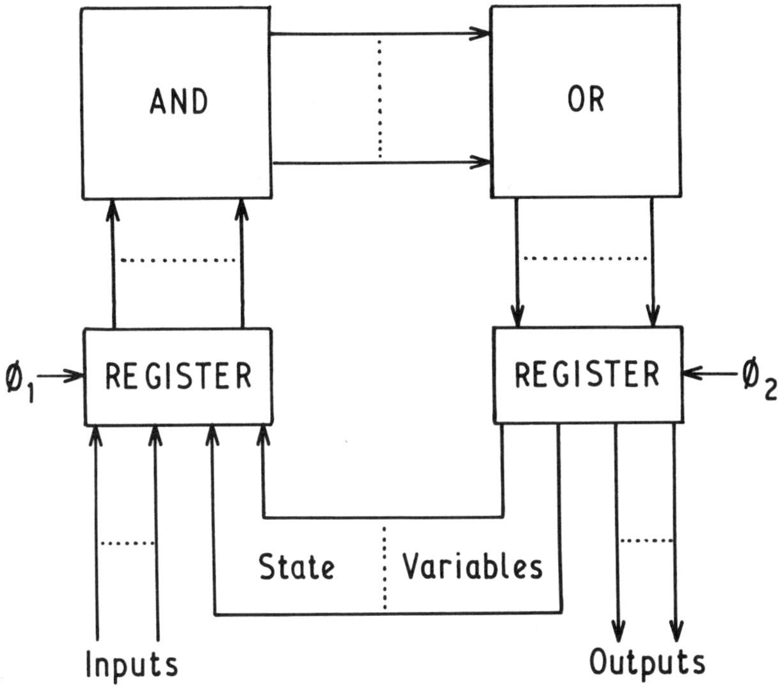

Fig. 4.14 PLA implementation of a Moore machine.

stored in the first register during clock phase 1, and the outputs of the combinational logic are stored in the second register during clock phase 2. The primary output and the feedback state variables defining the machine's next state appear on each clock cycle, phase 1 then phase 2. Note that the clock period must be long enough to ensure that signals can propagate through the combinational logic within the period. Such an implementation generates the basic level of sequencing control.

Since the Moore machine is a finite state automaton, the finite set of its states, the possible state transitions and the corresponding sets of inputs and outputs may be represented by a state diagram or, equivalently, in a state transition table.

The systolic array, important for many image processing applications, is a regular system of Moore machines with nearest neighbour connections. The representation of data and control flow through these systems is a subject to which we shall return in Section 4.8.2.

4.7.3 Performance constraints

The notion of a general purpose image processing computer, capable of the requisite versatility and running at circa 10^8 to 10^{10} operations per second, is as yet ill defined and in practice may prove to be highly elusive for some time to come. To date, cost-effective solutions to serious image processing problems have required a significant degree of very specific functional specialisation in the implemented computer architecture.

As previously discussed, the advent of readily accessible VLSI technology and design aids has now made it possible to provide sophisticated but inexpensive computational capability by implementing a system consisting of a large number of identical processors organised in a simple and regular structure, so minimising the need for application-specific constraints. In order to exploit this incipient computational capability usefully it is important to employ efficient parallel algorithms running on computationally optimal machines. This section is concerned with briefly discussing some of the measures which can be usefully applied to computational architectures in order to evaluate performance, and hence establish the constraints which result.

The common theme which lies behind the search for new computational architectures, in general and for image processing in particular, is the drive to abandon the classical von Neumann machine architecture with its characteristic bottleneck in the single memory-processor communications channel. Unfortunately, what is perhaps most characteristic of this search is the sheer diversity of solutions which have been proposed.

The important and difficult question of how to evaluate and use performance figures of merit, and hence constraints, for given algorithms running on particular architectures is not an easy one to answer. Further, we would also like to be able to establish the performance constraints imposed by various implementation strategies for a given algorithm implemented on a particular class of architecture. In practice, as one might expect, the situation is complex. As the characteristics of one component of the system design improve, overall performance improves until some other component becomes the bottleneck in the system's performance. Certainly, there is no general framework yet available for investigating these questions.

At the simplest level, the performance of parallel processors may be measured in terms of the speedup achieved over a corresponding sequential machine and the degree of processor utilisation. Consider first a k-stage pipelined vector processor. Let n be the number of program instructions, and N_i the number of vector operands in instruction i. Then it is easy to show that the ratio of the total execution times required in the pipeline processor and the equivalent serial computer is given by

$$S^{k\text{-pipe}} = \frac{T_{\text{serial}}}{T_{\text{pipe}}} = k \sum_{i=1}^{n} N_i \Big/ \left(\sum_{i=1}^{n} N_i + (k-1)\, n \right)$$

$S^{k\text{-pipe}}$, the 'speedup', is a commonly quoted performance metric, as is the efficiency e, given by

$$e = \frac{S^{k\text{-pipe}}}{k}$$

As k, the number of pipeline stages, becomes greater than the average vector length, the speedup becomes insensitive to the value of k, and the efficiency of processor utilisation becomes poor. The moral is an obvious one: pipelining favours long vectors.

Similarly, the performance of an SIMD array of n processing elements may be discussed in terms of the speedup ratio

$$S^{n\text{-array}} = \frac{T_{\text{serial}}}{T_{\text{array}}}$$

and the efficiency of its utilisation

$$e = \frac{S^{n\text{-array}}}{n}$$

Hockney and Jesshope (1981) have reviewed this characterisation of performance and subsequently introduced two parameters r_∞, the maximum or asymptotic rate of producing results, and $n^{1/2}$, the vector length which produces half maximum performance. The significance of these two parameters is that while r_∞ is primarily technology dependent, $n^{1/2}$ is a measure of the amount of parallelism inherent in the architecture. They then define the 'specific performance', $r_\infty/n^{1/2}$, which they claim to be more characteristic of particular types of computer architectures than either of these parameters alone. A high value of this ratio is regarded as desirable in principle because less parallelism is needed in order to achieve a given performance. Computers for image processing will then generally tend to have large values of $n^{1/2}$ and hence relatively low specific performance ratios. For further detailed discussion see Hockney and Jesshope (1981).

Hon and Reddy (1979) have looked at this question, in the context of some specific image processing algorithms, in a fairly general way and heuristically established a number of global and fairly intuitive rules of thumb. These relate mainly to the performances of primary and secondary memory and appropriate trade-offs and do not attempt to quantify performance constraints in any way.

The potential advantages, or conversely the constraints, which arise from a consideration of the underlying design and fabrication technologies are various, complex and frequently interrelated.

The general significance to image processing systems of advances in VLSI fabrication technologies has already been discussed in Chapters 1 and 3 and will be raised again in Chapter 7. We refer the reader to these sections and will only refer in passing here to their implications in terms of how they relate to actual design limitations for highly concurrent image processing engines.

It is still a truism that computing is little more than storage, communication and computation. Yet, as systems grow in size and complexity, communication becomes by far the most dominant factor and the most serious potential stumbling block. Unfortunately, the physics of device scaling (Mead and Conway, 1980) makes this a problem which tends to be exacerbated, rather than ameliorated, by advances towards ever smaller feature sizes on ever larger chips. This is because long range interconnection delays scale as α^2, with the concomitant gate-relative delays scaling as α^3, where α is the geometrical scaling factor. Further, the phenomenon of metal migration may well impose some fundamental constraints on the global distribution of clock, power, control and data signals in highly concurrent structures, such as systolic and SIMD arrays. This means that there is a strong requirement to enforce short range ordering, with special provision for inherently long range signal communication such as another, coarser, *ad hoc* level of design rules for distributing these signals.

Another means for avoiding undue communications delay is in pipelining the communications flow, buffering data at various nodes in the network. In an extreme case this becomes indistinguishable from packet switching. The Burroughs BSP design used this technique quite successfully.

In the end, to obtain peak performance, a composite approach will need to be used – structuring both the data and the architecture. Hierarchically structured systems are a good example of such an approach. Towards the top of the hierarchy, communication is characterised by large amounts of information sent infrequently with large latencies, while towards the bottom small amounts are sent frequently with low latency. There may be a number of levels in the hierarchy between these two levels, as dictated by the application, but with the frequency × delay product kept approximately constant so as to ensure efficient communications overlapping, layer to layer. This approach does not work with algorithms involving the global communication of unstructured data, such as matrix multiplication or inversion, where global communication is a basic property of the algorithm. Of all the constraints that will face the systems designer in the future, it seems that global communication is going to be the most difficult.

A design consideration which is closely related to communication is that of memory organisation. Here too hierarchy is an important concept, with registers, cache, primary (often interleaved) and secondary memories being almost universal in practice. It is worth noting the fact that the actual partitioning of memory to cope with a given problem's requirements can have a very significant effect on memory bandwidth requirements (Hockney and Jesshope, 1981 – Section 6.2.3). Again algorithmic and architectural concepts become closely and inextricably intertwined. Mead and Conway (1980, Chapter 8) have considered, *inter alia,* the impact of VLSI technology on the design of hierarchical, recursively defined memories and their performance. They conclude with the comment that they consider the greatest impediment to harnessing concurrency effectively is our present inherently sequential mode of thinking about computer organisation, and our greatest need is for a unifying theory of computation.

4.8 MAPPINGS

The theme of modern design for image processing, as for many other applications, is to exploit fully the use of parallelism, both in software and hardware, in the interest of maximising computational throughput. Applications in which data have a natural block structure, and for which the same computations are repeated with different variables or parameters, are eminently suitable for this approach. The motivations for exploiting parallelism in image processing are then:

(a) the requirement of real-time performance (circa 10^8 to 10^{10} operations/s)
(b) the regularity of image processing algorithms and data structures (e.g. arrays of pixels)
(c) the ability of VLSI technology to realise specialised parallel architectures cheaply and efficiently.

The efficiency of an algorithm depends on the particular parallel architecture chosen for its execution (see Hon and Reddy, 1979, for a comparative performance evaluation of an edge detection algorithm on three machines, the C.mmp, Cm* and HARP). Ideally, the structure of the algorithm should match that of the machine on which it is implemented. But how is this structural mapping to be achieved? To begin, we need a classification of parallel algorithms corresponding to that of parallel architectures. A first attempt has been made by Kung (1980).

A parallel algorithm may be considered as a number of independent communicating processes operating under an overall control, which supervises the interactions among the concurrent processes to achieve correct execution. Kung's classification of concurrency controls of parallel architectures is shown in Fig. 4.15.

Parallel algorithms may be further characterised in terms of their granularity, or amount of computation a typical process may achieve before communication with other processes is required in order to progress. Clearly, if an algorithm requires heavy interprocess communication, efficiency is increased when the communication structure of the architecture on which the algorithm is to be run matches well the interprocess communication paths. Indeed, communication within a parallel structure is one of the most important influences on design. A classification of communication geometries is given in Fig. 4.16 (Kung, 1980).

Chiang and Fu (1983) extend this characterisation to include data movement and memory space as further attributes of a parallel computation. The data movement between subtasks of the algorithm is characterised as regular or irregular; the connection between subtasks may be variant or invariant in time. Memory space is referred to as local or global, depending on the absence or presence of secondary memory. If the required memory space does not depend on the input data, it is referred to as constant.

Recall from Section 1.4.2 and 4.7.2 that machine architectures can be

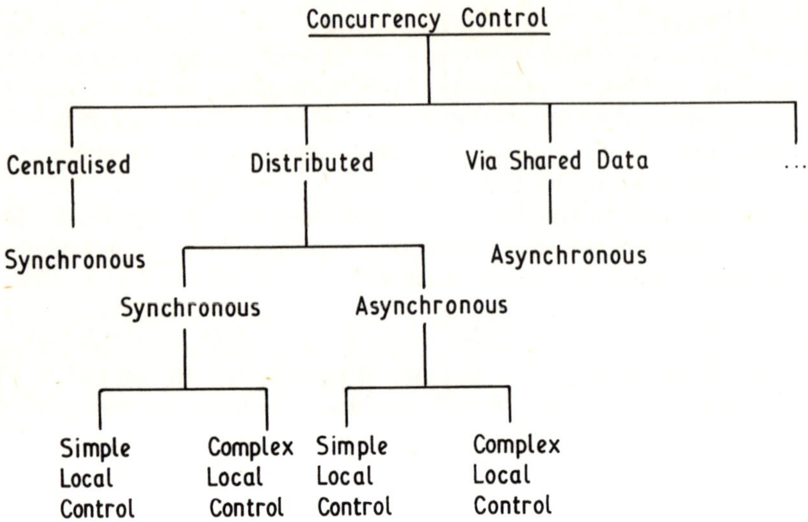

Fig. 4.15 Classification of concurrency controls.

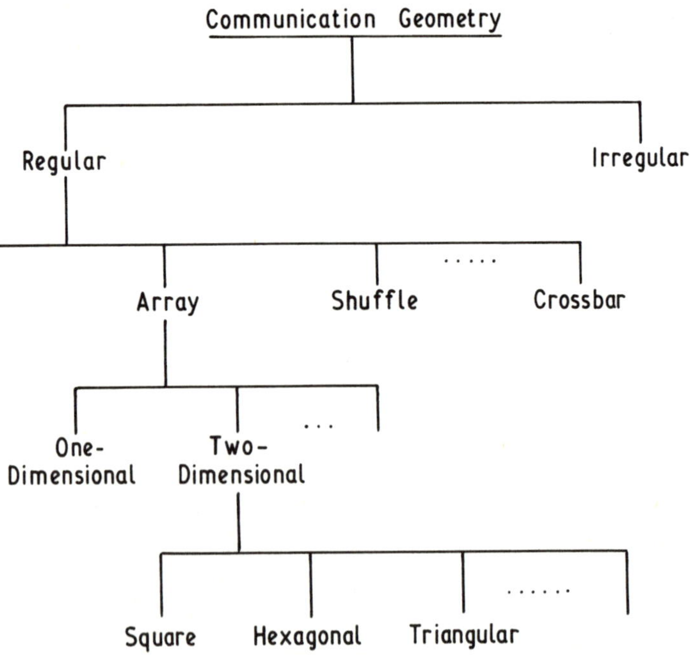

Fig. 4.16 Classification of communication geometries.

classified broadly according to their patterns of data and control flow. A SIMD (single-instruction multiple-data stream) machine consists of a number of processing elements, each with local memory. Identical instructions from the central control unit are executed in lockstep fashion. The MPP and GRID (see Section 7.2.2) are examples of VLSI implementations of SIMD architecture.

MIMD (multiple-instruction multiple-data stream) machines consist of an array of processing elements operating asynchronously on independent instruction streams. The Cosmic Cube machine (see Section 7.2.2) falls into this class.

Systolic machines (see Section 1.4.2) are networks of processors through which a regular flow of data is pumped periodically. In contrast to the programmable SIMD and MIMD machines, these are special purpose architectures designed for algorithms with frequent and regular interactions among subtasks.

With this background, we can now characterise broadly the classes of parallel algorithms matching the systolic, SIMD and MIMD classes of architecture. Synchronous algorithms with small constant granularity using distributed control through simple local control correspond to systolic machines. Those synchronous algorithms requiring centralised control match SIMD machines, whereas asynchronous algorithms correspond to MIMD architectures. The characteristics of parallel algorithms suitable for implementation on the systolic, SIMD and MIMD classes are given in Fig. 4.17.

	Granularity	Concurrency control	Communication geometry	Memory space	Data movement
SIMD	small constant, small or large	centralised, synchronous	regular	small, local	regular, variant or invariant
MIMD	large	shared data, asynchronous	irregular	large, global	irregular, variant
Systolic	small constant	distributed, synchronous, by simple local	regular	small, local	regular, invariant

Fig. 4.17 Characteristics of parallel algorithms.

As we have already mentioned, the choice (or design) of architecture for image processing applications should reflect the particular structures of data and algorithm. Processor arrays (SIMD) reduce computation time by mapping the data structure directly onto an array of processing elements. If the number of processing elements is greater than or equal to the number of pixels, the throughput rate is independent of image size. Decreasing the number of PEs, however, may incur the price of a greater memory size per element (Rosenfeld, 1981). Connections between PEs may be extended beyond nearest neighbour at the expense of a higher degree of design complexity.

At the opposite end of the scale lie machines whose architecture more exactly mirrors algorithmic structure. Systolic systems are examples of this latter class; these special purpose architectures are designed so that the communication paths match the algorithmic data dependencies. Systolic array designs for filtering, correlation, convolution and the discrete Fourier transform may be found in Kung (1980).

The question of algorithmic and data partitioning for multiprocessor systems, process synchronisation and interprocessor communication are important areas of research. The reader is referred to Cantoni and Levialdi (1982) who benchmarked the SIMD array and pipeline processor architectures by an analysis of the time performances of a number of image processing tasks (point operations, local operations, image statistics and two-dimensional transforms).

The next four sections explore in more detail the problems which underlie any attempt to establish mappings between computational algorithms and implementation architectures.

4.8.1 Topological maps: static considerations

To begin with, we can look at the algorithm-architecture mapping problem in topological terms as the matching of algorithmic communication geometry and hardware interconnection network. To minimise communication overheads it is desirable that processes which communicate with each other should, as far as possible, be assigned to directly connected processors. A good example of this strategy is the shuffle-exchange network (Stone, 1971), which directly realises the parallel FFT algorithm (Pease, 1968). The choice of assignment comes down to the maximisation of the number of pairs of communicating processes that fall on pairs of directly connected processors, and is formulated mathematically as a graph matching problem (Bokhari, 1981).

An algorithm to be matched onto an array of processors may be represented by a graph $G_p = (V_p, A_p)$ where V_p denotes the set of processes making up the algorithm. Processes $x, y \in V_p$ are connected by an arc $(x,y) \in A_p$ if the two processes communicate. The processor array can also be represented by a graph $G_a = (V_a, A_a)$, where V_a is the set of processors and the set of arcs A_a forms the interconnection network of the array.

The graph G_p (and similarly for G_a) can also be regarded as the set of vertices V_p and a membership function $M_p{:}V_p \times V_p \rightarrow \{0,1\}$ with the properties:

$$M_p(x,y) = M_p(y,x) = 1 \text{ for } (x,y) \in A_p$$

$$= 0 \text{ for } (x,y) \notin A_p$$

$$M_p(x,x) = 0$$

for all $x, y \in V_p$.

Suppose now that $|V_p| \leq |V_a|$ where $|V|$ denotes the number of elements of the set V. If $|V_p| < V_a|$, the necessary number of 'fake' vertices may be inserted to ensure that $|V_p| = |V_a|$.

We are looking then, for that function among the $|V_p|$! possible bijections (mappings which are both one-to-one and onto) $f{:}V_p \rightarrow V_a$ which maximises the overlap of process on processor graph arcs. That the number of such overlaps is given by

$$|f| = \tfrac{1}{2} \sum_{x,y \in V_p} M_p(x,y) M_a(f(x),f(y))$$

is easy to see. There is a contribution to the sum only if both factors in the summand are 1. $M_p(x,y){=}1$ if x and y are connected by an arc in the algorithm graph. $f(x)$ and $f(y)$ represent the processors onto which processes x and y are mapped, and M_a $(f(x), f(y)){=}1$ only if these processors are connected. Finally, a symmetry factor of 1/2 is needed to prevent overcounting.

The problem of maximising $|f|$ for arbitrary G_p and G_a is equivalent to the problem of determining whether two graphs are isomorphic (Bokhari, 1981), which is an unsolved problem! In the absence of exact procedures for solving the topological mapping problem in general, heuristic procedures have been explored for generating mappings onto specific processor arrays (Kung and Stevenson, 1977; Bokhari, 1981). An example of the mapping of the data flow graph for the factorial function onto an hexagonal array will be given in Section 4.8.4. Such heuristically found solutions may not, of course, be optimal. One such heuristic procedure is that of simulated annealing, discussed in Chapter 2.

4.8.2 Data and control flow maps: dynamic considerations

The data and control flow aspects of algorithms and architectures have been separately considered in Sections 4.6.2 and 4.7.2. We are now concerned with how these dynamic aspects of computation can be efficiently represented and mapped.

A formal representation linking a mathematical expression to be calculated to a network architecture designed to perform the computation has been given by Cohen (1978) and further developed by a number of other authors (e.g. Johnsson and Cohen, 1981). The flow of data along directed paths through a computational network can be described with data and control flow treated within the same formalism. Given an expression to be computed, the representation allows transformation between implementations with the same output but different throughput characteristics.

The description of the flow of data and control requires a treatment of the distribution of the computation both in space and in time. For this purpose, Cohen introduces a delay operator Z modelling a storage element. Consider a sequence of data elements $\{x(i)\}$, where $x(i-1)$ is the element which arrived at a particular point in the network one time unit before element $x(i)$. The delay operator is then defined by

$$Z[x(i)] = x(i-1) \tag{4.1}$$

or, recursively:

$$Z^n[x(i)] = Z[Z^{n-1}[x(i)]] = x(i-n) \tag{4.2}$$

A graphical representation of Equation 4.1 is shown in Fig. 4.18.

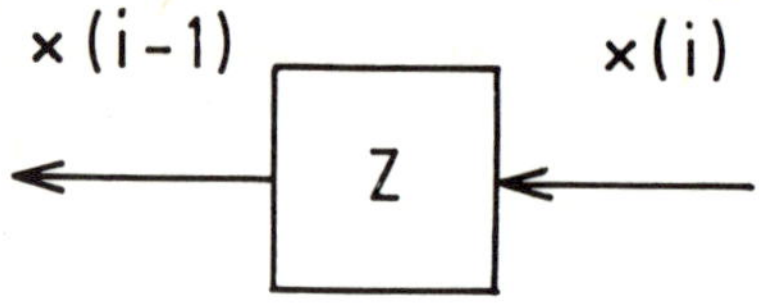

Fig. 4.18 A delay element.

Expressions defining calculations to be performed may be formally manipulated to generate alternative computational networks evaluating the same function with different performance characteristics. These manipulations rest on the following properties of the delay operator:

$$Z^n[F(x,y)] = F(Z^n[x],Z^n[y]) \tag{4.3}$$

or, graphically, Fig. 4.19, and, for constants C,

$$Z^n[CX] = CZ^n[X] \tag{4.4}$$

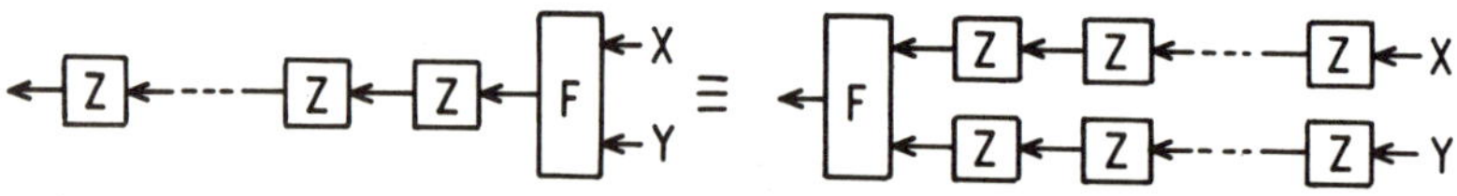

Fig. 4.19 A diagrammatic representation of equation 4.3.

A simple example of the manipulation of network implementations is provided by a finite impulse response filter (Johnsson and Cohen, 1981), defined by

$$y(i) = \sum_{n=0}^{N-1} a(n)\, x(i-n) \tag{4.5}$$

This expression may be rewritten as

$$y(i) = \sum_{n=0}^{N-1} a(n)\, Z^n[x(i)] \tag{4.6}$$

or, alternatively, since $\{a(n)\}$ are time-independent coefficients:

$$y(i) = \sum_{n=0}^{N-1} Z^n[a(n)x(i)] \tag{4.7}$$

To each expression in the formalism there corresponds a network implementation. The networks given by equations 4.6 and 4.7 are shown in Figs. 4.20 and 4.21 respectively. The pipelined implementation, Fig. 4.21, achieves a higher throughput rate than the initial implementation Fig. 4.20. More complicated examples, involving explicit control signalling, have been analysed by Johnsson and Cohen (1981).

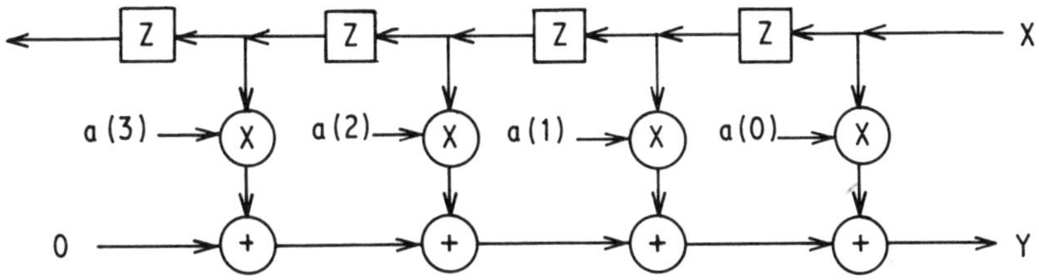

Fig. 4.20 Implementation of equation 4.6. Taken from 'A Mathematical Approach to Modelling and Flow' by Johnsson and Cohen, *VLSI Systems and Computations*, edited by Kung *et al.*, with the permission of Carnegie Mellon University and the publisher Computer Science Press, Inc., 1803 Research Blvd., Rockville, Maryland, 20850, USA.

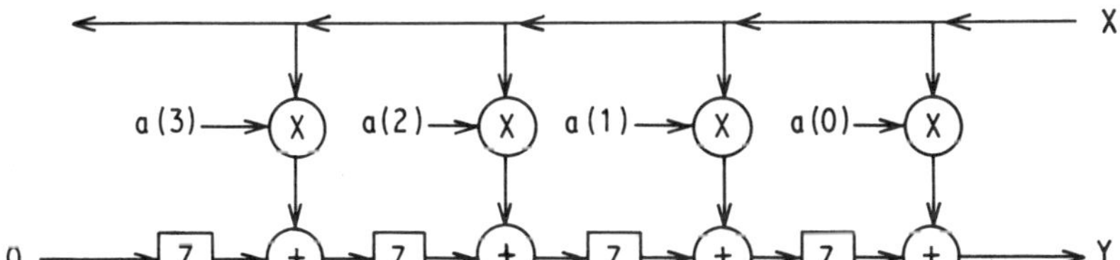

Fig. 4.21 Implementation of equation 4.7. Taken from 'A Mathematical Approach to Modelling and Flow' by Johnsson and Cohen, *VLSI Systems and Computations*, edited by Kung *et al.*, with the permission of Carnegie Mellon University and the publisher Computer Science Press, Inc., 1803 Research Blvd., Rockville, Maryland, 20850, USA.

The delay formalism enables exploration of design alternatives in a controlled and methodical fashion amenable to machine aid through automatic symbol manipulation. Functionally equivalent expressions may be analysed to compare computation rate and delay (Cohen, 1978; Jagadish *et al.*, 1984). Indeed, this functional approach has been used to derive a number of systolic array designs significantly better in performance than those proposed by Mead and Conway (1980). Furthermore, designs can be verified formally since they correspond to transformations of the expression to be computed.

The progression of 'wavefronts' of data streams through a computational network is described by the application of operators to ordered sets of data elements $x(i,j)$ where i and j are position and time indices respectively. These operators may be used to delay, rotate or shift wavefronts. The combination of two input wavefronts to produce an output datastream may also be represented in the formalism. An example of the use of such wavefront transformations in the design of a network performing the discrete Fourier transform is given by Johnsson and Cohen (1981).

Extensions of the Petri Net model (Brauer, 1980) have been proposed for the graphical representation of the dynamics of control. In its simplest form, a Petri Net **N** is a directed bipartite graph, in which the nodes are of two distinct types: places **P** and transitions **T**. Places are represented by circles and transitions by bars. Arcs **A** may only join places and transitions, i.e. $\mathbf{A} \subseteq (\mathbf{P} \times \mathbf{T}) \cup (\mathbf{T} \times \mathbf{P})$. Places may be occupied by tokens, which flow through transitions along arcs from place to place according to a set of 'firing rules' enabling the transitions. System states are modelled by the places, and events by the transitions, in a Petri Net model of the coordination or synchronisation of system events. The technique is applicable to the modelling of both software and hardware control flow and can be used for a very qualitative assessment of the suitability of an algorithm for a particular architecture (Chiang and Fu, 1983).

4.8.3 Performance and optimisation

In order to assess the quality of any proposed map or matching between algorithm and architecture, we require suitable measures of 'goodness of fit'. Here, yet again, is an open and difficult area of research: a major problem is that any one measure may quantify particular aspects of performance whilst neglecting others, so that design trade-offs cannot be adequately assessed. A simple example suffices to make the point.

Siegel, Siegel and Swain (1982) have compared a number of proposed measures on an algorithm for the computation, using $N \geqslant B$ processing elements, of the B-bin histogram of an $m \times m$ image. To each PE is assigned an $mN^{-1/2} \times mN^{-1/2}$ sub-image. The PEs each compute, in parallel, a B-bin histogram of a sub-image, and the histogram of the complete image is computed from these by a recursive doubling technique.

As a performance measure we might first consider the speed, or number of data points processed per unit time. Figure 4.22 shows the logarithm of the speed, appropriately normalised, as a function of the logarithm of N. The speed increases with m for given N, a reflection of the decreased effect of the overhead incurred in PE masking operations in the communication of data between processing elements. For large m, the number of pixels processed in unit time is N, the theoretical maximum speed. In the intermediate range of m, the speed increases at a rate which decreases with

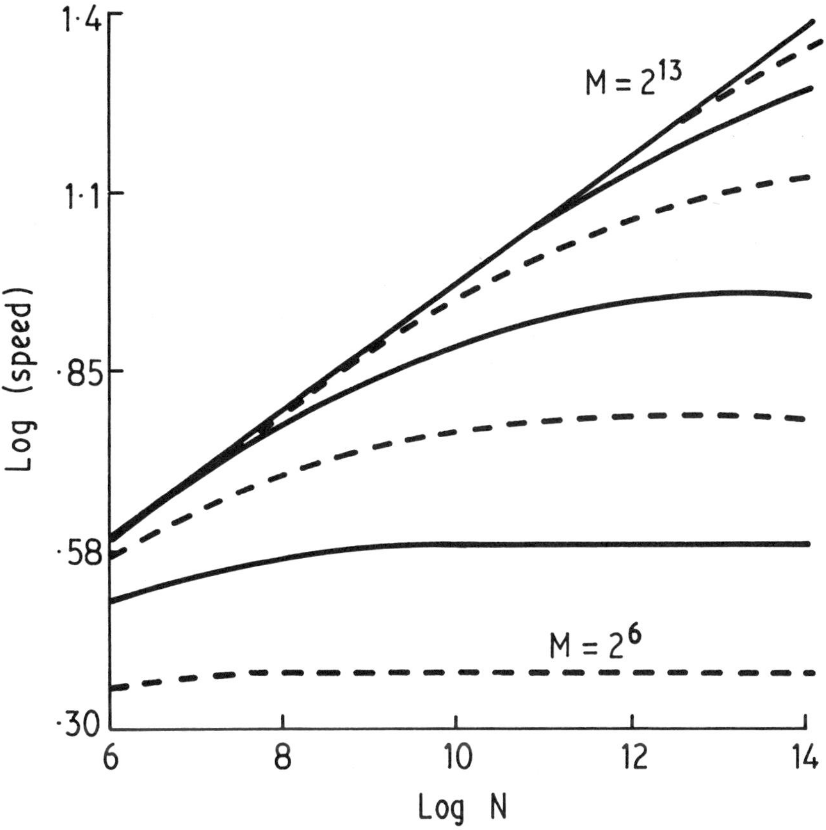

Fig. 4.22 Log of speed of a histogram algorithm on an *N* x *N* processor array as a function of log *N*. Adapted from Siegel, Siegel and Swain (1982).

N. Finally, for small *m*, the speed is roughly constant in *N*.

Alternatively the 'utilisation', or fraction of time during which the processing elements are executing the algorithm, may be taken as a performance measure. The utilisation is unity when overhead is negligible. According to Fig. 4.23, utilisation decreases with *N* for all values of *m*, though at a slower rate for large *m*. For very large *m*, utilisation is high throughout the range of *N* and relatively independent of *N*. On the other hand, taking speed as performance measure would dictate that *N* should be large for large *m*. For intermediate values of *m*, the utilisation is quite sensitive to *N*, and the value of *N* chosen to maximise utilisation is smaller than that which maximises computational speed. Quantification of speed or execution time does not indicate the efficiency of use of processing resources.

More generally, we need a better understanding of what is meant by matching an algorithm to an architecture efficiently. Much intensive work in the areas we have discussed in the previous sections will be required in order to partition algorithms effectively into processes, map them appropriately onto processors and to analyse quantitatively design trade-offs (e.g. communication *vs.* time) and optimisation.

The application of modern VLSI complexity theory should be very useful in analysing the cost of processor-to-processor and processor-to-memory communication in probing the relation of algorithms to communication network topology. Lower bounds on the time required for execution of an algorithm on architectures composed of a set of p processors with no shared memory can be derived (Gannon and van Rosendale, 1984) using Thompson's 'fooling' technique (recall Section 4.3). Within the context of a simplified model in which the problem to be solved is uniformly distributed over the set of processors, upper bounds on speedup performance are available, which depend only on the problem to be solved and the communication topology. As an example, a comparison of bounds for the FFT for a variety of network connection graphs is given in Fig. 4.24. Analysis of particular algorithms shows that some of these bounds are exact, as functions of the problem size n, up to constant terms.

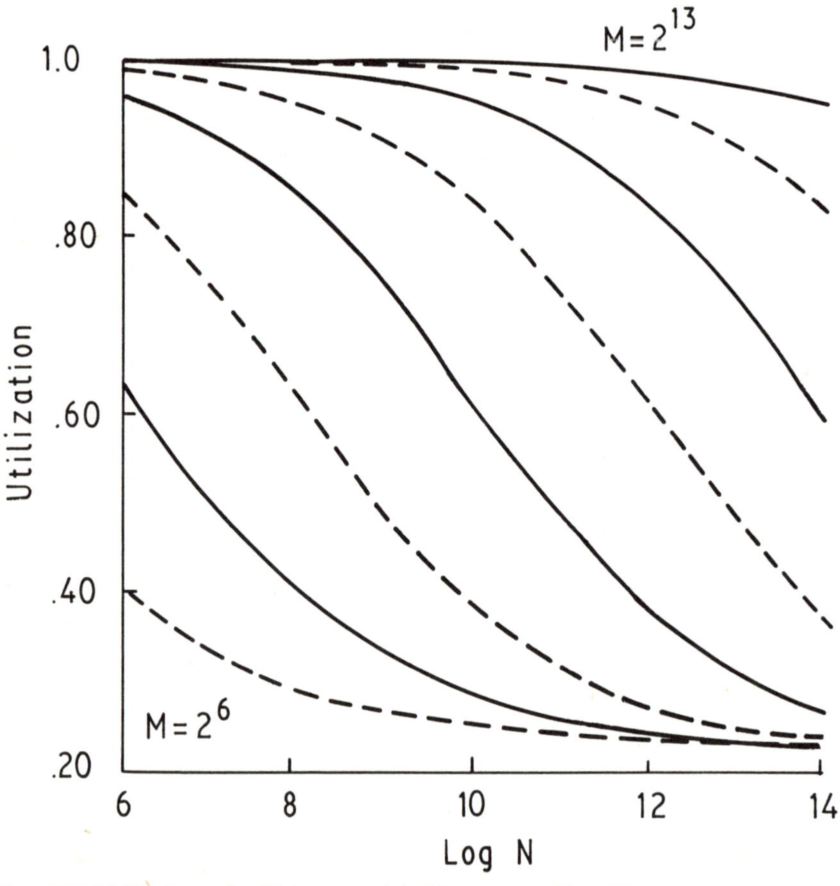

Fig. 4.23 Utilisation of a histogram algorithm on an $N \times N$ processor array as a function of log N. Adapted from Siegel, Siegel and Swain (1982).

TREE	$= \dfrac{P}{1 + cr\,\dfrac{P}{\log(n)}}$
RING	$= \dfrac{P}{1 + cr\,\dfrac{P}{\log(n)}}$
MESH	$= \dfrac{P}{1 + cr\,\dfrac{P^{1/2}}{\log(n)}}$
SHUFFLE	$< \dfrac{P}{1 + \dfrac{cr}{\log(n)}}$

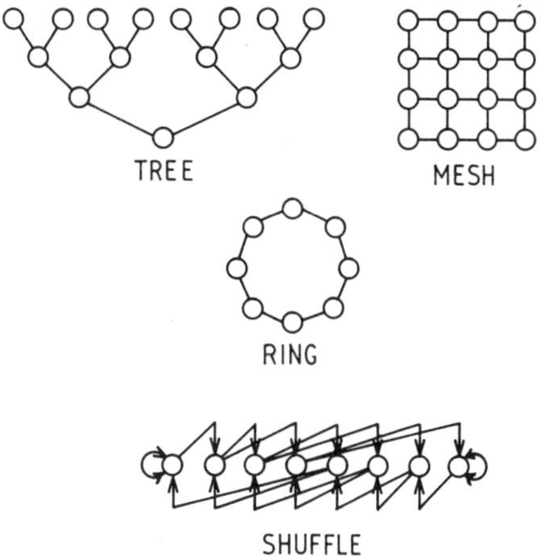

Fig. 4.24 Speedup upper bounds for the FFT on some interconnection topologies. *r* and *c* are constants. Taken from Gannon and van Rosendale (1984) – copyright © 1984 IEEE.

4.8.4 Examples of simple mappings

We have seen that, with the delay formalism, transformations are made at the algorithmic level to provide a direct mapping of the transformed expression or algorithm onto the architecture. Another approach to the mapping problem is to give a model of the algorithm, for example in terms of a data flow graph or an extended Petri Net, and to map that representation onto an appropriate architecture.

An example of a simple mapping onto a processor array using this strategy has been given by Koren and Silberman (1983). The idea here is to express the algorithm as a data flow graph (Davis and Keller, 1982) which is mapped directly onto the array.

The example given employs a hexagonal array of processing elements, each of which is capable of performing the basic data flow operations. Recall from Section 4.6 that, in data flow graphs, nodes represent operations; tokens carrying data objects flow along the arcs, and an operation may be executed when all its operand tokens have arrived at the node. The basic operations are arithmetic, logical and conditional, along with a number of operators specifically controlling the flow of tokens:

(a) The ‘copy’ operator **C**, which simply outputs two copies of its input data token.
(b) The ‘merge’ operator **M** is used when a token may arrive from two different arcs, and is to be merged into a single output arc. This operator is also capable of producing a Boolean token whose value depends on the input arc which supplied the token for the output.

(c) The 'router' operator **R**, which receives two inputs: a logical value and a data token. The data token is copied into exactly one of two output registers, according to the value of the logical input.
(d) The 'gate' operator **G** transfers the incoming token to an output register if a second token is present at another input register.
(e) The 'self-iterating' operator **L** is employed when the result produced by the processing element is immediately used as argument to the next operation.

In terms of these operators, the factorial functions, for example, may be expressed as a data flow graph as shown in Fig. 4.25. The results of a

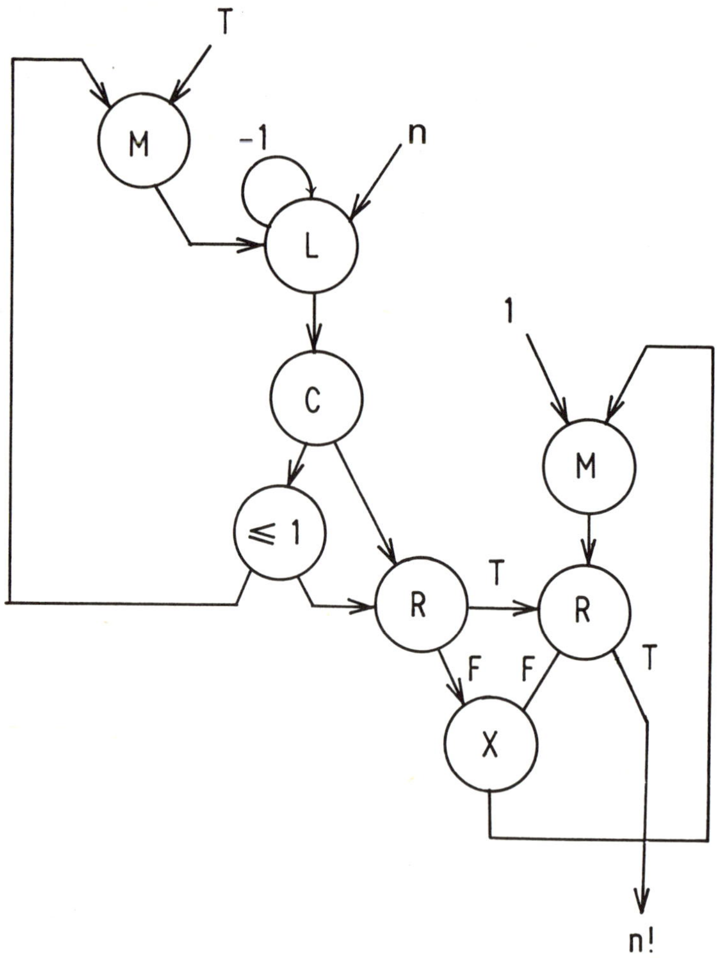

Fig. 4.25 Data flow graph for the factorial function. Taken from Koren and Silberman (1982) – copyright © 1983 IEEE.

mapping algorithm, executed by an external host, is presented to the array for the distribution of the data flow graph. This algorithm is dependent on the array topology chosen, in this example hexagonally connected. The result of this mapping for the factorial function is shown in Fig. 4.26(a). Finally, a reduction procedure is applied to compactify the mapping; the result is Fig. 4.26(b).

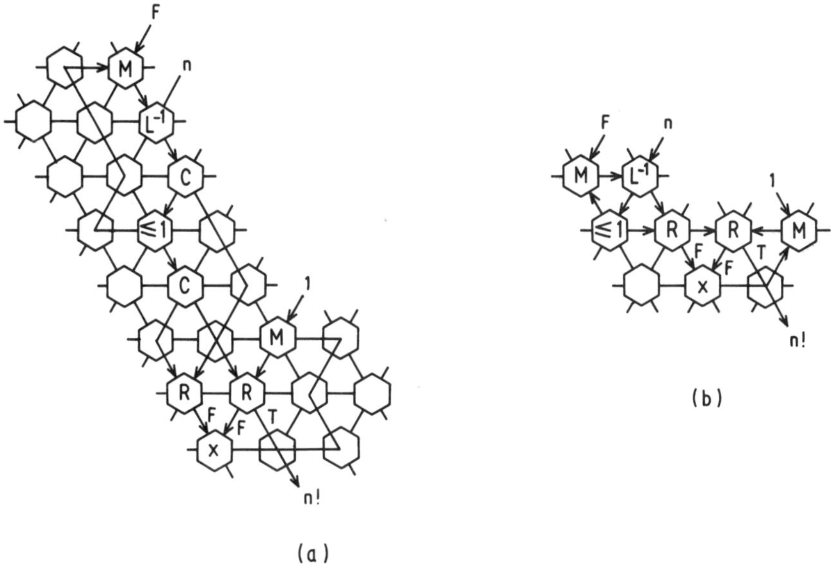

Fig. 4.26 (a) Initial and (b) reduced mappings of the factorial function. Taken from Koren and Silberman (1983) – copyright © 1982 IEEE.

Despite the interest of this technique, its usefulness in practice remains to be demonstrated. The mapping algorithm given by Koren and Silberman is rather inefficient; we must again face the problem raised in Section 4.8.3: how to optimise the mapping.

A method for optimisation, in terms of minimising a complexity measure such as computation time T or a space-time measure AT^2, has been discussed recently by Li and Wah (1983). In this example a procedure is given for optimising the mapping of linear recurrences:

$$z^k\ (i,j) = f[z^{k-\delta}\ (i,j), x(i,k), y(k,j)] \tag{4.8}$$

where $\delta = \pm 1$, onto systolic architectures. Here, f is a function to be evaluated, $z^{k-\delta}\ (i,j)$ is the result of the $k-\delta$ step of the computation at position (i,j), and $x(i,k)$ and $y(k,j)$ are iteration independent linear functions. The multiplication $\boldsymbol{C} = \boldsymbol{AB}$ of $n \times n$ matrices $\boldsymbol{A}$ and $\boldsymbol{B}$, for example, may be performed by the recursion

$$c^0{}_{ij} = 0,\ c^k{}_{ij} = c^{k-1}{}_{ij} + a_{ik}\, b_{kj}$$

where $1 \leqslant i,j,k \leqslant n$. Recurrence algorithms are characterised by the spatial distribution of data, period of computation and velocity of data flow.

Recall that, in systolic processing, streams of data arrive at a processing element, which performs a calculation and passes the result to another PE in the systolic network for further calculation. The idea behind this technique is to optimise a cost function for the algorithm (such as computation time) under constraints expressing the simultaneity of arrival of the operands at a PE for a computational step of the recurrence. The optimisation problem for the systolic recurrence mapping is thus a non-linear integer program.

The unit of distance is fixed as the separation between two neighbouring directly connected PEs. Data items can move to neighbouring PEs or buffers in one clock cycle, which is taken as the unit of time. The velocity of flow of data item x is defined as the vector distance moved by x within a clock cycle, $\mathbf{x}_d$. Since PEs may be separated by buffers, $|\mathbf{x}_d| \leqslant 1$.

The distribution of data item x over the two-dimensional systolic array must also be parametrised. The vector separation between $x(i,j)$ and $x(i+1,j)$ is called the row vector of x, $\mathbf{x}_{is}$; similarly, $\mathbf{x}_{js}$, the column vector of x, is the vector separation between $x(i,j)$ and $x(i,j+1)$; see Fig. 4.27. In the case of the array index k, $\mathbf{x}_{ks}$ is the row (or column) vector if k is the row (or column) index.

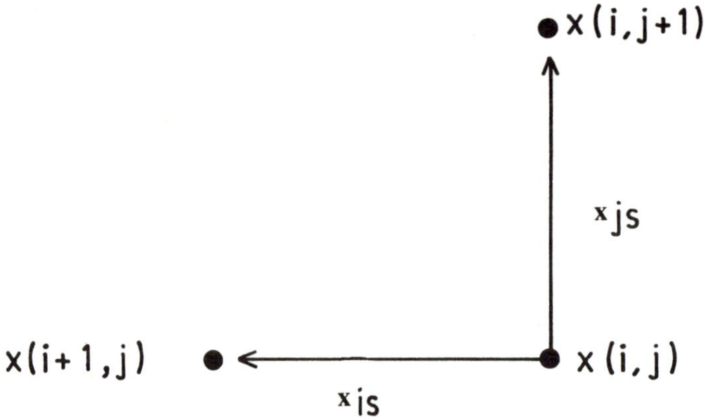

Fig. 4.27 Data distribution row and column vectors; heavy dots represent PEs.

We shall also need the notion of periods of computation. Denote the time at which a computation takes place by τ. Then the periods of i and j are defined as

$$t_i = \tau\ (z^k\ (i+1,j)) - \tau\ (z^k\ (i,j))$$
$$t_j = \tau\ (z^k\ (i,j+1)) - \tau\ (z^k\ (i,j))$$

The period of iterative computation t_k is defined as

$$t_k = \tau\ (z^{k+1}\ (i,j)) - \tau\ (z^k\ (i,j))$$

Finally, the period of x or y with respect to k, denotes t_{kx} or t_{ky}, is defined to be the difference in time between consecutive accesses of data items x or y. These are related to t_k by $|t_{kx}|=|t_{ky}|=|t_k|$ (see Li and Wah, 1983, for an elaboration of sign conventions).

With these definitions, it is easy to prove the following 'theorem of systolic processing', which expresses the relationship between timing and the distribution of data in a systolically performed recurrence algorithm:

Theorem: The velocity, data distributions and periods of a recurrence computation of the form represented by equation 4.8, performed on a systolic array, obey the following equations:

$$\begin{aligned} t_{kx}\,\mathbf{x}_d + \mathbf{x}_{ks} &= t_{kx}\,\mathbf{z}_d \\ t_{ky}\,\mathbf{y}_d + \mathbf{y}_{ks} &= t_{ky}\,\mathbf{z}_d \\ t_i\,\mathbf{x}_d + \mathbf{x}_{is} &= t_i\,\mathbf{y}_d \\ t_i\,\mathbf{z}_d + \mathbf{z}_{is} &= t_i\,\mathbf{y}_d \\ t_j\,\mathbf{y}_d + \mathbf{y}_{js} &= t_j\,\mathbf{x}_d \\ t_j\,\mathbf{z}_d + \mathbf{z}_{js} &= t_j\,\mathbf{x}_d \end{aligned}$$

On the basis of this theorem, the optimisation procedure is:

(a) Write the recurrence formula for the problem to be solved.
(b) Write the corresponding systolic processing equations using the theorem of systolic processing and constraints on the parameters.
(c) Write the cost function based on the design requirements in terms of the systolic processing parameters and problem size.
(d) Find the parameter values which minimise the cost function by solving the non-linear program.
(e) Design a basic cell for the systolic algorithm and find a possible interconnection of the cells based on the parameters obtained.
(f) Finally, eliminate cells that do not perform any useful computations.

The technique gives a systematic approach for the optimised mapping of algorithms which can be represented as linear recurrences onto systolic arrays. Li and Wah (1983) have used this procedure to generate a number of systolic array designs for image processing applications.

4.9 TOWARDS A SYNTHESIS

Throughout this chapter we have really only assembled and briefly discussed the various elements which are necessary for developing a methodology capable of coping with the complexity of VLSI designs incorporating considerable degrees of concurrency. Progress towards such a methodology is bound to be difficult and protracted, but already some key elements can be identified. Hierarchical structuring is crucial if the design is to be managed in a systematic and ordered way. Adequate formal means of representing and analysing the concurrent aspects of both algorithms and architectures, and the mappings between them, must be available. A realistic VLSI complexity model should underpin any attempt to implement an algorithm in silicon in a way that is efficient in area and computation time.

Most formal VLSI design work to date has concentrated on the chain of transformations from an abstract functional view of architecture to physical realisation, i.e. the right-hand side of Fig. 4.1; we can characterise this as the 'silicon compilation' approach. Here, the design process involves successive steps of refinement from initial functional specification through floorplanning, logic design and layout to final implementation. If each level can be provided with a formal representation, then direct transformation is in principle possible between the level descriptions and the mapping between representations can be rigorously analysed. Each level of design being made up of components whose composition is mathematically understood, using the algebraic properties of the description language, hierarchical verification can be used to demonstrate the behavioural equivalence of design and specification; verification then becomes an integral part of the design process. Further, design alternatives can be compared by manipulation of formal descriptions to give alternative designs with equivalent function. It should be emphasised, though, that in the silicon compilation approach the formal linkages to algorithms and their underlying computational theories is still very weak.

A first step in this direction has been taken by Sheeran (1984) with the algebraic VLSI design language μFP. This is an extension of the functional language FP (Backus, 1978), to encompass the notion of state. Circuit cells are described algebraically by functions manipulating signals: an expression in the language is a function relating time sequences of data objects and captures the behaviour of a circuit by specifying the relationship between its input and output traces. The forms of the functions have geometric interpretations, so that a μFP expression describes both circuit behaviour and floorplan. A number of primitive functions and combining forms are provided, in terms of which hierarchical circuit descriptions can be constructed. Different μFP expressions may be behaviourally equivalent but correspond to different floorplans. Floorplanning can be improved by using the algebraic properties of μFP to manipulate the original specification. Since μFP is a finite state machine description, an obvious area of application is to synchronous arrays, and the first realistic example analysed has been a systolic correlator chip designed at the GEC Research Laboratories, Wembley, UK (McCabe *et al.*, 1982).

Uniform formal notations for both abstract behavioural specification and lower level architectural description, and the transformations between them, have been explored by Cardelli (1982) and Subrahmanyan (1983).

Several tools are under development to support hardware design through verification by formal proof rather than simulation (Barrow, 1983, Gordon, 1983, Mishra and Clarke, 1983). Mishra and Clarke describe a hierarchical verification system for asynchronous circuits. Circuit design is verified incrementally, building the correctness proof of complex modules upon the correctness of their submodules. Temporal behaviour is specified using a propositional temporal logic. The circuit to be verified is modelled by its state transition graph, and a model checker analyses the truth of the specification in all states. Barrow's VERIFY is a Prolog-based system for proving the equivalence of structural and behavioural descriptions of a design. The design is ordered as a hierarchical set of modules, which are modelled as finite-state machines. Modules interact by communicating signals through input and output ports. The behaviour of primitive modules (e.g. registers) is specified by equations relating the output signals and new internal states to the inputs and current internal state. The behaviour of larger modules can then be defined in terms of the specification of the submodules and their internal interconnections. VERIFY attempts to prove equivalence of the specified behaviour to that deduced from the structural description and the behavioural specifications of the modules; the problem is to determine the correspondence between two finite-state machines. The system has been used to verify a simple computer, as has Gordon's LCF-LSM system.

Less is known, and certainly no general theory exists, of the relationship between algorithmic and architectural design. Most approaches have been concerned with mapping particular classes of algorithms, e.g. those describable as recurrences, onto systolic arrays (for a review see Fortes *et al.,* 1984). Algorithms involving repetitive execution with regular data access are matched onto a regular pattern of PE interconnections and pipelined data flow. This matching is accomplished by deriving, in a more or less *ad hoc* fashion, a meaning-preserving transformation between a chosen representation or model of the algorithm and a functional representation of the systolic structure. Systolic designs have been achieved in this way, for example, for finite impulse response filters, convolution, matrix-vector multiplication, DFT and FFT.

We have attempted to give an overview of the mapping problem and to describe some representative attempts to achieve suitable mappings. It has to be admitted that we still lack a systematic, formalisable methodology for deriving such mappings and optimising them in terms of, e.g. power, area, time and pin count. Although most approaches to date have been rather *ad hoc,* the regularity of the communication patterns and data structures in image processing algorithms encourages our belief that development and unification of the strands discussed here will bear important fruit for future image processing applications.

BIBLIOGRAPHY

Aho, A.V., Hopcroft, J.E. and Ullman, J.D. (1983) *Data Structures and Algorithms,* Addison-Wesley.

Amamiya, M. and Hasegawa, R. (1984) 'Data-flow Computing and Eager and Lazy Evaluation', *New Generation Comp.,* **2**, 105–129.

Cantoni, V. and Levialdi, S. (1982) 'Matching the Task to an Image Processing Architecture', *Proc. 1982 Int. Conf. on Parallel Processing* (K.E. Batcher, W.C. Meilander and J.L. Potter, eds.), Computer Science Press, 254–257.

Hwang, K. and Briggs, E.A. (1984) *Computer Architecture and Parallel Processing,* McGraw-Hill.

Meyer, B. (1985) 'On Formalism in Specifications', *IEEE Software,* Jan., 6–26.

Seitz, C.L. (1984) 'Concurrent VLSI Architectures', *IEEE Comp.,* **C–33,** No. 12, 1247–1265.

Ullman, J.D. (1984) *Computational Aspects of VLSI,* Computer Science Press.

VLSI Design, Oct. 1984.

REFERENCES

Aho, A.V., Hopcroft, J.E. and Ullman, J.D. (1983) *Data Structures and Algorithms,* Addison-Wesley.

Arvind and Iannucci, R.A. (1983). 'Two Fundamental Issues in Multiprocessing', *MIT Report: MIT/LCS/TM–241.*

Ayres, R.F. (1983) *VLSI Silicon Compilation and the Art of Automatic Microchip Design,* Prentice-Hall.

Backus, J. (1978) 'Can Programming be Liberated from the von Neumann Style?' *Comms. of the ACM,* **21**, No. 8, 613–641.

Barr, A. and Feigenbaum, E.A. (1981) *The Handbook of Artificial Intelligence,* Vol. 1, Pitman.

Barrow, H.G. (1983) 'VERIFY: A Program for Proving Correctness of Digital Hardware Designs', submitted to *Artificial Intelligence.*

Baudet, G.M. (1981) 'On the Area Required by VLSI Circuits', in *CMU Conf. on VLSI Systems and Computations* (ed. H.T. Kung, R. Sproull and G. Steele), Computer Science Press, 100–107.

Bokhari, S.H. (1981) 'On the Mapping Problem', *IEEE Trans. on Comps.,* **C–30**, No. 3, 207–214.

Brauer, W. (1980) 'Net Theory and Applications', *Lecture Notes in Computer Science,* **84**, Springer-Verlag.

Bryant, R. (ed.) (1983) *Third Caltech Conference on Very Large Scale Integration,* Springer-Verlag.

Cantoni, V. and Levialdi, S. (1982) 'Matching the Task to an Image Processing Architecture', *Proc. 1982 Int. Conf. on Parallel Processing* (K.E. Batcher, W.C. Meilander and J.L. Potter, eds.), Computer Science Press, 254–257.

Cardelli, L. (1982) 'An Algebraic Approach to Hardware Description and Verification', *Univ. of Edin. Dept. of Comp. Sc. Tech. Rep.,* CST–16–82.

Chazelle, B. and Monier, L. (1981) 'A Model of Computation for VLSI with Related Complexity Results', *Proc. 13th Annual ACM Symp. on the Theory of Comp.,* 318–325.

Chazelle, B. and Monier, L. (1981) 'Optimality in VLSI', *VLSI 81* (ed. J.P. Gray), Academic Press.

Chiang, Y.P. and Fu, K-S. (1983) 'Matching Parallel Algorithm and Architecture', *Proc. 1983 Intern. Conf. on Parallel Proc.,* Computer Society Press, 374–380.

Cohen, D. (1978) 'Mathematical Approach to Iterative Computational Networks', *Proc. Fourth Symp. on Comp. Arithmetic,* 226–238.

Davis, A.L. and Keller, R.M. (1982) 'Data Flow Program Graphs', *Comp.,* **15**, No. 2, 26–41.

Feng, T-Y. (1981) 'A Survey of Interconnection Networks', *Comp.,* **14**, No. 12, 12–27.

Folts, H.C. and desJardins, R. (eds.) (1983) Special Issue of *Proc IEEE,* **71**, No. 12.

Fortes, J.A.B., Fu, K-S. and Wah, B.W. (1984) 'Systematic Approaches to the Design of Algorithmically Specified Systolic Arrays', *Purdue University Technical Report,* TR–EE–84–39.

Gannon, D.B. and van Rosendale, J. (1984) 'On the Impact of Communication Complexity on the Design of Parallel Numerical Algorithms', *IEEE Trans. on Comps.,* **C–33**, 1180–1194.

Gordon, M. (1983) 'Proving a Computer Correct', *Univ. of Camb. Comp. Lab. Tech. Rep.,* No. 42.

Gray, J.P. (ed.) (1981) *VLSI 81: Very Large Scale Integration,* Academic Press.

Gurd, J. and Watson, I. (1980) 'Data-Driven System for a High-Speed Parallel Computing–Part 1: Structuring Software for Parallel Execution', *Comp. Des.,* June, 91–105.

Havens, W. and Mackworth, A. (1983) 'Representing Knowledge in the Visual World', *Comp.,* **16**, No. 10, 90–96.

Hicks, P.J. (ed.) (1983) *Semi-Custom IC Design and VLSI,* Peter Peregrinus.

Hockney, R.W. and Jesshope, C.R. (1981) *Parallel Computers,* Adam Hilger Ltd.

Hofstadter, D.R. (1980) *Goedel, Escher, Bach: An Eternal Golden Braid,* Vintage Books.

Hon, R.W. and Reddy, D.R. (1979) 'The Effect of Computer Architecture on Algorithm Decomposition and Performance', *High Speed Computer and Algorithm Organisation* (D.J. Kuck, D.H. Lawrie and A.H. Sameh, eds.), 411–421, Academic Press.

Hopcroft, J.E. (1984) 'Turing Machines', *Sc. Am.,* **250**, No. 5, 70–80.

Jagadish, H.V., Kailath, T., Newkirk, J.A. and Mathews, R. (1984) 'On Hardware Descriptions from Block Diagrams', *Proc. IEEE Internat. Conf. on Acoustics, Speech and Signal Proc.,* San Diego.

Johnsson, L. and Cohen, D. (1981) 'A Mathematical Approach to Modelling the Flow of Data and Control in Computational Networks', *VLSI Sys. and Computations* (H.T. Kung, R. Sproull and G. Steele eds.), Computer Society Press, 213–255.

Johnson, S.C. (1984) 'VLSI Circuit-Level Design Reaches to the Level of Architectural Description', *Electronics,* May 3, 121–128.

Kennaway, J.R. and Sleep, M.R. (1983) 'Novel Architectures for Declarative Languages', *Software and Microsystems.* **2**, No. 3, 59–70.

Koren, I. and Silberman, G.M. (1983) 'A Direct Mapping of Algorithms onto VLSI Processing Arrays Based on the Data Flow Approach', *Proc. 1983 Conf. on Para. Proc.* (eds. H.J. Siegel and L. Siegel), Computer Society Press, 335–337.

Kung, H.T. and Stevenson, D. (1977) 'A Software Technique for Reducing the Routing Time on a Parallel Computer with a Fixed Interconnection Network', *High Speed Comp. and Alg. Org.* (D.J. Kuck, D.H. Lawrie and A.H. Sameh, eds.), Academic Press.

Kung, H.T. (1980) 'The Structure of Parallel Algorithms', *Advances in Comps.,* **19**, 69–112.

Leighton, F.T. (1983) *Complexity Issues in VLSI: Optimal Layouts for the Shuffle-Exchange Graph and Other Networks,* MIT Press.

Leiserson, C.E. (1983) *Area-Efficient VLSI Computation,* MIT Press.

Lerner, E.J. (1984) 'Data Flow Architecture', *IEEE Spec.,* Apr., 57–62.

Li, G.J. and Wah, B.W. (1983) 'Optimal Design of Systolic Arrays for Image Processing', *Workshop on Comp. Arch. for Pattern Analysis and Image Database Management,* 134–141.

Li, G.J. and Wah, B.W. (1985) 'The Design of Optimal Systolic Arrays', *IEEE Trans. on Comp.,* **C–34**, 66–77.

Marr, D. (1982) *Vision,* W.H. Freeman.

McCabe, M.M. *et al.* (1982) 'New Algorithms and Architectures for VLSI', *GEC J. of Sc. and Tech.,* **48**, No. 2, 68–75.

Mead, C.A. and Rem, M. (1979) 'Cost and Performance of VLSI Computing Structures', *IEEE J. of Solid-State Circuits,* **SC–14**, No. 2, 455–462.

Mead, C.A. and Conway, L.A. (1980) 'Introduction to VLSI Systems', Addison-Wesley.

Melliar-Smith, P.M. and Schwartz, R.L. (1982) 'Formal Specification and Mechanical Verification of SIFT: A Fault-Tolerant Flight Control System', *IEEE Trans. on Comps.,* **C–31**, No. 7, July, 616–630.

Mishra, B. and Clarke, E.M. (1983) 'Automatic and Hierarchical Verification of Asynchronous Circuits using Temporal Logic', *Carnegie-Mellon Univ. Dept. of Comp. Sc. Tech. Rep.,* CMU–CS–83–155.

Nilsson, N.J. (1980) *Principles of Artificial Intelligence,* Tioga, Palo Alto, California.

Parker, I.N. (1984) Private Communication.

Patel, N.H. (1981) 'Performance of Processor-Memory Interconnections for Multiprocessors', *IEEE Trans. on Comps.,* **C–30**, 771–780.

Pease, M.C. (1968) 'An Adaptation of the Fast Fourier Transform for Parallel Processing', *J.ACM,* **15**, 252–264.

Post, E.L. (1943) 'Formal Reductions of the General Combinatorial Decision Problem', *Am. J. Math.,* **65**, 197–268.

Potter, J.L. (1978) 'The STARAN Architecture and its Application to Image Processing and Pattern Recognition Algorithms', *Proc. 1978 Nat. Comp. Conf.*

Preparata, F.P. (1984) 'VLSI Algorithms and Architectures', Mathematical Foundations of Computer Science 1984', eds. M.P. Chytil and V. Koubek, *Lecture Notes in Comp. Sc.* **176**, Springer-Verlag.

Proceeding of a Seminar on 'Custom and Semi-Custom VLSI', (1983) Gnostic Concepts and McGraw Hill.

Rabiner, L.R. and Gold, B. (1975) *Theory and Application of Digital Signal Processing,* Prentice-Hall.

Reeves, A.P. (1984) 'Parallel Computer Architectures for Image Processing', *Comp. Vision, Graphics and Image Processing,* **25**, 68–88.

Rosenfeld, A. (1981) 'Cellular Architectures: from Automata to Hardware', *University of Maryland Rep.* TR–1048.

Sheeran, M. (1984) '*μFP, a Language for VLSI Design',* Symposium on LISP and Functional Programming, Austin, Texas, 104–112.

Siegel, H.J. (1979) 'Interconnection Networks for SIMD Machines', *Comp.,* **12**, No. 6, 57–65.

Siegel, H.J. (1979) 'A Model of SIMD Machines and a Comparison of Various Interconnection Networks', *IEEE Trans. on Comps.,* **C–28**, 907–917.

Siegel, L.J., Siegel, H.J. and Swain, P.H. (1982) 'Parallel Algorithm Performance Measures', *Multicomputers and Image Processing: Algorithms and Programs'* (eds. J. Preston Jr. and L. Uhr), Academic Press, 241–252.

Stefic, M. *et al.* (1981) 'The Partitioning of Concerns in Digital System Design', *Xerox PARC Rep.,* VLSI–81–3.

Stone, H.S. (1971) 'Parallel Processing with the Perfect Shuffle', *IEEE Trans. on Comps.,* **C–20**, No. 2, 153–161.

Subrahmanyam, P.A. (1983) 'Synthesizing VLSI Circuits from Behavioural Specifications: a Very High Level Silicon Compiler and its Theoretical Basis', *VLSI 83* (eds. F. Anceau and E.J. Aas), North-Holland Publishing Company.

Tarjan, R.E. (1978) 'Complexity of Combinatorial Algorithms', *SIAM Review,* **20**, 457–491.

Thompson, C.D. (1979) 'Area-time complexity for VLSI', *Proc. 11th Ann. ACM Sym. on the Theory of Comp.,* 81–88.

Treleaven, P.C. and Lima, I.G. (1984) 'Future Computers: Logic, Data-Flow, . . ., Control Flow?', *Comp.* **17**, No. 3, 47–57.

Ullman, J.D. (1984) *Computational Aspects of VLSI,* Computer Science Press.

Waterman, D.A. and Hayes-Roth, F. (1978) *Pattern-Directed Inference Systems,* Academic Press.

Chapter 5
Towards Computer Vision

H. BUXTON and J. WIEJAK

5.1 INTRODUCTION

The previous chapters have discussed image processing algorithms and VLSI architectures, and the mappings between them. This chapter is concerned with the goals of computer vision, that is of determining what useful information should be extracted from images and how. The algorithms and architectures provide a way of implementing the devised strategies.

It is best to make it clear straight away what is meant here by the term 'computer vision'. In computer vision, the input is in the form of images; perhaps a single image, images from different viewpoints, or more generally, a time series of images. The output is a useful interpretation of the image data in terms of a symbolic description of the scene content. Rather than achieve this by solving specific problems in particular application domains, computer vision focuses on understanding the process of vision (Brady, 1982), in particular by concentrating on the identifiable functional modules of the human visual system. Much of the current computer vision research concerns the extraction of three-dimensional information from images, using techniques based on a computational theory of visual perception. This emphasis is reflected in the work described in the following sections of the chapter. These general techniques can then be applied to particular tasks, rather than having to start from scratch to build a solution to each problem. The term 'image understanding' is also used to cover this field of research.

The human visual system provides a demonstration of an enormously powerful and flexible vision system. For a computer vision system, it is not necessary to try to copy all the details of the human visual system, but it is important at least to try to understand what lies behind its power and flexibility. In fact, because the human visual system can perform so effortlessly such a wide range of tasks, it is difficult at first to appreciate just how difficult the problems it is solving are. It is only when trying to solve the same problems with computer algorithms that the true complexity becomes clear.

The first step in the process of computer or human vision is the capture of the raw images. At this stage, the information content is represented by some sort of spatial array of light intensity values. Because of past computational limitations, most computer vision work has dealt with static, grey-level images, largely ignoring the additional complexities of colour or time varying scenes. However, as discussed in the following sections, it is now becoming feasible to implement some of the motion and optical flow algorithms in real time. As a result, this area of research is likely to expand over the next few years. Indeed, the study of time varying images may well characterise a new direction in computer vision as current research is opening many interesting new avenues.

A grey level computer image is simply a regular two-dimensional array of intensity values, e.g. 512 × 412 pixels each with a grey-level value in the range 0 to 255. To begin to appreciate the difficulty of computer vision, it is useful to consider the difficulty of interpreting such a matrix of values with the ease of interpretation of the image intensity pattern. The information content is the same but while a biological vision system automatically processes the intensity to see the content, the matrix remains uninterpreted as it is not in the right input format. It remains just as it is, the set of grey level values at each pixel. This illustrates the need for the computational machinery of the human visual system to be implemented in an artificial system to extract or 'make explicit' the information about the content of the image that is hidden or only 'implicit' in the local intensity values.

The next problem to be faced in computer vision, then, is how to make accessible to the computer the vast amount of information that is available to the human system about the appearance of objects in the world and what is expected in particular domains. As well as representing this knowledge, there is the problem of how and when to use the information. What is required is a control structure which allows the effective interaction between 'bottom up' processing, i.e. using only knowledge of general principles of the physics of image formation to determine the set of parameters that best account for the intensity values observed, and 'top down' processing, i.e. using knowledge specific to the particular domain or *a priori* knowledge. This problem is discussed further in the section on model based vision.

The other input to a computer vision system is some description of the visual task to be performed. This task description should determine what output is required. The visual task may itself be part of a much larger task, for example, the visual location of an object as part of an assembly workcell operation. The computer vision system takes these three inputs, images, knowledge and task, and delivers an appropriate output. The kind of output that will be useful really depends on the task being performed.

As an example of an application that could make use of such a computer vision system, consider the underwater inspection of pipes and structures for the oil and gas industries using a remotely operated vehicle (ROV). Such an inhospitable environment is, at best, dangerous for divers, and below about 700 m a ROV is the only option. Typical image capture facilities on a ROV would include a fixed stereo pair of cameras. It may be sufficient just to do some video-rate image processing on each image stream to enhance and possibly code the images for transmission to the surface. There, the enhanced images are reconstructed and displayed for fusion and interpretation by an operator. Any actions which must be taken are then relayed back down to the ROV control system.

On the other hand, it is possible to envisage a future system which is almost completely autonomous, finding its own way around using computer vision and other sensors such as sonar and touch as necessary to make inspections and perform routine maintenance using its manipulators. This would probably need to be done with the aid of stored models of the environment. It would thus be an example of a sophisticated and flexible computer vision system performing a variety of inspection and navigation tasks. Such a system is beyond the state of the art and needs a great deal of further research and development. However, there are several more immediately realisable situations such as the automatic computer generation of a quantitative three-dimensional picture of the environment using stereopsis algorithms. This three-dimensional information could then be compared automatically to a model of the environment to highlight likely structural defects. This would constitute an intermediate stage that extracted the useful visual information but did not act upon it. The final coordination of the visual information to guide the ROV and perform the maintenance, however, is still quite a long way off.

To understand the process of vision, it seems to be useful to split it into three basic stages as proposed by the late David Marr, who has contributed greatly to the computational theory of visual perception. Many *ad hoc* solutions that try to go directly to a three-dimensional object based description from the two-dimensional image description founder as the real world invariants are three-dimensional surface or volumetric features, not their two-dimensional projections. The general framework for visual processing is given below, together with a brief description of the various stages. These stages are described in much greater detail in the subsequent sections.

Stage	Purpose	Primitives
Image	Represents intensity	Intensity value at each point
Primal sketch	Represents intensity changes and two-dimensional organisation	Zero crossings, edge segments, blobs terminations
2½-D sketch	Represents orientation and depth of visible surfaces plus discontinuities	Local surface orientation and distance from viewer and discontinuities
3D model representation	Describes shapes and three-dimensional spatial organisation using modular hierarchical representation that is volumetric	Three-dimensional hierarchical models with volumetric or surface shape primitives

The visual system starts out with just the intensity map, so the first stage is to pull out important intensity changes (edge detection) in order to get the raw primal sketch which is a purely two-dimensional description of the data. Next the information in this primal sketch is used to obtain the surface description. Note that only the visible surfaces are available in this 2½-D sketch and to get this level of description it is necessary to use one of the 'shape from' methods, e.g. stereo and optic flow. At this stage only a viewer-centred coordinate system is accessible, so distance from the observer and orientation with respect to the observer is represented. The surface description can then be used to obtain the full three-dimensional representation of the scene. It is worth computing the description using object-centred coordinates at this stage since viewer dependence would mean storing multiple views of the objects. This level of representation may use volumetric primitives, e.g. generalised cylinders organised in a modular hierarchical manner, and can be used for full three-dimensional spatial reasoning about the scene.

The visual system can be regarded as an unpicking of the image formation process in that the three-dimensional scene that gave rise to the two-dimensional image(s) must be recovered. However, vision is an undetermined problem in the sense that a lot of information is lost when the three-dimensional world is projected down into two dimensions. This means that many sets of parameter values could have given rise to the local intensity values, edges, surfaces, etc. The problem then, is that the data must be interpreted to get a fairly unique description of the original scene. This can only be done by building in knowledge, either very general knowledge about the nature of surfaces, edges, etc., which are referred to as 'assumptions' or 'constraints', or domain specific knowledge in the form of three-dimensional models at the higher levels of the visual system. It is clear that in human vision a lot of knowledge about the appearance of objects is stored, but it is not as evident that low–level assumptions about the nature of the scene are being used. However, as discussed below it is possible to find clues to the powerful constraints being used by the human visual system.

In human vision the system is sometimes fooled, resulting in 'illusions'. This occurs when some assumption is being violated, e.g. the moon appears to sail across the cloud or the train next to yours pulls out but it seems as if it is your train that moved. The assumption or constraint used by the visual system is usually true, i.e. there is a high probability mapping of the image feature onto the scene interpretation. In the above examples, only relative motion is available so in some sense it is arbitrary whether you say the cloud is moving left or the moon is moving right or whether the train you are standing in or the one next to it is doing the moving. The assumption being used is probably that the visually 'smaller' object is in motion unless there is further evidence to the contrary. There are many examples of this kind; e.g. the Ames room where systematic distortions fool the eye into thinking there is a regular room. These examples give rise to the concept of mimicry and the 'general viewpoint assumption' (Cowie, 1983) which help in thinking and reasoning about ambiguity and deriving parsimonious interpretations of the image data.

A major part of the discussion in this chapter concerns the physical principles involved in extracting useful information from the various representations using powerful constraints. It is the proper posing of these constraints and the physical principles of the mapping of one level onto another that constitute the computational theory which allows an understanding of both human and machine vision and principled solutions to the processes of visual perception (Marr, 1982). The computational theory can motivate many algorithms using a variety of techniques as discussed in the main body of this chapter. The algorithms must in turn be carefully matched to the machine architecture or hardware at the implementation stage. This is the theme of this book.

The following sections of this chapter present techniques for processing images and extracting useful information from them. The sections follow the processing from the early stages at the image level (Section 5.2), through surface level descriptions (Section 5.3), to model-based interpretation (Section 5.4). Although these three sections present logically sequential processing stages which act on the representations produced by previous stages, it should be emphasised here that the control structure of any vision system should be fully dynamic, not rigidly bottom up. It is envisaged that the stages could communicate in both directions, allowing effective top down processing strategies. In addition, information need not always go through all these transformations – the output of any stage might be sufficient for a particular task. The final section (5.5) looks at some implementation issues, and considers future directions for work on computer vision.

5.2 IMAGE DESCRIPTIONS: EARLY STAGES OF PROCESSING

The previous section introduced the idea that implicit information in an image or image sequence is made explicit by processing in a series of stages. The output descriptions of one stage may be used in the next, possibly with feedback, to build a complete computer vision system. Also, the introduction stressed the importance of understanding the fundamental mechanisms and functional modules of the human visual system to help in the development of visual processing algorithms. The way in which these ideas work in practice will become clearer in this section when the early stages of visual processing are discussed. Methods to obtain rich descriptions of the image data are described in the following subsections on edge detection, optical flow extraction, texture and colour determination.

5.2.1 Edge detection

The process of image formation consists of a series of transformations which confound the effects of many parameters describing the lighting direction, viewing conditions, orientation of surfaces, etc. In addition, the geometry of the imaging process leads to a loss of information due to projection from the three-dimensional scene onto the two-dimensional image plane. The strategy in computer vision, therefore, is to concentrate on features of the image, most commonly edges, that allow a reconstruction of the scene geometry in a way that is not crucially dependent on other imaging parameters. Thus, the first stage of extracting visual information usually involves processing the image intensity, more strictly 'irradiance' (Horn and Sjoberg, 1979), $E(x,y)$, to find the significant intensity changes in the image.

This process of edge detection has attracted a great deal of research effort since the resulting image descriptions will limit the quality of subsequent stages. Significant intensity changes correspond to a variety of events in a scene, such as depth discontinuities between surfaces, shadow boundaries and changes in reflectivity, orientation and texture of a surface. Also, it is important to note that not all such scene structure is apparent. For example, not all object boundaries are marked by intensity changes in the image.

Leaving aside these problems of interpretation for the moment, the detection and localisation of intensity change itself poses some problems due to the presence of noise. This image noise is a combination of quantisation and sampling effects and variations in sensor sensitivity over time and across the image coordinates, which carries no useful information but is confounded with the scene-related intensity changes. Both noise and edge features tend to be high spatial frequency events, i.e. intensity is changing rapidly over the image. It is inevitable, then, that the detectability of scene-related edges will be affected by their signal-to-noise ratio. Threshold and local context methods as well as multiple scale correspondence methods have been used to assess the evidence for significant edges whilst discarding noise.

It is well known that an edge, where the intensity changes from one level to another, corresponds to a large value of the gradient of the image intensity and to a zero crossing in the second derivative (gradient of the gradient) of this map. Many local edge detection operators have been developed to approximate these first and second derivatives for computer vision systems. These can be represented as simple 'masks' which take weighted differences of the intensity map values in the neighbourhood of each point (pixel) in the image map. Note that first difference operators tend to produce a large value over a broad region and thus do not give accurate localisation, whereas the zero crossing of a second difference operator is highly localised. Taking differences, however, amplifies any error component, so second difference operators tend to be much noisier than first difference operators. Most differencing operations are therefore used in conjunction with image smoothing to minimise this problem. The operators are convolved with the image, that is, a weighted average is found by summing neighbouring points multiplied by the mask values. The masks take differences in systematically varying directions and the convolution signals are used to obtain estimates of both the magnitude and direction of the detected edges.

A lot of further processing is often needed to combine and clean up the output of small convolution masks such as the Sobel and Navatia-Babu operators before the data is useful; for example, thinning, thresholding and linking of the edge points. These processes tend to need optimising with respect to the particular image being analysed and can be very complex and time consuming as well as being rather arbitrary with respect to the goal of a general vision system.

In the mammalian visual system, 'centre-surround' cells with varying receptive field sizes have been identified (Hubel and Wiesel, 1966) and are thought to be fundamental to edge detection. These cells seem to be connected to a neighbourhood of receptor cells with a roughly Gaussian distribution of strengths with a larger inhibitory neighbourhood than excitatory, giving the characteristic 'DOG' (difference of Gaussian) function. Marr and Hildreth (1980) in their influential theory of edge detection proposed the use of the Laplacian of a Gaussian, $-\nabla^2 G(x,y)$ operators at a range of scales for computer vision. This operator closely approximates the DOG function (See Fig. 5.1).

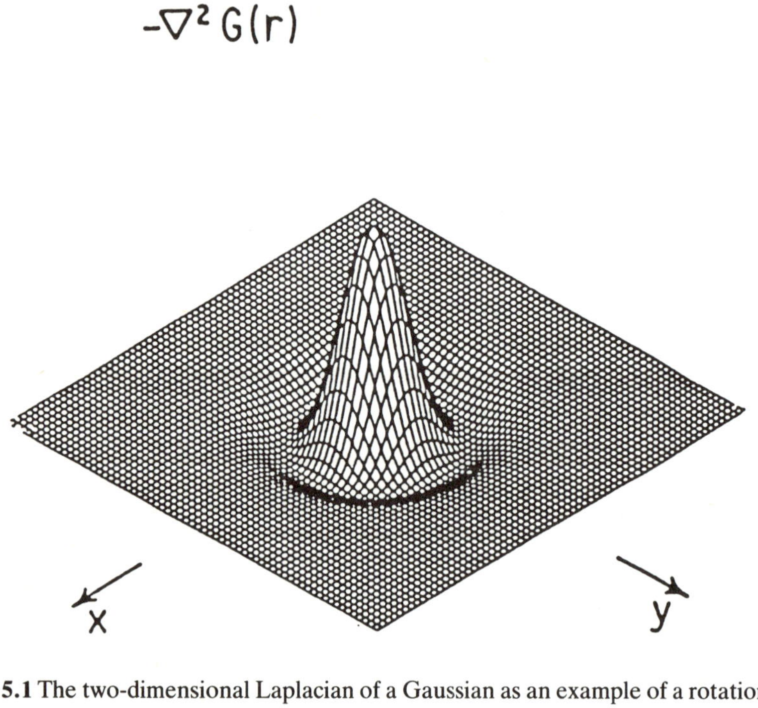

Fig. 5.1 The two-dimensional Laplacian of a Gaussian as an example of a rotationally symmetric second difference operator.

The $-\nabla^2 G(x,y)$ operator has many advantages. For example, the Gaussian smoothing tends to give a much larger operator and thus may give better noise suppression than the small operators. Sobel, etc., mentioned above, edges are detected simultaneously and without bias in all directions due to the rotational symmetry, the zero crossings of the resulting signal are sharply localised and form connected contours. The output of the different spatial frequency channels (different scale or extent of Gaussian smoothing) varies systematically with the image events as has been described by Witkin (1983). A full theory of the combination of these outputs and the behaviour of zero crossings has not yet been developed and will probably require a detailed knowledge of 'topology', i.e. the way in which the convolution surfaces must map onto one another.

The problem of accurate localisation together with good noise suppression is also not fully solved; the Marr–Hildreth operator does tend to produce zero crossing contours that wander off into the noise. To overcome this, Canny (1982) has combined adaptive thresholding with edge localisation in the smallest channel for which there is enough evidence (signal-to-noise ratio). The interpretation of the convolution signal in the larger channels then takes into account the predicted behaviour of edges detected in the smaller channels. This method has produced very encouraging results. In general, 'optimal' edge detection depends on the set of criteria and new operators are being investigated which explicitly mark step discontinuities in the image.

Other approaches to edge detection involve explicit edge models and yield a description of the edge shape parameters as well as magnitude and direction. Herskovitz and Binford (1970) found 'step', 'roof' or 'peak' changes in image intensity corresponding to depth and reflectance boundaries, and concave or convex surface orientation discontinuities respectively. This useful information would be available if the edge type can also be found in the processing. Surface fitting and parameter estimation is also used in the 'sloped facet model' developed by Brooks (1978) and Haralick (1980). Indeed, Haralick, Watson and Laffey (1983) have suggested that a topological primal sketch has considerable advantages over an edge based description, giving a richer, more invariant labelling set of ridge, ravine, peak, pit, flat, etc. It remains to be seen whether this type of processing will be useful for computer vision.

It must be remembered that image descriptions in terms of the edges or the shape parameters are not an end in themselves except for image encoding and reconstruction. Further processing exploiting either general or domain specific knowledge is needed to interpret representations at this low level, as is discussed in the sections on visible surface descriptions and model based vision.

5.2.2 Optical flow extraction

Edge detection was considered only for a single image in the discussion above, but image intensity usually changes over time as well as space. Low level processing of this time varying pattern, $E(x,y,t)$, can allow moving features to be detected and a measure of the instantaneous velocity to be derived. Spatio-temporal processing has two main advantages over processing of a single view. First, there is a larger amount of data available which can allow more reliable detection of intensity changes by smoothing out the time varying image noise. Second, and more important, information about the relative depth, motion and the orientation of surfaces in the scene can often be inferred from the velocity measures.

The optical flow, $\dot{\mathbf{r}} = (\dot{x},\dot{y})$, which is the instantaneous two-dimensional motion of image points, can be measured in various ways given the image intensity pattern. Two basic methods can be distinguished. The first uses an extension of the static edge detection process to measure the velocity directly and the second uses features matching over time to obtain a disparity measure. The second method tends to give fairly sparse data and is more difficult as it entails solving the 'correspondence problem', i.e. finding the best global match when there are many possible local matches. Recent success has been reported in following the motion of corners (Nagel, 1983) and edge features (Lawton, 1981).

Correspondence is discussed further in the next section under the heading of 'stereopsis' as the matching of views separated in space is very similar to the matching of views separated by time when there is relative motion of observer and scene. Note that distance moved from one view to another, $\Delta\mathbf{r}$, equals the product of the instantaneous image velocity, $\dot{\mathbf{r}}$, and the time interval, Δt, so that, in a fundamental sense, optical flow and stereopsis can be used to determine the same kind of information.

The rest of this discussion will concentrate on intensity based methods to extract the optical flow field, leaving the recovery of scene structure and motion until the next section. These methods assume that the only changes in image intensity are due to motion and can thus exploit the relationship that the change in intensity over time can be explained as the product of the optical flow and the change of intensity over space. This is equivalent to saying that a sharp edge moving slowly can give the same intensity change over time as a shallow edge moving quickly.

The main problem for these intensity based methods is that the full optical flow cannot be computed at a point without additional information, as only the steepest gradient in space can be determined and this is not necessarily in the direction of the motion. To determine the full flow vector involves the solution to the 'aperture problem', shown schematically in Fig. 5.2. This is equivalent to trying to find the component of motion, $\dot{\mathbf{r}}_{//}$, along a line. Unless there is a kink, curvature, end point, features in common motion or other source of information, this will not be possible.

The component perpendicular, $\dot{\mathbf{r}}_{\perp}$, to the line can be observed but the scene parameters depend on the full flow, $\dot{\mathbf{r}}$. The methods developed by Horn and Schunk (1980), Hildreth and Ullman (1982) and Buxton and Buxton (1983) differ in their treatment of the aperture problem as well as in

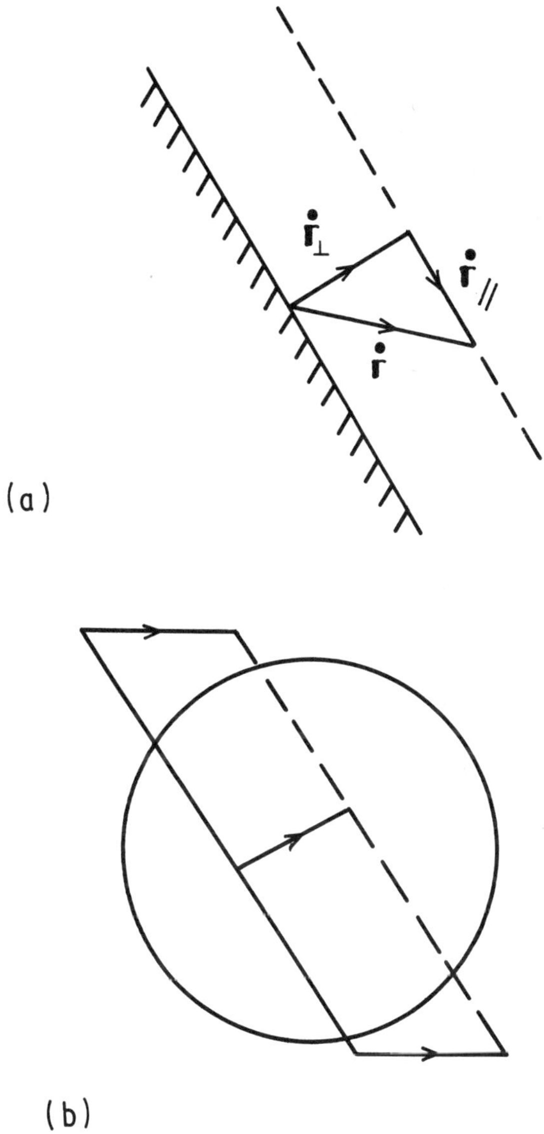

Fig. 5.2 The aperture problem; (a) the component of motion along the line (b) cannot be observed.

the initial measurement of the perpendicular component of optical flow. The method of Horn and Schunk computes optical flow by differentiating the intensity distribution with respect to time. They assume the case of a flat surface with uniform incident illumination and smoothly varying reflectance. The intensity at a point **r** at time t does not change so $\mathrm{d}E(\mathbf{r},t)/\mathrm{d}t = 0$. It follows from the chain rule that the component of the image motion perpendicular to ∇E, $\dot{\mathbf{r}}_\perp$, can be approximated by the ratio of the gradient in time and the gradient in space, i.e. $\dot{\mathbf{r}}_\perp \approx -(\partial E/\partial t)/\nabla E$, but in practice, quantisation errors, noise, etc., mean that an error term Er_1 must be introduced. To compute the full flow field requires an extra constraint and Horn and Schunk derive a measure of the departure from smoothness of the flow extracted by taking the square of the gradient of the flow $Er_2 = (\nabla\dot{x}^2 + \nabla\dot{y}^2)$. The combined errors $Er = \alpha Er_1 + Er_2$ are minimised using an iterative scheme to yield an estimate of the full flow field over the entire image plane. The solution is improved on iteration over frames, assuming uniform motion.

The method of Hildreth and Ullman assumes that edge contour motion in each frame has already been computed, e.g. by convolving the images with the $-\nabla^2 G(\mathbf{r})$ operator and computing the flow normal only at the edge zero crossings. They use a smoothing constraint around these contours to obtain the full flow field, again using an iterative scheme. The estimated flow corresponds closely to that observed for human vision, with similar illusions. Both the methods described so far assume that it is necessary to obtain the full flow field for further processing. However, it is possible to derive interpretations using only the measured normal component if this is derived accurately (see Section 5.3.2).

Finally, then, the method of Buxton and Buxton depends on a full spatio-temporal operator in which the scale in time is made equivalent to that in space by multiplying by a scaling velocity u; $\Delta r \equiv u\Delta t$. The space-time convolution signal has zero crossings in the centre frame at the location of moving edges. The gradients of this convolution signal in both time and space will be large, enabling an accurate measure of the normal component of the image velocity to be obtained provided that the edge velocity is not too great. The aperture problem described above is given a three-dimensional rather than a two-dimensional treatment by postponing any estimation of the flow parallel to the edge until the combined interpretation of structure and motion in the scene (Buxton, Buxton, Murray and Williams, 1984).

Spatio-temporal processing means that more information is available for reliable edge detection (better signal-to-noise ratio). In addition, the integration over time can be traded against that over space to change the operator characteristics while keeping the support (number of pixels used in the convolution) constant. Averaging over a long time but small area gives high spatial resolution but averaging over a short time but a large area gives high temporal resolution. This is in accord with evidence from human psychophysics where it is evident that sensitivity to high spatial frequencies entails low temporal sensitivity and *vice versa* (Kelly, 1979), although recent work indicates that this difference may not be as great as first thought.

5.2.3 Texture

Texture is a compelling visual cue for segmentation and recognition of surface type in natural scenes. It is possible to distinguish regions of an image that are grass, pebbled paths, etc., and terrain classification of satellite images depends, amongst other things, on texture measures. In addition, gradual change in texture is a cue for changes in depth and orientation of surfaces, as is discussed in Section 5.3.4 on visible surface descriptions. However, although many articles have been written on the subject of computer recognition and description of texture, no precise definition has been agreed. Many texture discrimination techniques are, therefore, rather *ad hoc*.

In the literature, statistical and structural methods for texture classification and description often conflict. Many accounts of texture for computer vision use the idea that the distinguishing characteristic is the regularity of primitive structural elements called *texels*. These elements have invariant properties and consist of aggregates of pixels forming, for example, ellipses or linear features. A description may have a hierarchical structure as there is always a scale associated with the texture elements or measures. For example, a regular brick pattern may be evident at one scale, but at a smaller scale the texture of each brick may be visible.

Several structural models for the placement of texels have been proposed. These consist of shape or tree grammars and graphs (e.g. Stiny and Gips, 1972, Zucker, 1976, Lu and Fu, 1978). However, the regularity of elements is not exact and is more often modelled statistically. In fact, statistical methods could be used to yield the entire description of texture without regard to any structural elements. The main problem with such an approach is that it may fail to discriminate perceptually distinct textures if inappropriate measures are taken. Fractal descriptions (Pentland, 1983) have been used with some success both to reconstruct the shape of natural objects and to segment images.

The simplest features to extract are first-order statistics that describe the distribution of individual pixel features such as intensity. Second-order statistics give the probability of co-occurrence of features for pairs of pixels. Various combinations can yield many features for texture discrimination. Higher order statistics may also be used but involve more computation and do not necessarily give better results. Indeed, for human visual perception, Julesz (1975) proposed that two textures are not discriminable if their second-order statistics are identical, although recently counter examples have been found.

It seems that although it may be possible to account for pre-attentive texture discrimination using simple statistical features, perceptual systems may learn to extract more complex features where appropriate for the particular task. Complex features are usually modelled using explicit texel descriptions, as discussed earlier. However, when automatic feature selection is better understood (see the discussion in the concluding section), a computationally more efficient method may be devised. A promising approach, using proximity codes which capture structural-relational features of the texture (Fox and Mayhew, 1978), could account for the immediate low level segmentation and may be a useful representation for further processing that extracts more complex features.

One further point should be made here. It is probably more useful to take the statistical or structural-relational measures on the local primitive features, as in Marr's primal sketch, rather than on the image pixels. Marr (1976) argues that features derived from this representation, which makes explicit the two-dimensional relationships between local primitives, can explain texture discrimination by humans. This account also naturally incorporates the notions of scale and hierarchy mentioned earlier as the local primitives are always extracted at a particular scale with the relationship between zero crossings in scale space giving a hierarchial tree structure (see Witkin, 1983). In addition, the low order statistics of such features would automatically depend on the high order statistics of the image, which might explain the counter-examples to Julesz' original proposal.

5.2.4 Colour determination

The additional information provided by colour, in full colour images, can be very effective, especially for low level segmentation (Ohlander, 1975). This is also suggested by the intuitive feeling that colour makes visual perception easier, and by its importance in graphics and display work. One example of the use of colour information is that shadow boundaries on a surface can be simply identified since the spectral balance stays roughly constant while the intensity changes. Infrared and other sensors can also be useful as fairly direct sources of information for special tasks; in particular, terrain classification uses a great deal of multispectral information.

Colour images are produced by viewing the scene through colour filters which transmit primarily red, green and blue wavelengths (*RGB* images). This coding is sufficient to reproduce colour images but alternative encodings (see Foley and van Dam, 1982) with an organisation more like that of the human visual system have been devised, as colour perception is not related in a simple way to colour measurement. The sensation of colour starts with the differential sensitivity of three types of neurochemical sensors in the retina to different parts of the visible spectrum. Combinations of the three principal wavelengths can account for most perceived colours and 'colour spaces' are often defined using these three co-ordinates.

Psychologists usually describe colour quality by 'hue', the colour name which depends on average wavelength, by 'saturation', the purity of the colour, and by 'intensity', the overall brightness of the colour. In the corresponding '*IHS*' scheme for computer vision, standard formulae using the *RGB* values define these dimensions.

$$\text{Intensity} = R + B + G$$
$$\text{Hue} = \cos\left((R-G) + (R-B) \right) / \left((R-G) + (R-B)(G-B) \right)$$
$$\{ \text{ or if } B > G, \cos(2\pi - \text{Hue (as above)}) \}$$

Saturation = 1 − (3 min (R, G, B) / Intensity)

Kender (1976) pointed out some problems for the *IHS* system in that there may be indeterminate encodings for the non-linear combinations used to define saturation and hue.

In human colour vision, it seems that, although there are three 'filters', these are organised as an opponent process system. This is reflected in colour naming – e.g. 'reddish-blue' has a sensible interpretation but 'reddish-green' does not. In addition, it has been suggested that opponent colour sensors can account for luminance judgements as well as colour perception. The '*YIQ*' coding used in commercial television transmission has a superficial resemblance to such a system. It requires a luminance component (Y) for monochrome sets and two chrominance components (IQ) which take weighted differences of the RGB values.

$$Y = 0.30R + 0.59G + 0.11B$$
$$I = 0.60R - 0.28G - 0.32B$$
$$Q = 0.21R - 0.52G + 0.31B$$

Further understanding of the way in which human colour and luminance processing is organised could well lead to a more efficient scheme for computer vision, although it must be realised that some potential spectral information is not extracted by our visual system. For example, a uniform mixture of red-yellow wavelengths is not distinguished from a mixture of just two wavelengths with the same average. However, an artificial system could detect this difference even with broad spectral responses of the filters, by taking variously weighted differences of the signals. On the other hand, it may be that spectral composition in nature is evenly spread around a central value so that this type of processing would not normally yield any useful information. Human colour discrimination is amazingly sensitive even when the illuminating light consists of only two wavelengths (Land, 1977) so it seems to have evolved well adapted to the environment and to use relative colour signal levels.

5.2.5 Segmentation

Segmentation is essentially the complement or 'dual' of edge finding since it involves finding regions that do not contain significant changes in intensity or some other feature measure. A great deal of effort has been expended on segmentation or 'region analysis', as it is a form of data compression where there are fewer regions than pixels in an image. Each region consists of a single cluster of simply connected pixels that satisfy some rule, such as having a feature value or set of feature values within a tolerance range, while adjacent regions satisfy a different rule. The difficult problem, as in edge detection, is finding a suitable definition of 'significant change' so that the segmented regions correspond to perceptual entities in the image or, even more ambitiously, scene.

The main weakness of region segmentation is that uniform regions of the image do not always correspond to portions of visible surfaces. Segmentation, then, is an ill defined operation as any violation of this correspondence will lead to errors in interpretation. However, it has important uses for encoding and two-dimensional image reconstruction. Also, many authors have pointed out that region finding is no more or less difficult than edge finding (Haralick, 1979 and Binford, 1981).

There are two basic approaches to region segmentation; 'region growing' and 'region splitting'. Region growing consists of starting with many small regions, such as single pixels, and iteratively merging similar regions until no further merging is possible. The remaining regions will then be internally uniform and have reached significant boundaries. Region splitting is the opposite and consists of starting with a single large region, such as the whole image, and splitting it into pieces until only uniform regions remain.

Each of the processes for measuring two-dimensional features described in this section (5.2) can be adapted for segmentation. For instance, if every intensity change found by a Marr–Hildreth edge detection operator is classed as significant, the closed contours of zero crossings correspond to regions. Indeed, a refinement of this approach has been found useful for automatic image segmentation and encoding (Bartliff, 1983). The segmentation cannot be easily used for higher level processing, however, as is discussed below.

It is clear from work on shape from shading (Horn, 1975), that there are considerable changes in intensity on curved surfaces. It is only planar sufaces that give rise to more or less uniform intensity values and it seems likely that much of the early segmentation work, based on intensity measures, only looked promising due to the use of limited 'blocks world' scenes. Changes in orientation give rise to changes in irradiance (Horn and Sjoberg, 1979) so the flat faces of blocks were easily segmented but the result cannot generalise to more complex scenes.

Many surfaces have perceptually constant texture, colour and certainly move as a whole. Also, as mentioned in the previous subsection, spectral composition can be uniform even if there is significant variation in intensity. It is difficult, in general, to devise measures that capture this uniformity, as is obvious from the fact that changes in measures of intensity, texture and optical flow can be used to recover changes in depth of a surface (see Section 5.3). The image measures are based on simple two-dimensional patterns but these are the result of three-dimensional scene constancies or 'invariance'. Thus, for instance, the projection of even simple translational motion of a plane tilted with respect to the image plane will be greater for the nearer points than the more distant ones. Segmentation, in general, needs to be delayed until the three-dimensional information is available. It is an intimate part of the interpretation process and cannot easily precede it (Buxton and Murray, 1985). Ohlander, Price and Reddy (1975, 1978) experimented with multispectral descriptions including hue, saturation and brightness and found considerable success for images that consisted of large regions that were uniform in one of the measures used. However, the process was fairly easily fooled (Tennenbaum and Barrow, 1977).

An alternative approach to improving low level segmentation is to use domain-specific knowledge to drive the processing. This should help as the goal of region segmentation is to obtain meaningful objects or parts of objects, such as the 'sky', 'tree', etc., from the image. Here the research becomes extended pattern analysis or classification using semantic information and is used with success in essentially two-dimensional situations, such as aerial image labelling (Selfridge and Sloan, 1981) and in medical images of cells, etc. Perceptual segmentation, in general, is complex and seems to occur at the stage where three-dimensional objects are recognised and interpreted as such in human visual processing. An approach to three-dimensional perceptual segmentation using both general constraints derived from the imaging process and allowing for domain specific knowledge to be incorporated as additional constraints is described in the paper by Buxton and Murray (1985). Low level segmentation, based on two-dimensional image descriptions, necessarily violates the 'principle of least commitment' (Marr, 1976, 1982) as it attempts an interpretation before all the information is available.

5.3 VISIBLE SURFACE DESCRIPTIONS: INTERMEDIATE PROCESSING

In this section, functional modules to recover information about the depth, orientation, curvature and motion of the visible surfaces from various two-dimensional descriptions of the image are discussed. These 'shape from' methods could be combined to produce a very powerful set of constraints for interpreting the visual information. This is the idea behind the 2½ D sketch level of representation which allows explicit assertion of information from the different sources. A great deal more work needs to be done on this combination of knowledge, as discussed in Section 5.3.6 and the final section on future directions (5.5). The two major shape recovery modules discussed in the following subsections are binocular stereo and optical flow interpretation, which both allow a reconstruction of the depth map for surfaces in view. In binocular stereo, the depth is determined from the disparity measures for features in the two eyes' views. The nearer objects give rise to larger disparities. The same relationship holds for measures of the optical flow field when a sensor is in motion through a static environment. The nearer objects give rise to larger velocity vectors. The translational and rotational motion of the sensor can be recovered from the flow field in addition to the surface slant, tilt and relative depth. Other modules considered are shape from shading, from texture and from contour. The concluding subsections consider interpolation of the surface information recovered and techniques for combining information both within a module and from different sources.

5.3.1 Stereopsis

In stereopsis, three-dimensional structure is recovered from a pair of images of a scene taken from slightly different viewpoints. The resulting disparity between corresponding points is measured and, from the viewing parameters the depth/range data is computed. Note that the stereo pair of images may be obtained in different ways depending on the application. There may be a pair of sensors in some fixed relative geometry, with axes verging or parallel, or a sensor which is moving and recording images at different times and hence from different views. On the other hand, for example in electron microscopy, it is the sample under examination which is rotated by a small amount to give a stereo pair of electron micrographs. The technique thus has much wider application than simply to a fixed pair of cameras staring straight ahead.

There are now several stereo algorithms at various stages of development with varying degrees of performance. Rather than trying to catalogue them all, some of the fundamental considerations will be discussed, followed by a description of one current algorithm which is based on a computational theory of stereopsis developed at Sheffield University by Mayhew and Frisby and recently extended by Pollard.

The first question to consider is, at what level of description should the matching of corresponding primitives between the pair of images take place. Julesz (1960) demonstrated with random dot stereograms that human stereopsis can proceed without extensive monocular processing, in other words it can operate on low level descriptions at an early stage of the visual processing. Possible low level descriptions which could be used are the original grey level images, or representations of the intensity changes in the images. Area correlation algorithms use small areas of the original grey level images as the matching primitives, and are therefore most effective for textured or smoothly varying scenes such as those found in terrain following or mapping tasks.

On the other hand, in edge-based algorithms such as those of Marr, Poggio and Grimson at MIT (Grimson, 1981), Baker and Binford at Stanford (Baker, 1981), and Mayhew, Frisby and Pollard at Sheffield (Pollard *et al.*, 1985), various operators are used to detect and locate intensity changes in each image, and it is these primitives which are matched. This edge based approach is potentially more efficient and capable of greater accuracy than area correlation. The first step is to compute an image description in terms of intensity changes, using the techniques discussed in the previous section. In fact, stereopsis provides a tough challenge for any edge detection scheme since it rests on the assumption that the edges detected correspond to the same spatially localised discontinuities in the real world, such as discontinuities in depth, surface orientation, surface reflectance or illumination. In current stereo algorithms, the edge description is often based on the zero crossings in the output of a Marr–Hildreth operator, for reasons already discussed.

When it comes to solving the correspondence problem, that is matching the selected primitives, without prior information on the viewing parameters, an item in one image might match an item anywhere in the other image. To reduce the combinatorial problems of the search, the viewing parameters need to be determined. This could be achieved either with measurements from a calibrated camera control system or by analysing the images themselves. In the simplest case, the images are arranged to be registered, usually by a simple horizontal translation of one camera with respect to the other, so that the matching items occur in corresponding rows.

Even having determined the viewing parameters, the correspondence problem is far from solved since it is still necessary to select the correct global match for an item amongst other matches which locally appear plausible. In order to resolve the ambiguities it is necessary to identify constraints from the geometry of the projection of objects in the real world, and build the algorithms around these constraints. The constraints which apply to stereo projection have been discussed previously in detail by Mayhew (1983). Briefly, they can be summarised as follows:

Figural continuity. Surface edges and markings are usually continuous, giving rise to continuous intensity changes in the images and a smoothly varying disparity. Thus having found a match for one edge element, there is a constraint on how neighbouring edge elements which lie along the same boundary can be matched.

Ordering. The order and adjacency relationships of intensity changes are usually preserved. For registered images this means that the order of intensity changes will be the same along corresponding rows in the images. In those cases where the ordering does not hold, the human visual system also has difficulty in fusion, e.g. when looking past a thin object close up to the face it is not possible simultaneously to fuse the close object and a distant scene.

Spatial uniqueness. A given location in space can hold only one object at any one time. Thus any particular location in space should give rise to similar symbolic descriptions in each image in terms of the direction and scale of intensity changes since the description will refer to the same physical object.

Subject to the above constraints, stereo algorithms try to minimise the number of unmatched points since points on objects will normally be visible in both views. However, the possibility that an object point is visible in one image and occluded in the other view cannot be ruled out.

To show how these constraints are incorporated, the stereo algorithm of Pollard, Mayhew and Frisby (Pollard, 1984), referred to as PMF, is described. The first issue is the choice of matching primitives. PMF uses an edge based approach, taking zero crossings of a high spatial frequency Marr–Hildreth operator as the matching primitives. Initially, the set of plausible local matches consists of those zero crossings of similar contrast sign, orientation and scale, lying within very loose disparity constraints along corresponding rows of the images (the images are assumed to be registered).

PMF allows a plausible match of a primitive in one image to gain support from a neighbouring primitive in the same image only if that neighbour itself has a potential match such that when considered as a pair the two matches lie within a disparity gradient limit. The disparity gradient is the ratio of the change in the disparity to the distance between the pair of points. Clearly this ratio is related to the orientation of the line between the pair of points with respect to the observer. PMF thus attempts to preserve surface continuity. Burt and Julesz (Burt, 1980) demonstrated that the disparity limit for humans is about unity.

The disparity gradient limit is implemented by a cooperative, iterative technique. The support score for a match is the sum of the best support scores of neighbours within a circular neighbourhood. A match is included in the sum if it is within the disparity gradient limit, and is weighted by the inverse of its distance from the match under consideration, since the effect of the limit is weaker the further away the match is. Also the sum score is expressed as a ratio of the maximum support that its neighbourhood could have given if there were no disparity gradient limit, so as not to penalise unfairly sparse areas of the zero crossing map. The scores are thus updated iteratively, with local support for matches propagating across the image to provide global support. The figural continuity and ordering constraints are implicitly implemented in this algorithm because matches which break these constraints will also tend to violate the disparity gradient limit.

Other attractions of this algorithm are that the iterative updating equation is simple and allows a parallel implementation. However, the algorithm has some problems at densely textured depth discontinuities which exceed the disparity gradient limit. To overcome this, it will be necessary to allow a match to survive even if it receives support only from one side of a boundary. Another problem, common to all edge based approaches, is that the output of the stereo matching is usually rather sparse. Possibly ways to overcome this problem are to increase the resolution of the edge detection or to fill in the map using area based primitive matching guided by the initial sparse edge map, following the approach of Baker and Binford (Baker, 1981). Other further work on PMF will involve a formal analysis of some of the mathematical properties of such an optimisation problem, and examination of parallel implementation issues.

Nearly all the work to date on stereopsis has dealt with pairs of static images. What happens if each image is varying with time? A naive approach would be to solve the correspondence problem in real time for each pair of frames independently, but surely this is not the best way to do it, since it throws away the time continuity element. This powerful constraint could be used at various stages of the processing. At the lowest level, use of the spatio-temporal operator described in Section 5.2.2 instead of simply a spatial operator (e.g. Marr–Hildreth) would impose a continuity constraint on the image description. At a higher level, the present surface description could be used to guide the correspondence search on the future image frames, thereby giving preference to a time continuous solution. However, any such approach would still have to allow for discontinuous events such as new surfaces suddenly coming into view.

The work on motion and optical flow described in the next section has the same goal as stereopsis in the sense that both techniques aim at producing a three-dimensional interpretation of a scene. It does seem likely that for the case of time varying images they may well complement each other, although exactly what the mechanism should be is the subject of further research.

5.3.2 Motion and optical flow

The optical flow field is the motion of points in the image plane due to motion of the corresponding object points relative to the sensor. The sensor can be regarded as a camera with origin, 0, of the coordinate system at the nodal point of the lens with effective focal length, l (see Fig. 5.3). Points in the image plane $\mathbf{r} = (x,y,-l)$ are related to object points $\mathbf{R} = (X,Y,Z)$ by the perspective transform, $\mathbf{r} / l = \mathbf{R} / Z$. In general, the coordinates $\mathbf{R}$ can be time dependent due to motion of the sensor or objects in view giving rise to optical flow, $\dot{\mathbf{r}} = (\dot{x},\dot{y},0)$, in the image plane. Differentiating the perspective equation and using the equation describing the motion of object points relative to the sensor, $\mathbf{R} = \mathbf{V} + \mathbf{O}\wedge\mathbf{R}$, the optical flow field may be written as:

$$\dot{\mathbf{r}} = -l\mathbf{V}/Z - \mathbf{r}(\hat{\mathbf{z}}\cdot\mathbf{V})/Z + \mathbf{O}\wedge\mathbf{r} + \mathbf{r}[\hat{\mathbf{z}}\cdot(\mathbf{O}\wedge\mathbf{r})]/l$$

where $\mathbf{V}$ stands for the translational velocity, $\mathbf{O}$ stands for the rotational velocity of the object point at a particular moment in time, and $\hat{\mathbf{z}}$ is a unit vector in the direction of the optic axis.

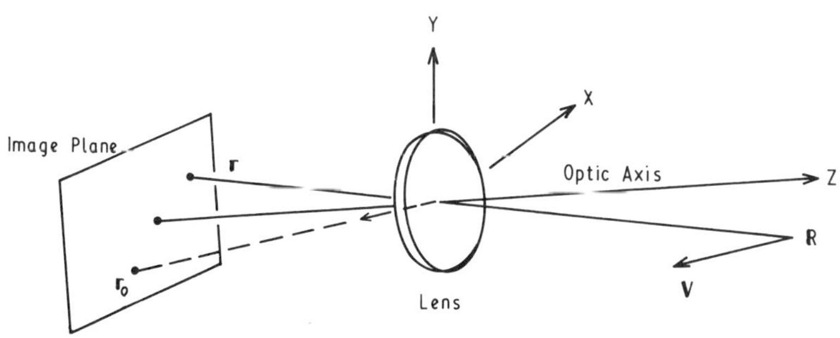

Fig. 5.3 The perspective sensor geometry.

It can be seen from the equation above that only the translational part of the flow field, dependent on $\mathbf{V}$, involves the depth, Z and, therefore, contains information about the three-dimensional structure of the scene (Longuet-Higgins and Prazdny, 1980). Some authors (e.g. Prazdny, 1980) therefore try to decompose the optical flow field into its translational and rotational components before attempting to obtain the structural information required. This is frequently done by locating the focus of expansion, $\mathbf{r}_0$, in the translational part of the flow field (Jain, 1983, Jerain and Jain, 1984) as the translational part of the flow field may be written as $\dot{\mathbf{r}}_{trans} = -(\mathbf{r}-\mathbf{r}_0)(\hat{\mathbf{z}}\cdot\mathbf{V})/Z$. Thus, once $\mathbf{r}_0$ is known, the depth Z can be determined up to a factor involving the speed V. In fact, it is the 'time to contact', $T= V/Z$, or relative depth that can be determined if the motion is unknown (Lee, 1980).

If the motion parameters, **V** and **O**, are known *a priori*, determining the translational part of the flow field and the focus of expansion and hence the depth, Z, is straightforward. This is also true even when only 'edge motion', the normal component of the optical flow (see Section 5.2.2), is available. However, the solutions are more complex if these motion parameters are to be recovered in addition to the scene structure as discussed below.

Motion parallax effects, which depend only on the translational part of the flow field, can be used to find the focus of expansion directly (Reiger and Lawton, 1983). To see this, consider the optical flow field on either side of a discontinuity in the flow field where the depth of the visible surface changes abruptly from Z_1 to Z_2 due to occlusions, the 'motion parallax' or difference in flow is given by: $\dot{\mathbf{r}}_1 - \dot{\mathbf{r}}_2 = -(\mathbf{r}-\mathbf{r}_0)(\hat{\mathbf{z}}\cdot V)(1/Z_1 - 1/Z_2)$, and is proportional to the distance from the focus of expansion, $(\mathbf{r}-\mathbf{r}_0)$. Determining the motion parallax at two points, therefore, enables the focus of expansion to be found. However, discontinuities in the flow field can occur whenever the motion parameters change as well as at occlusions. Application of this idea, in general, implies that the image has been segmented, which, as argued in Section 5.2.5, is something that should be delayed until three-dimensional interpretation. Indeed, the visual motion itself should be used as a cue for segmentation and object recognition (Buxton and Buxton, 1983).

An approach that avoids this presegmentation is to solve simultaneously the 'structure from motion' and the 'aperture problem' (see Section 5.2.2). That is, to infer the range, orientation and motion of the visible surfaces in a scene and, at the same time, reconstruct (implicitly) the missing information about the optical flow parallel to the edges in the image. On its own, edge motion does not contain enough information to solve both these problems, as a simple equation counting argument shows (Buxton, Buxton, Murray and Williams, 1984). Either some of the information must be known *a priori* or extra assumptions or constraints must be introduced.

The most important case where *a priori* information may be available is the passive navigation problem when the motion of the vehicle to which the sensor is attached may be accurately known. It is then easy to obtain the depth of object points directly from the motion of edge features in the image as the focus of expansion and rotational part of the flow field (independent of depth) can be calculated from the motion parameters, **V** and **O**. Reconstructing the visible surface, however, still requires simplifying assumptions and the use of several image points. The simplest assumption is that of locally planar surface facets. Then, just three points are needed to obtain the local surface normal, **N**, describing its depth, slant and tilt. The structure from known motion algorithm for obtaining this solution, the 'SM 3 point algorithm', is described in the paper by Buxton, Buxton, Murray and Williams (1984).

In the more general case, where the motion parameters, **V** and **O**, must be inferred in addition to the local surface description, **N**, there seem to be nine quantities (the three components of each vector) to be determined. However, there are only eight independent quantities that can be determined in this way as there is a fundamental ambiguity in optical flow fields since it is impossible to tell whether the flow is due to a small object close by moving slowly or a large object far away moving quickly. An absolute measure of the depth cannot, therefore, be obtained: only the time to contact, $T = Z/V$, is available without further information. The structure from unknown motion algorithm, then, determines eight not nine parameters and this 'SM 8 point algorithm' is also described in the paper by Buxton *et al.* (1984).

Both the SM 3 point algorithm and the SM 8 point algorithm fail in cases where the information in the flow field is deficient in some way. For example, there will clearly be no chance of inferring the depth if there is no translational motion, **V**. Also, there must be some edge motion, that is, the motion must not be only parallel to visible edges in the image and the translational component of the optical flow must not be parallel to the edge features. Other indeterminacies can occur when, for example, the edge features lie on lines through the focus of expansion and result in a failure to determine the relative depth of a set of image points. Even when the relative depth can be found, the recovery of the slant and tilt can fail due to rank deficiencies in the matrix (set of simultaneous equations) used to determine the solution. This happens when the image points used are collinear so that no information about the orientation of the surface perpendicular to this direction is available.

The failures described so far are common to both the 3 point and 8 point algorithms and correspond to precisely defined mathematical conditions. This analytical approach to the failures is possible due to a well developed computational theory of the interpretation of the optical flow field. Only performance criteria could be used if the algorithms had been developed *ad hoc*. Since it is impossible to test algorithms with all possible inputs and since ignorance of the limitations of algorithms can be dangerous, it is essential to formalise problems in this way for computer vision. This leaves complete freedom to explore different algorithms and implementations (see Section 5.5).

When the motion parameters, **V** and **O**, are unknown, there are additional indeterminacies in the solution when the image points lie on certain smooth curves. This problem has been studied by Waxman and Wohn (1984) for an algorithm similar to the SM 8 point algorithm. They give the conditions for failure and show that this includes the class of conic sections, but more general solutions satisfy the condition. The general conditions for failure of the SM 8 point algorithm are discussed in Buxton, Buxton, Murray and Williams (1985).

The numerical stability of the SM 3 point and 8 point algorithms has been tested on simulated data (Buxton, Buxton, Murray and Williams, 1985). The instabilities arise from the fact that these structure from motion algorithms, like most others to date (e.g. Ullman, 1979), rely on small second-order geometrical effects in the flow field to determine a solution. There are many ways the algorithms could be made more robust to the failures just described above and to the instabilities nearby. One way to eliminate such ill conditioning is to 'regularise' the solution, i.e. impose a prior potential (extra constraints). This is essentially what Hildreth and Ullman did with their smoothness constraint (see Section 5.2.2). The result is then a balance between the data and the expected solutions (Yuille, 1984). The concept is more general than this, however, and can include high level knowledge as well as low level constraints that embody knowledge of how contours, surfaces, volumes, etc., change with the parameters used in the solution. Recent work (Ullman, 1984) indicates that temporal consistency assumptions are likely to prove most powerful in constraining the process to reasonable solutions.

There are other important problems that remain concerning the interpretation of the data. One of the main reasons for working on optical flow is the belief that it can be a very strong cue for segmentation. Examples of the ease with which camouflaged creatures such as stick insects or a chameleon can be spotted when they move, reinforce this idea, as do simple demonstrations with interpenetrating random dot patterns moving relative to each other. The SM 3 point and SM 8 point algorithms can be extended to produce motion segmentation (Buxton and Murray, 1985). The important points are, first, that the segmentation is based on the scene surface and motion parameters, not on properties of the image patterns *per se*. Second, that the least squares processing used in these algorithms provides a natural opportunity to perform multiple regression in which new surface and motion parameters can be introduced in turn to explain variations in the visual motion data at different scales and resolutions. In principle, the scale and resolution chosen could be made to depend on the task in order to generate useful hierarchical descriptions of the objects in a scene.

5.3.3 Shape from shading

The input to 'shape from shading' is the unprocessed grey level values of the image rather than some edge or flow map as in stereo or motion surface reconstruction. The output is, however, a representation of the shape such as the local surface normal or orientation. One example requires the 'gradient space' parameters, p and q, the change in depth in the x and y directions in the image, dz/dx and dz/dy (see Fig. 5.4).

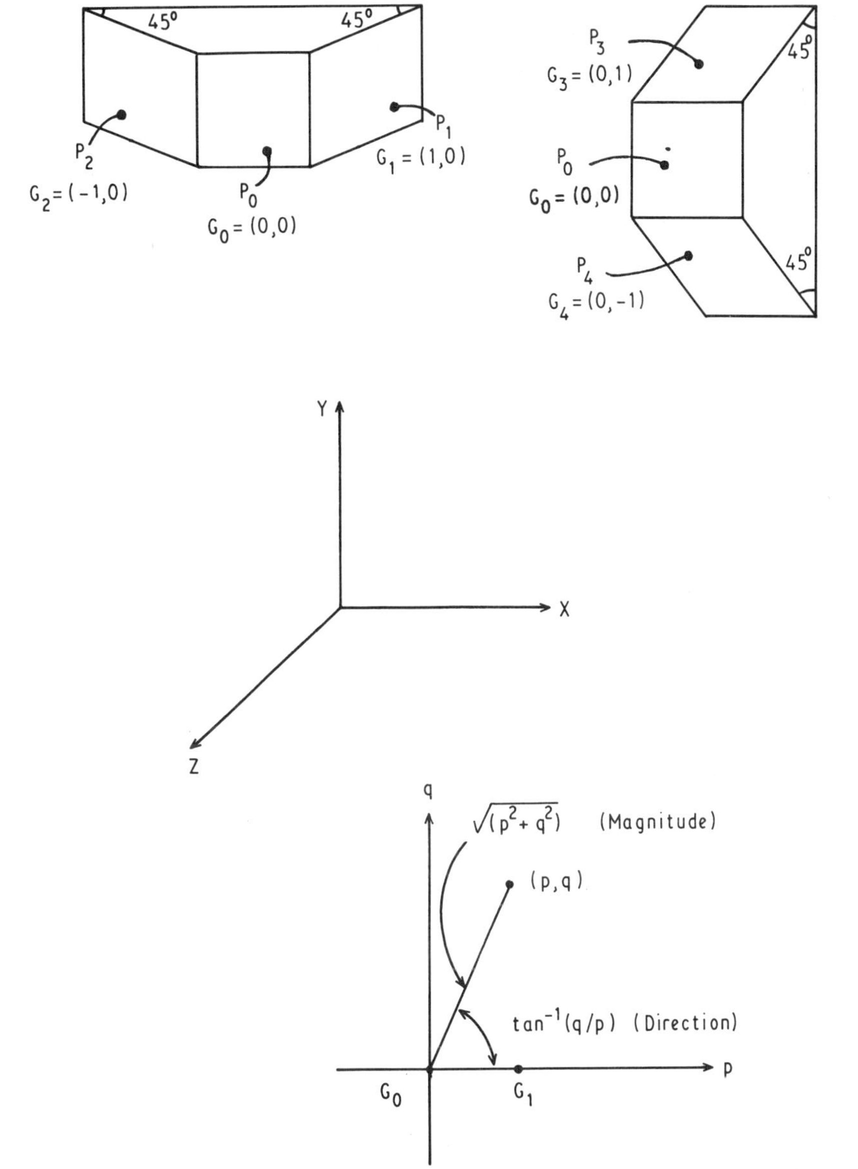

Fig. 5.4 Gradient space where p is the component of the change in depth along the x axis and q is the component along the y axis.

The image intensity at a point (x,y) can be represented as a function, $R(p,q)$. The 'image irradiance equation', $I(x,y) = R(p,q)$, is a non-lincar first-order partial differential equation for p and q. The function, R, depends, in general, on the reflectance characteristics of the surface and on the relative positions of the viewer and light sources (Horn, 1977). For a fixed distribution of light sources and given reflectance characteristics, the image irradiance equation assigns the intensity values to particular surface orientations, i.e. the 'reflectance map'. Note that this is not a one-to-one mapping.

There have been a variety of algorithms proposed to exploit this relationship. The original work by Horn (1975) used a standard technique for solving partial differential equations. This 'characteristic strip method' can be used to reformulate the image irradiance equation as a set of five ordinary differential equations. These form the basis of an iterative computation which starts by assuming that the surface gradient (p,q) at one image point (x_0,y_0) is determined and then stepping through the image across isobrightness contours, assigning the gradient values according to the reflectance map.

The problems with this method concern the choice of starting point (x_0,y_0). First, this point must be such that the surface gradient (p,q) is determined uniquely by the intensity data and second, the algorithm depends on an assumption that the underlying surface is convex at this point. Although, in general, there is only one convex solution to the image irradiance equation (Bruss, 1981), other solutions can exist so further research was needed to define the class of images and starting points for which the algorithm could find the correct solution.

At the occluding contours or 'boundaries' in the image, the surface normal is uniquely determined since it must be perpendicular to both the line of sight and the contour direction in the image. The gradient space parameterisation of local surface normal cannot represent these, however, as at least one of the gradients is infinity at such occluding boundaries. The stereographic projection has been used together with a smoothness constraint (Ikeuchi and Horn, 1981) in an iterative solution that uses the boundary values of the occluding contour of an object to fill in the orientation of interior points. The stereographic space parameters, f and g, are related to p and q by the equations:

$$f = 2p(\ 1+p+q\ -1\ )/(p+q)$$
$$g = 2q(\ 1+p+q\ -1\ )/(p+q)$$

The recurrence relation used by Ikeuchi and Horn is fairly robust to errors in the placement of light sources, the surface orientation at the boundary and its reflectivity function, although there is no formal proof that it will converge nor of how many boundary points are necessary. Progress on the formulation of the boundary condition functions and refinement of the iterative process continues (Horn and Brooks, personal communication) and can be directed to interpretations that describe the appropriate model for the discontinuity consistent with the data, in addition to determining the orientation from such a description.

A related method of computing shape from shading is 'photometric stereo' (Horn, Woodham and Silver, 1978, and Woodham, 1981). This uses multiple images of a scene lit from different, but known, positions. Although the process is not exploitable for human vision, it can be used to solve industrial problems such as bin picking. The first intensity value, $I_1(x,y)$, restricts the surface orientation to a particular isobrightness contour in the reflectance map. The second, $I_2(x,y)$, provides another constraint defining the intersection, and finally the third, $I_3(x,y)$ can uniquely determine the solution. The method involves a calibration phase on a known shape such as a sphere, and can then be both fast and accurate when a few quantised triples of intensity values are used to reconstruct the surface shape of unknown objects (Silver, 1980).

In conclusion, shape from shading can be used to obtain the local surface orientation. It seems to be used in human vision as a strong cue to curved surfaces. Indeed, artists use density of shading to represent the occluding boundaries of curved objects such as vases. Simple contours are interpreted as flat without such shading, although filled in as silhouettes they are usually given a three-dimensional interpretation (Marr and Nishihara, 1978) which is evidence for volumetric primitives in three-dimensional representations.

5.3.4 Shape from texture and contour

Gibson (1950) drew attention to the fact that changes in texture or texture gradient could be used to determine changes in depth of a surface. However, much of the computer vision work that relates texture to shape is more recent, for example, important studies include Kender (1980), Stevens (1980) and Witkin (1981). The last of these can also be classed as shape from contour and is related to the work of Kanade (1980) and Brady and Yuille (1983) which are discussed later in this section.

Stevens (1980) assumes a ground plane and that two-dimensional texture features on this plane are computed ready for a two-step process. First, 'characteristic directions' in which there is no change in depth are found, and second, depth is computed from the slant and tilt assumed perpendicular to these directions. Some of the work on shape from texture has close parallels to the work on shape from shading (Kender, 1980). In this approach, primitive texels are assumed to be already extracted from the image. Just as the intensity at a pixel varies with surface orientation, so does the appearance of a texel in the image, so each texel is assigned a set of possible scene parameters in the generalisation of the reflectance map called 'normalised texture property maps'. The computation assumes surface continuity and uniqueness to propagate constraints and find the globally correct solution.

Witkin (1981) has implemented a useful approach based on the statistical method of maximum likelihood. The method relies on the probability density function for the orientation of edges in the image plane for a given surface orientation. For a surface aligned to the image plane, all directions are equally probable but foreshortening due to slant and tilt of the surface bias the distribution. Given an irregular contour or set of contours in the image plane, the process will pick the interpretation that maximises the 'compactness' using the assumption of uniform orientation of edge segments in the scene. In effect, this will equalise the ratio of the virtual axes in the figures. This method fails for regular figures, e.g. a chessboard, since the edge distribution is strongly biased and violates the statistical model.

Two-dimensional contours are given a three-dimensional interpretation by the human visual system but there remains the problem that while many three-dimensional shapes could in principle have given rise to the two-dimensional projection, people are rarely aware of such multiple possible interpretations. The computation of the quantitative values of the perceived slant and tilt is discussed below.

Kanade (1980) introduced 'skewed symmetries', which are two-dimensional linear or 'affine' transforms of real symmetries. Such skewed symmetries correspond to the set of projections of symmetric shapes that lie in planes oriented with respect to the image plane. Kanade, therefore, proposed the heuristic assumption that a skewed symmetry is interpreted as an oriented real symmetry. The calculation of the slant and tilt of the oriented plane is naturally formulated in gradient space where the set of values, (p,q), satisfying a skewed symmetry in the image plane lie on a hyperbola. Kanade further proposed that the 'natural' interpretation, i.e. matching human perception, is the point on the hyperbola nearest the origin. This corresponds to the least slanted interpretation consistent with the image contour.

The minimum slant proposal above is not always correct, however, although skewed symmetries are indeed always interpreted as oriented real symmetries. The cases where it fails are for high symmetry contours, e.g. a square or ellipse. Brady and Yuille (1983) show that an extremum principle that determines three-dimensional surface orientation from a two-dimensional contour links the work of Witkin and Kanade described above. They propose that the measure to be maximised is $M = \text{Area} / (\text{Perimeter})^2$ which forces an interpretation in which both symmetry and compactness are maximised. The principle can deal with both regular figures and irregular figures; in the latter case it approximates to the maximum likelihood method.

Even the extremum principle above is insufficient to account for perceptual interpretation of figures in varying contexts. Higher level structure can then be apparent and the interpretation chosen will be in some sense the most 'parsimonious', or that which requires minimum information to specify. The ambiguity of an isolated two-dimensional contour can be reduced when considered as part of a more global percept. A formalisation of the work of Gestalt psychologists on grouping phenomena and its extension to three-dimensional description of recurrent elements using the principles of maximum likelihood should be possible in the near future. This would then automatically deliver structured interpretations of the image or scene for known organisations of data.

5.3.5 Interpolation and discontinuity description

The output of a stereo algorithm such as PMF is a fairly sparse (and usually noisy) map, with disparity or depth values assigned to those primitives which have been matched. The next question is, what to do with this output to make it useful for later stages of processing. The answer to this question depends on the application. In some limited circumstances the sparse depth map may be sufficient as it is, but usually it will be necessary to organise the information in some way by identifying surfaces bounded by discontinuities in depth or surface orientation, and explicitly describing these surface regions and discontinuity contours. For example, identifying an occlusion boundary and specifying which surface is occluded and which surface is doing the occluding; information which is not explicitly provided in the output of the stereo algorithm.

One possible strategy for achieving this, which has been explored by Grimson (1981), and more recently by Terzopoulos (1982), is to interpolate over the sparse depth map. Grimson's approach was to find a surface which fits the sparse stereo data subject to a surface consistency constraint. The surface must be such that it does not give rise to any zero crossings other than those which occurred in the original convolved image. In other words, the absence of zero crossings constrains the possible surface shapes. This constraint has informally become known as 'no news is good news' since it says that if there is no zero crossing then the surface cannot be varying very radically. By examining the image irradiance equation, it is shown that the best surface to fit is that which minimises the quadratic variation in surface orientation and passes through the known points. To implement this constraint, Grimson used an algorithm that satisfied three criteria of biological feasibility, namely parallelism, local support and uniformity; these criteria are also appropriate for computational implementation.

Unfortunately the iterative technique chosen by Grimson required a very large number of iterations since it was applied solely to the finest resolution output of the stereo algorithm, obtained by assimilating the outputs from several different spatial frequency channels. Terzopoulos avoided this problem by keeping separate the sparse representations at the different levels of resolution. He developed a multi-level surface reconstruction algorithm, allowing inter-level relaxation and bidirectional inter-level local interpolation with a vastly improved efficiency. The surface most consistent with the sparse depth constraints is characterised as the solution to a variational problem, that of minimising a functional subject to constraints. In this case it corresponds to the physical situation of minimising the energy of a thin flexible plate which is attached by springs to pins at certain heights. A discrete approximation to this problem is then posed, and solved by a local parallel computation.

This technique could be extended to include different sources of constraints such as depth from motion or surface orientation from texture or shading. One problem with the technique at present, however, is that the algorithm currently takes no account of depth discontinuities – it tries to fit one smooth surface over the whole area. Clearly this is not desirable, and one way suggested by Terzopoulos to avoid the problem would be to break the interpolation procedure at places where high strain is generated, and localise the interpolation to the regions between these boundaries.

Recently, Blake (1984) has argued that the thin plate model assumptions are not appropriate for this problem, and anyway this model does not lead to a viewpoint independent surface shape. He suggests using as an energy function an integral of surface curvature over the surface area. This approach may overcome the problems, but much further work remains to develop these ideas into a working solution.

Even having interpolated over the depth map, there still remains the task of grouping the three-dimensional points into curves and surfaces, and interpreting them. An alternative strategy suggested by Mayhew (1983) is to try to extract a three-dimensional structural interpretation directly from the cues in the original depth map by grouping the depth points into curves and analysing the curve terminations, intersections and discontinuities, and making the assumption that the scene is being observed from a general viewpoint. This is similar to the approach taken by Lowe and Binford (Lowe, 1981) who analyse image curves (rather than three-dimensional space curves) to obtain three-dimensional structure as part of the ACRONYM project (Brooks, 1981).

On the other hand, the information so derived concerning the surface discontinuities could be used as an input to an interpolation process, informing it of the positions of knots and breakpoints corresponding to surface orientation and depth discontinuities respectively. A robust working system may thus combine both approaches.

5.3.6 Combining information

Two aspects of combining information can be distinguished and will be discussed below. The first is using complementary sources of information to resolve any ambiguity in the interpretation of the scene. The functional role of the 2½ D sketch is just this bringing together of constraints to build up the full description of the visible surfaces and discontinuities. The second is what techniques to use when combining information, whether it be from single or multiple sources. Here, two methodologies can be distinguished: either using explicit cost function that relates the input data to the interpretation, or providing a mapping of the data onto the set of parameters that constitutes the interpretation as outlined below.

Very little research has actually been done on the details of combining diferent sources of information at the surface description level. For instance, it is known that stereo and motion, although they can operate as separate modules, could together provide a much more robust and flexible mechanism for obtaining the surface and motion parameters. Current research is exploring the details of such a scheme. The whole family of 'shape from' methods can help in building up this level of description. Ambiguities in shape from texture could be resolved by either stereo or optical flow data, for instance, or shading information could be integrated with stereo motion to characterise regions of high curvature. There is potential for a great deal of research in this area as the principles behind the individual modules are now quite well understood. However, as discussed in the last section (5.5) there are questions of control and flow of information in the system that also need to be worked out.

Techniques for combining the information are described in some detail in Chapter 2 in the context of image processing algorithms. Here, the discussion focuses on some advantages and disadvantages of these methods. The basic idea is that once the relationships between the levels of description are known and the constraints that should be applied are decided, a cost function can be constructed that will define the set of parameter values that best interprets the data. For any complex cost function, it will be necessary to perform an iterative search for some kind. If the function is guaranteed to be convex and have only a single minimum, simple relaxation can be used. However, this is not usually the case and on convergence, the gradient descent methods cannot detect whether a local or a global minimum has been found, so the solution depends critically on the starting point. It is sometimes possible to do multi-scale relaxation to smooth out the local minima to overcome this problem. A more powerful search can be performed by stochastic optimisation, however, which can deal with complex cost functions where there are many conflicting constraints. The simulated annealing proposed by Kirkpatrick, Gelatt and Vecchi (1983) involves a temperature function which adds a decaying amount of noise to the search step decisions so that large disadvantageous moves can be accepted initially to get out of the local minima but gradually the average step size is reduced to reach the stable state.

Another way of organising the combination of data is to use some generalisation of the Hough technique which maps the data into parameter space. This was developed for detecting straight lines and curves in noisy images (Duda and Hart, 1972) but has been generalised as a strategy for detecting arbitrary shapes (Ballard, 1981) and can be adapted for arbitrary mappings of data onto a set of parameters. The parameter space is divided into an interval array and all data consistent with a particular set of these values 'votes' for that interpretation. The best interpretation is detected as a peak in the array and the values can be accurately found by interpolation around the distributed peak. The array can be extremely large for multidimensional interpretations and the effective search therefore very slow, but it is usually possible to decompose the problem into a number of stages using higher level features. The levels can then usually be made to reinforce each other. Ways of speeding up the process on serial machines have been described by Brown, Curtiss and Sher (1983) using 'culling' of empty cells in the array as the processing continues together with look-up tables.

Using peaks in the array, that is the mode, gives an advantage in that the interpretation is not overly affected by outlying data points in the histogram-type methods above. The same effect can be achieved in the cost function search by changing the weights of the data points, but this takes time on iterations of the algorithm. On the other hand, there are problems of choosing the right scale that are harder to deal with in the histogram methods than in the cost function search where the scale can usually be changed by a weighting of the cost of continuity constraints. For instance, in line detection the spatial relations between the points voting for a particular set of parameters are lost in the histogram so there is a possibility of spurious peaks.

In conclusion, then, there is clearly a long way to go before the best way of using these methods is fully understood. However, they can be used as generic techniques to implement, at the algorithmic level, the set of constraints or mapping decided at the computational level of theory.

5.4 THREE-DIMENSIONAL MODELS AND MODEL BASED VISION

So far this chapter has dealt with the description of images and of visible surfaces in images. This bottom up processing involves knowledge of general principles from mathematics, e.g. calculus and geometry, and physics, e.g. image formation, but is largely independent of knowledge specific to the objects to be found in the scene. Furthermore the resulting descriptions are viewer centred, in other words they depend on the particular viewing parameters. The subject of this section is the interpretation of these descriptions in terms of instances of objects, that is making explicit the correspondence between the image or surface description primitives and the object primitives. To do this will require, in addition to the descriptions already discussed, models or representations for the objects, as well as mechanisms and strategies for the prediction and verification of comparisons between the models and the descriptions. In some cases it is also necessary to introduce other intermediate representations.

5.4.1 Computational requirements for three-dimensional models

A vision system which must reason about a variety of complex objects, whose positions and orientations are not known even approximately, requires object models which are viewpoint independent. It is not usually appropriate, even if feasible, to try to store multiple viewpoint dependent models which attempt to cover views from all possible positions and orientations. This is to be contrasted with the case of simple objects with only a few stable positions, and hence few two-dimensional projections, as might occur for instance on a conveyor belt in an industrial setting. Viewpoint independence suggests that the models must provide three-dimensional structural information about the objects, from which the vision system itself can compute the required views.

There are several general requirements of such models. Clearly they must be capable of representing efficiently the (possibly large) class of (possibly complex) objects which the vision system must recognise. Since these models will be used for making predictions which will be compared to the image or visible surface descriptions, the information required to support this prediction phase must be readily accessible in the models, if not available explicitly. The models should also support representations at different levels from coarse to fine, that is from wholes to parts and subparts, with correspondingly greater detail. Furthermore, the models may need to serve several different applications, such as object recognition and manipulation and trajectory planning.

Modelling techniques for solids and surfaces are already in use in computer aided design (CAD) and computer graphics, so it is worth seeing what they have to offer for computer vision. It is also worth considering these techniques from the point of view that in future fully computer integrated manufacturing systems, a CAD database will have been created with models of the objects to be manufactured; these models will thus already exist and be available for use in a computer vision/robotic system (see Section 7.3.2).

5.4.2 Solid modelling

Many modelling systems for CAD now satisfy the requirements for viewpoint independent three-dimensional representations. Their purpose is to allow the interactive design of a model for a specific part, and the automatic generation from this model of information for particular applications, for example instructions for the part's manufacture. However, such systems have not been designed with the interface to a vision system specifically in mind, thus the information required for visual tasks is not always readily accessible, although for display purposes it is necessary to synthesise images from the models using techniques such as ray casting and shading.

Some important formal properties that are considered in these solid modelling representation schemes are (Requicha, 1980):

Validity. Any representation which is syntactically correct corresponds to a sensible real solid. In this way a simple parser can detect whether a nonsense object has been created.
Completeness. Any valid model corresponds to a single object so that any of its physical properties (e.g. mass, moments) can be computed.

Uniqueness. Any object has a single representation so that equality of objects can be assessed.

Other important, but less formal considerations addressed include:

Conciseness vs. redundancy. It may be useful to explicitly store redundant information which is frequently used, rather than having to compute it from a rather concise model; this is basically a space *vs.* time trade-off, and the choice will depend on the target application.
Editing. This refers to the ease with which models can be created and interactively modified.

Various representation schemes which are used for solid modelling are listed below. An excellent survey giving more details of these schemes and of their properties can be found in the paper by Requicha (1980).

Pure primitive instancing. A generic primitive is used to represent a class of objects; it can be instanced to a particular object by assigning values to the associated parameters. This can be useful when there are several objects of basically similar shape but varying dimensions.
Cell decomposition. A model represents an object as the union of disjoint primitives. In the simplest case these cells may be identical cubes, so that the scheme reduces to simple spatial enumeration.
Constructive solid geometry (CSG). A frequently used representation in which a solid is modelled by a hierarchical Boolean combination of components. A CSG model is a binary tree in which non-terminal nodes are either rigid transformations (translation, rotation, scaling) or generalised set operations (union, intersection, difference), and terminal nodes are simple primitive solids or values of parameters for the transformations. Unfortunately this scheme does not necessarily give a simple description of an object in terms of a hierarchical model. It may even lead to complex structures for fairly simple objects, especially if the set of primitives is not sufficiently rich. (See Fig. 5.5(b).)
Boundary representations. A solid is represented by the combination of a finite number of enclosing faces or patches. An object will be described by a graph structure in which the nodes correspond to the faces, boundaries and vertices, and the arcs correspond to their connectivity and adjacency relationships; both the nodes and arcs are three-dimensional constructs. This scheme is closest to that required for visual tasks since it is based on the external surfaces of an object, i.e. what can be seen. (See Fig. 5.5(c).) However, there are problems concerning the segmentation into patches for objects with complex curved surfaces. Some progress has been made in this direction and is discussed in the section on differential geometry. Special techniques for modelling complex, free form doubly curved surfaces have been developed for the field of computer graphics and will be described in the next section.
Sweep representations. An object is described by the volume swept out when either a planar cross section is swept along a spine curve (translational

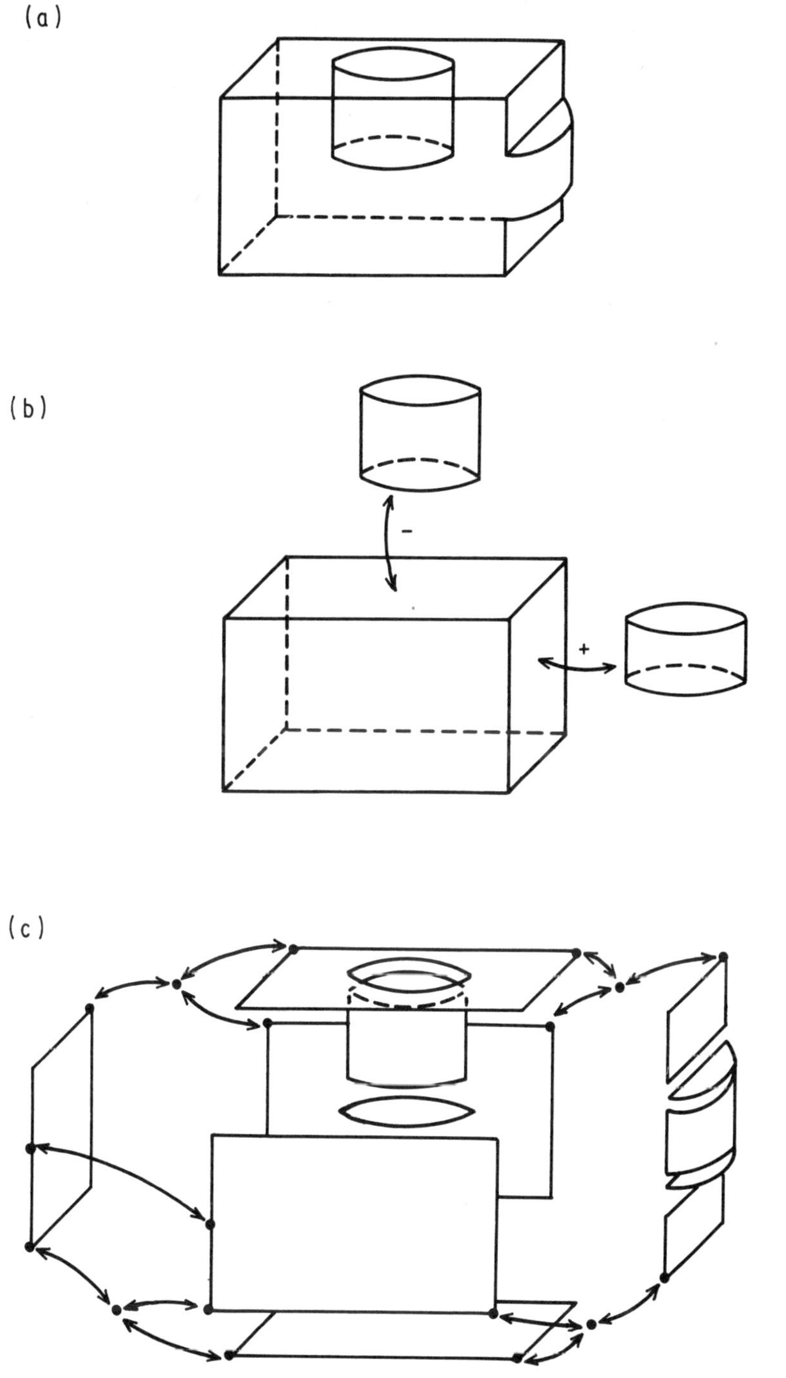

Fig. 5.5 (a) A simple three-dimensional object; (b) its CSG representation; (c) the boundary representation model.

sweeping) or a planar curve is rotated about an axis (rotational sweeping). These schemes are appropriate for elongated or rotationally symmetric objects, for example produced by extrusion or turning operations respectively. (See Fig. 5.6.)

Much work has also gone into the problem of converting between the various representation schemes, since a system which must support a variety of applications efficiently probably requires several consistent representations within it. Of particular significance is the conversion from CSG to boundary representations, for the purpose of computer graphics display. Many

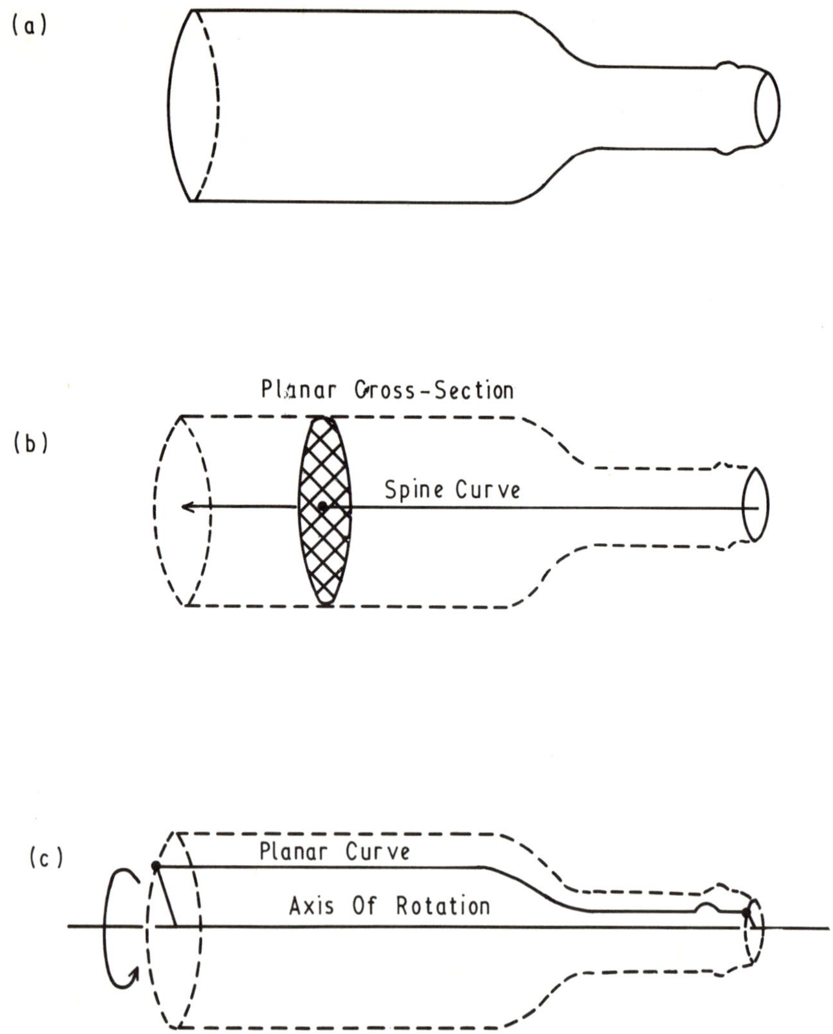

Fig. 5.6 (a) A simple three-dimensional object; (b) its translational and (c) its rotational sweep model.

modelling systems in fact use a combination of the above approaches; an example of such a hybrid system is now described.

TIPS-1. This geometric modelling system for CAD/CAM was developed at Hokkaido University in Japan (Okino, 1976). A mechanical part is modelled as the union of 'segments', where each segment is defined by the intersection of an 'element' with a 'domain'. Elements can be standard primitives (e.g. cylinder, rectangle, cone), sweeps of primitive planar shapes (e.g. circle, ellipse) along a coordinate axis, or non-standard, meaning defined by a free form surface or a surface given by a mathematical function. A domain is basically a set of half-spaces which are used to bound the element. For the case of a free form surface specified initially by a number of three-dimensional point coordinates, the data points are replaced by mesh grid points to which bicubic patches are fitted. In addition to simple segments, segments can be generated by operating on existing segments with operators such as intersection or revolution. The models are subsequently used for producing orthogonal projection views, computing volume and mass properties, finite element stress or thermal analysis, or generating programs for manufacturing operations such as milling.

Consideration is now being given to the extension of these CAD geometric modelling systems to support vision tasks efficiently, by introducing redundancy in the models. Bolles *et al.* (1983) has proposed such a system, called 3DPO (Part Orientation), based on an extended CAD model (such as PADL-2 (Brown, 1981)). This model will contain consistent CSG and boundary representations to which have been added pointers that link related features, for example each surface has a list of pointers to its boundary elements, and dually each edge has pointers to the adjacent surfaces; it will also have lists of particular classes of features such as surface elements that have the same surface normal.

Thus, attempts are already being made in this direction to close the gap between the features that conventional solid modelling supports and the requirements of computer vision.

5.4.3 Surface modelling

CAD surface modelling systems have been used mainly for the design and display of single complex surfaces, for example the UNISURF software package used by Renault for body-panel design, rather than the combination of surfaces into solids. Representing such complex surfaces mathematically raises problems of computational complexity and numerical instability. Similar well known problems occur in the two-dimensional case of fitting curves to data points, and techniques developed for this curve representation have been extended to deal with surfaces. Two basic concepts underly attempts to overcome the difficulties:

Piecewise approximation. Rather than trying to fit a single high order function over a complex surface, it is much better to join together simple surface patches with certain constraints on the continuity of the surface position and gradient imposed at the patch boundaries.

Parametric representation. A vector valued parametric representation avoids any axis dependencies and allows surfaces which are multi-valued with respect to the Cartesian axes. Thus a surface is given by a parametric equation of the form $\mathbf{r} = \mathbf{r}(u,v) = (f(u,v),g(u,v),h(u,v))$ where f, g and h are simple functions of u and v, and the parameters u and v usually cover the range $0 \leqslant u \leqslant 1$, $0 \leqslant v \leqslant 1$. (See Fig. 5.7.)

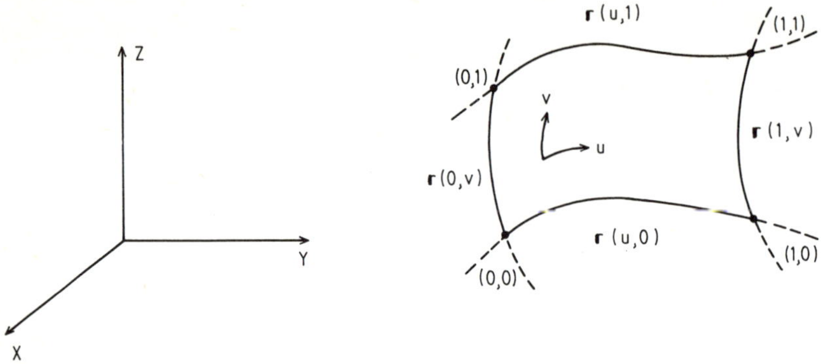

Fig. 5.7 Definition of a parametric surface patch.

Polynomial functions are normally used to approximate the surface patches since they can be easily differentiated to give the curvature and other geometrical measures. Typically cubics are chosen as a reasonable compromise between higher order polynomials, which are intuitively and numerically difficult to handle, and lower order polynomials (e.g. linear functions), a large number of which would be required to produce a reasonable description. Cubics also have a sufficient number of non-zero derivatives to allow reasonably smooth continuity constraints to be applied.

A good introduction to the large number of techniques used for surface and curve modelling can be found in the excellent text of Faux and Pratt (Faux, 1979).

One problem with these techniques is that although the representation is complete in the sense that the surface can be constructed at any point, basic qualitative information such as whether or not the surface is planar, is only contained implicitly. A very detailed surface description is in fact probably overkill for many vision tasks.

5.4.4 Differential geometry

Another problem with the surface modelling schemes described above is that such representations are established with respect to an externally defined coordinate system and mesh, and are not unique; thus there is no canonical description. For visual tasks it is preferable to determine locally a coordinate system and webbing based on the intrinsic properties of the surface itself, e.g. curvature. Visually important cues for surface shape are the boundary contour around the surface, and contours that lie across the surface. The visual interpretation of such contours, and the constraints applied by the human visual system to resolve the inherent ambiguities, have been discussed by Stevens (1980). It seems that a webbing derived from an analysis of the surface curvature may provide a natural description of the surface shape. As noted above, a very detailed description of the surface is probably not necessary for many visual tasks; a qualitative description will often suffice.

The analysis of surface curvature is conveniently framed in terms of differential geometry. Some of the relevant basic notions of differential geometry are briefly outlined.

Suppose a surface is given as a function of two variables $z = f(x,y)$. The local shape of this function can be described in terms of its first and second partial derivatives. (These determine the first few terms in a series expansion of the function):

$$\text{gradient} \equiv \left(\frac{\partial f}{\partial x}, \frac{\partial f}{\partial y} \right)$$

$$\text{hessian} \equiv H = \begin{pmatrix} \dfrac{\partial^2 f}{\partial x^2} & \dfrac{\partial^2 f}{\partial x \partial y} \\ \dfrac{\partial^2 f}{\partial y \partial x} & \dfrac{\partial^2 f}{\partial y^2} \end{pmatrix}$$

Consider the intersection of the surface with a plane which contains the normal to the surface at a given point and cuts the plane in a particular direction. The curvature of the resulting curve is the 'normal curvature' at that point in the given direction. Now the eigenvectors of the hessian matrix H give the directions of the maximum and minimum normal curvatures, the 'principal directions', and the corresponding eigenvalues give the 'principal curvatures' usually denoted by k_1 and k_2 (without loss of generality $k_1 \geqslant k_2$). A patch of surface can be characterised qualitatively by the different cases for the curvature values (k_1,k_2):

$k_1=0$ and $k_2=0$	$\Rightarrow$ planar
$k_1=0$ and $k_2 < 0$ or $k_1 > 0$ and $k_2=0$	$\Rightarrow$ developable
$k_1 > 0$ and $k_2 > 0$ or $k_1 < 0$ and $k_2 < 0$	$\Rightarrow$ elliptic/convex
$k_1 > 0$ and $k_2 < 0$	$\Rightarrow$ hyperbolic/saddle-shaped

The sign of the 'gaussian curvature', which is the product $k_1 \times k_2$, is thus a convenient qualitative descriptor. A surface could be segmented into regions where the gaussian curvature is of one particular sign, that is at parabolic lines. Haralick *et al.* (1983) suggested a classification of surface patches into categories such as peak, pit, ridge, ravine, saddle and flat, based on a local analysis of curvature, although his analysis was actually applied to the original image intensity surface, rather than a physical surface.

A 'line of curvature' is constructed by tracking one of the principal directions across the surface. An orthogonal webbing across the surface will be formed by the lines of the two principal curvatures (see for instance Fig. 5.1). In fact psychological evidence suggests that indeed surface contours are usually assumed to be lines of curvature for the purpose of inferring surface shape (Stevens, 1980).

A surface representation scheme in terms of 'curvature patches', using ideas from differential geometry, has been developed by Brady (1983) to support visual tasks. It involves parametrising the surface so that the webbing corresponds to lines of curvature (which form an orthogonal mesh). Then instead of describing the surface in terms of the values of the coordinates and their derivatives at the corners of the patch as in the computer graphics approach, it is interpolated using the variational calculus techniques already discussed. Having set up a webbing on the surface and formed the curvature patches, the next aim is to group connected sets of patches to form a concise description in terms of surface types, e.g. singly or doubly curved. This curvature patch representation is also being used in an investigation of a theory of three-dimensional object grasping in which it is suggested that objects are grasped along lines of curvature.

Questions which still need to be answered include how to deal with the different levels in a coarse to fine representation and what to do about extremal boundaries (limits of surfaces from a particular viewing position). Some of the questions will only be answered when there is a clearer understanding of the modules which are to operate on this description, that is when model based vision systems want to make use of this surface representation level.

5.4.5 Model based vision

In the model based approach to vision, it is specific knowledge about the objects expected in the scene which guides the analysis through a prediction/verification strategy. Model based vision systems can usefully be described according to the types of models (two- or three-dimensional) and the level of description (image or surface) at which they attempt interpretation. Model based vision will make increasing use of surface descriptions as the modules for computing these richer descriptions mature. Some model based systems in fact already use surface descriptions but derived from laser ranging (Kuan, 1983) or structured light (Bolles, 1983) data, rather than passively from vision. Several examples of model based vision systems are summarised to give some idea of what is involved.

WALKER. This is a program developed by Hogg at Sussex University, UK, which takes as input a time series of images of a person walking, and produces a description of the motion in terms of instances of the WALKER model (Hogg, 1983). The program 'generates' instances of the WALKER model and 'tests' them against the raw image data. The WALKER model defines spatial (positional) and temporal (movement) relationships between the major parts of the person, represented by elliptic cylinders, in the context of walking movements. It does not attempt to model in detail the local surface shape.

The range of possibilities to be generated for a particular frame is reduced by applying the dynamic constraints contained in the model to the decisions that have been made for the previous frames. This also ensures consistency between the interpretations of successive frames. For each frame in the sequence, a plausibility value (between 0 and 1) is assigned to each generated instance of the model. Trying to predict directly the image intensity values followed by area correlation is not a sensible strategy, since it requires detailed knowledge about the lighting and reflectance which would not normally be available. The more practical and robust scheme of comparing the predicted appearance of occluding surface boundaries with actual image intensity discontinuities (located by the Sobel edge operator) is in fact used. A plausibility value is obtained from a measure based on the difference in position and orientation between the predicted lines and image edges for a particular part. The plausibility values for the different parts are combined to give an overall plausibility value for the instance.

Hogg suggests possible extensions to this initial scheme; for example, checking that occluding boundaries actually end where they are supposed to by searching the appropriate parts of the image for edge terminations. This type of feedback mechanism between top down and bottom up processing is powerful, but can make control issues extremely complex.

The program has been tested on a sequence of 45 images of a person walking; after initial test runs on the first 15, it ran successfully first time on the whole sequence. Its performance is monitored by displaying the model instances superimposed in the original images.

ACRONYM. This system was designed to support vision tasks at the AI Laboratory at Stanford University (Brooks, 1981), and represents a comprehensive effort in the area of model based vision. A volume element is represented by a generalised cone, defined by the volume swept out along a space curve spine by a planar cross section.

A solid is then described in terms of an 'object graph' where the nodes represent such volume elements and the arcs their spatial and part/subpart relationships. ACRONYM also contains a base of algebraic and geometric rules for manipulating symbolic constraints and transformations. An object graph represents a generic class of objects, a particular subclass or specific object being specified as a restricted version by putting constraints on the three-dimensional model.

ACRONYM searches for model instances at the image level. An edge and line finder locates ribbons and ellipses, two-dimensional projections of generalised cone surfaces and ends respectively, in the image. To generate predictions of ribbons and ellipses, all boundaries of surface that might be visible are identified, first in the object's own coordinate system and then by symbolic transformations into the camera coordinate system. As a local match is made, constraints are propagated back into the three-dimensional model, with accumulated constraints for a cluster of ribbons required to be self-consistent. ACRONYM has so far been demonstrated on aerial images of planes at an airport. Its performance depends critically on the output of the ribbon finder.

In an extensive survey of model based vision systems, Binford (1982) suggests that the way ahead for a vision system such as ACRONYM lies in making interpretations at the volumetric level rather than the image level; in other words, the matching should take place between three-dimensional scene features and three-dimensional model primitives. This of course requires surface descriptions to be generated from the image.

RAPT and Verification Vision. RAPT is an 'object level' robot programming language developed at Edinburgh University for the off-line programming of assembly (Popplestone, 1978). Object level means that the task is specified in terms of the objects that are to be manipulated and spatial relationships between them, rather than as a program of individual actuator commands as in more conventional robot programming languages. RAPT outputs a sequence of planned positions for the objects during the assembly, but takes no account of the inaccuracies of its own geometrical model compared to the real environment, or of errors that may occur. It has now been augmented to allow the use of vision data for modifying the plan during run time (Yin, 1983). Verification vision is special in that the environment and task are heavily constrained, there being a large amount of knowledge available about the objects; the goal is just to check or refine the position of a particular object relative to its expected approximate position. To support verification vision, several statements have been added to RAPT. The LOOK statement specifies which feature (usually an edge) is to be verified, the INVIOLATE statement specifies a constraint on that feature in terms of a relationship between it and the world, and the TOLERANCE statement specifies the maximum error relative to the object's coordinate system. So far the system has been tested only on simulated vision data; the interface to a real vision system will present a considerably greater problem. Indeed the use of vision for precision measurements is itself difficult, requiring high resolution and good access for the camera. It may need to be supported by touch sensor information.

5.4.6 Conclusions

The gap between the modelling systems originally used in CAD and the requirements for computer vision is being slowly closed from both directions. On one front, CAD systems are now being extended to provide features that are useful specifically for vision tasks. It seems that there are two main components to this. One is the provision of a rich qualitative and quantitative description of the object's surfaces, boundaries and junctions, i.e. its visible characteristics, and a network describing the adjacency/connectivity relationships between these primitives. In contrast to traditional model based systems, these features and their relationships are all three-dimensional entities. This description could conceivably be derived directly from another representation such as CSG, which serves as a core representation for several tasks, only one of which is vision. The other component is an off-line analysis of the model which identifies key features of the object on which the hypothesis/verification stage should concentrate.

Some progress has been made in the two-dimensional case by researchers at SRI International (Bolles, 1982) using what they call a 'local-feature-focus' method. By detailed analysis of the models, they select clusters of simple features (corners and holes) in order to recognise and locate overlapping industrial parts such as hinges. However, for the three-dimensional case this problem is perhaps one of the hardest yet to be tackled, since for three-dimensional objects there will typically be many structurally different views. In other words, different clusters of surface features will be potentially visible from different viewing positions.

In the other direction, as discussed in the previous sections, progress is being made in the extraction of rich qualitative and quantitive three-dimensional descriptions of surface patches and discontinuity boundaries from images using stereopsis, optical flow and so on. These descriptions will be similar to the ones found in the models.

Although ideas on the representations that are appropriate for model based vision are becoming quite well developed, the mechanisms and strategies that should be used during the prediction/verification stage are by no means clear. So far the control issues have been considered mainly from a computational point of view, rather than from ideas on how the human system solves such problems. For example, programs such as ACRONYM and WALKER neither make much use of, nor contribute much to, an understanding of the eye-brain performance. The next section contains some ideas on how the human system might learn and store representations of objects, and use them in visual tasks.

5.5 FUTURE DIRECTIONS AND IMPLEMENTATION

In this final section, some major extensions for vision research are discussed, many of which have a bearing on the architectural issues covered in other chapters. The discussion starts by considering the combination of information, using more constraints and knowledge to tackle complex visual tasks. This leads on to the question of how to use knowledge at different levels and self-organising connectionist architectures. These architectures have recently been shown to support symbolic level reasoning and inference, as discussed below. Next, an example implementation using a SIMD parallel machine is described which involves measuring and interpreting the optical flow field, going from an image sequence to a parameterised description of the scene structure and motion. The problem of mapping higher level processing such as geometrical reasoning about surfaces on such architectures is then considered and questions about the ideal architecture for visual systems are raised. Finally, two important areas of expanding activity in computer vision research are highlighted. These are first, the characterisation of three-dimensional surfaces and three-dimensional matching and second, the questions of stereo motion integration and incremental elaboration of scene descriptions. In conclusion, the driving force of these computer vision applications towards new and more powerful parallel architectures is discussed.

First, then, it seems likely that as understanding of the principles behind visual processing advances, it will be increasingly important to know how to combine several sources of information and to balance possibly competing constraints for interpretation. As discussed in Section 5.3.6, the current methods are either some kind of structured modal filtering through parameter nets (Ballard, 1981) where the constraints are implicit in the mapping of one set of parameters onto another or some kind of complex cost function incorporating the various constraints explicitly. The best interpretation is found by detecting the peak in parameter space or by some iterative search, e.g. relaxation or stochastic optimisation, on the cost function, respectively.

There seem to be two advantages of the connectionist schemes for modelling complex visual processing, although these may not apply to testing out theories in the short term. First, the connectionist schemes allow a very natural extension to the case where knowledge is available at many levels in the processing hierarchy allowing the best interpretation to be selected by top down and bottom up reinforcement, and second, these schemes can incorporate algorithms to learn and improve the mappings (Hinton, Sejnowski and Ackley, 1984). It is essential to have learning for complex tasks as there is a cost associated with breaking each constraint and where the set of such constraints is large, exhaustive search for the best values becomes impracticable. There has been a great deal of progress recently in developing theoretical models of generalised learning both with regularisers and within neural network schemes, arising from the work of Hopfield (1982).

The neural network approach has been subject to criticism on the grounds that the structure of the processing encoded is implicit in the mechanism and thus unable to contribute to our understanding. However, much of the recent work has been directed towards understanding the mapping of explicit theory onto implicit mechanism. For example, Touretzky and Hinton (1985) describe a working implementation of a production system interpreter and show how two important symbolic reasoning operations, pattern matching and variable binding, can be emergent properties of neuron-like elements. In these distributed neural network architectures, each symbol is represented by activity in many units and manipulating symbols is not as easy as in conventional computer architectures where each unit can be regarded as a symbol. On the other hand though, there is evidence that the distributed approach is used in nature (see Ballard, Hinton and Sejnowski, 1983). The ultimate goal of such research is to design a symbolic to parallel machine interface that allows the user to specify the explicit symbolic level mappings between the levels and for these to be compiled automatically into the parallel machine mechanism. It is clear that parallelism is required for computationally demanding vision problems, but for most current parallel architectures (see discussion below), the user must do the mapping of the problem onto the machine to exploit best the parallelism.

To make the last point a little more concrete, the implementation of an algorithm for measurement and interpretation of optical flow from a single moving plane using a SIMD parallel machine will be described briefly (see Buxton, Buxton, Murray and Williams, 1985). The calculation can be split into a number of steps. First, locating the moving edge features and measuring the edge motion (see Section 5.2.2), second, local calculation of matrix coefficients, third, forming products of the transposed matrix, fourth, solving the least squares equations, finding the middle root of a cubic equation and computing the four possible interpretations in terms of the surface normal, translation and rotation vectors and fifth, backprojecting these parameter sets to resolve the ambiguity in the interpretation. The early steps involve calculations over the image space (x,y) whereas later steps involve only the small space of eight surface and motion parameters. In the former, therefore, it is most convenient to map the image pixels into the processor array. The vectors describing the edge position and edge motion are stored vertically below the processors in the memory planes and the coefficients are calculated and treated in the same way. The next step involves multiplication and then taking sums in the memory planes and follows on naturally from the first two steps. However, the fourth step, involving the calculation of the surface and motion parameters, is best suited to serial computation and was carried out on an attached processor. The final step of backprojecting the interpretations was again performed pixel-wise on the processor array.

The processing above is quite complex but runs in near real time with one-to-one mapping of pixels onto processors. The process essentially takes evidence in two-dimensional image space and calculates the 2½ D scene parameter description and then checks it against the image data as there are ambiguities in the instantaneous flow field that can be resolved by knowing that the surface must be facing the image plane. The problem that has to be faced is that this computation needs to be embedded in a complete vision system for any application and the higher level processing does not have any obvious mapping onto the processor array. There are plans to implement parallel versions of PROLOG on both SIMD and MIMD architectures which will mean automatic use of the parallelism and high level specification by the programmer. However, changing from one mode of programming to another is likely to impair communication between levels so the uniformity of the connectionist scheme described above is very appealing. However, the current network algorithms are usually slow and a lot of research will be needed before they form the basis of practical systems.

The emphasis in computer vision research, as outlined in this chapter, is on obtaining three-dimensional descriptions of the scene and a great deal of progress towards principled solutions has been made for the lower levels of the visual system. However, until very recently there has been relatively little work on characterising the surfaces. The work by Brady, Ponce, Yuille, and Asada (1985) represents progress in this area. Also, matching the surface descriptions to three-dimensional models has been rather neglected, perhaps because reliable data from stereo and motion algorithms is only just becoming available. A lot of work has been done on the specification of three-dimensional models in the graphics field, but there is still a gap between this and the computer vision research. Current model-based vision systems tend to take three-dimensional models and predict two-dimensional descriptions that are instantiated in the image. However, it is the three-dimensional features that are invariant in the real world, not their two-dimensional projections. In addition, analysis of the combinatorics of constraints for matching two-dimensional data to two-dimensional models and three-dimensional data to three-dimensional models by Grimson and Lozano-Perez (1983) supports the argument for developing full three-dimensional matching. There has been some research on three-dimensional matching (for example, Faugeras and Hebert, 1983) and this is expanding rapidly, so it should soon be used in state of the art robotics and industrial vision systems.

Another expanding area in vision research is the development of a full stereo motion algorithm. There is a great deal of evidence that in human visual processing there is an important interplay between these modules in determining the structure and motion of the scene. New research in this area should also soon be exploited by general vision systems. The growing awareness of the importance of the temporal course of visual perception is also beginning to motivate some research into the incremental characterisation of a scene. For example, the work of Ullman (1984) builds up the three-dimensional structure from visual motion data over time. The central idea is that there is continuity of the scene over time and that the description should be gradually elaborated rather than building up from scratch moment by moment. There is also evidence in the psychophysical literature that suggests that different visual codes are lost at different rates, symbolic codes persisting longer than the iconic codes. It seems likely that useful savings can be made by exploiting the predictability in the visual signal. This is especially relevant as the enormity of the computation performed by the human visual system has become more apparent as research on the principles of visual perception has advanced.

In conclusion, it seems that although new computer architectures are making an impact on what can be achieved, there is still a long way to go before artificial systems will compete with the speed and flexibility of the human visual system. The demands of the computation required to perform even very specific visual tasks in real time outstrip current parallel machines. This fact, together with the requirement for automation involving visual sensors, means that computer vision applications are a major driving force in the development of faster and more powerful parallel architectures.

REFERENCES

Baker, H. and Binford, T.O. (1981) 'Depth from Edge and Intensity Based Stereo', *Proc. IJCAI–81,* Vancouver, 631–636.

Ballard, D.H. (1981) 'Generalising the Hough Transform to Detect Arbitrary Shapes', *Pat. Rec.,* 13, 111–122.

Ballard, D.H. and Brown, C.M. (1982) *Computer Vision,* Prentice-Hall.

Ballard, D.H., Hinton, G.E. and Sejnowski, T.J. (1983) 'Parallel Visual Computation', *Nature,* 306, 21–26.

Bartliff, D.J. (1983) 'An Automatic Procedure for Image Segmentation', *Pat. Rec. Lets.,* 1, 435–442.

Binford, T.O. (1981) 'Inferring Surfaces from Images', *Art. Intel.,* 17, 205–245.

Binford, T.O. (1982) 'Survey of Model-based Image Analysis Systems', *Int. J. of Rob. Res.,* **1**, No. 1, 18–64.

Blake, A. (1984) 'Reconstructing a Visible Surface', *Proc. AAAI–84,* Texas.

Bolles, R.C. and Cain, R. (1982) 'Recognising and Locating Partially Visible Objects: the Local-feature-focus Method', *Int. J. of Rob. Res.,* **1**, No. 3, 57–82.

Bolles, R., Horaud, P. and Hannah, M. (1983) '3DPO: a Three-dimensional Part Orientation System', *Proc. IJCAI–83,* Karlsruhe, 1116–1120.

Brady, M. (1982) 'Computational Approaches to Image Understanding', *Comp. Surv.,* 14, 1–71.

Brady, M. (1983) 'Criteria for Shape Representations', *Human and Machine Vision,* Beck and Rosenfeld, eds., Academic Press.

Brady, M., Ponce, J., Yuille, A. and Asada, H. (1985) 'Describing Surfaces', *Int. Symp. Rob. Res.,* Hanafusa and Inoue, eds., MIT Press.

Brady, M. and Yuille, A. (1983) 'An Extremum Principle for Shape from Contour', *Proc. IJCAI–83,* Karlsruhe, 969–972.

Brooks, M.J. (1978) 'Rationalising Edge Detection', *Comp. Graph. and Im. Proc.,* 277–285.

Brooks, R. (1981) 'Symbolic Reasoning Among 3-D Models and 2-D Images', *Art. Intel.,* 17, 285–348.

Brown, C.M. (1981) 'Some Mathematical and Representational Aspects of Solid Modelling', *IEEE Trans. on Pat. Anal. and Mach. Intel.,* **3**, No. 4, 444–453.

Brown, C.M., Curtiss, M.B. and Sher, D.B. (1983) 'Advanced Hough Transform Implementations', *Proc. IJCAI–83,* Karlsruhe, 1081–1086.

Bruss, A.R. (1981) 'The Image Irradiance Equation: Its Solution and Application', *PhD dissertation,* AI Lab, MIT.

Burt, P. and Julesz, B. (1980) 'Modifications of the Classical Notion of Panum's Fusional Area', *Perception,* 9, 671–682.

Buxton, B.F. and Buxton, H. (1983) 'Monocular Depth Perception from Optical Flow by Space-time Signal Processing', *Proc. Roy. Soc. Lond.,* B218, 27–47.

Buxton, B.F., Buxton, H., Murray, D. and Williams, N.S. (1984) '3-D Solutions to the Aperture Problem', *Proc. ECAI–84,* Pisa, Italy, O'Shea, T. (ed.), 631–640.

Buxton, B.F., Buxton, H., Murray, D. and Williams, N.S. (1985) 'Machine Perception of Visual Motion' to appear in *GEC J. of Res.*

Buxton, B.F. and Murray, D. (1985) 'Optic Flow Segmentation as an Ill-posed Problem and Maximum Likelihood', submitted to *Image and Vision Computing.*

Canny, J.F. (1982) 'Finding Edges and Lines in Images', *AI Memo 720,* AI Lab, MIT.

Cowie, R.I.D. (1983) 'The Viewer's Place in Theories of Vision', *Proc. IJCAI–83,* Karlsruhe, 1952–1958.

Duda, R.O. and Hart, P.E. (1972) 'Use of the Hough Transform to Detect Lines and Curves in Pictures', *Comm. ACM 15,* 11–15.

Faugeras, O.D. and Hebert, M. (1983) '3D Recognition and Positioning Algorithm Using Geometrical Matching between Primitive Surfaces', *Proc. IJCAI–83,* Karlsruhe, 996–1002.

Faux, I. and Pratt, M. (1979) *Computational Geometry for Design and Manufacture,* Ellis Horwood.

Foley, J.D. and van Dam, A. (1982) *Fundamentals of Interactive Graphics,* Addison-Wesley.

Fox, J. and Mayhew, J. (1978) 'Texture Discrimination and the Analysis of Proximity', *Perception,* 8, 75–91.

Gibson, J.J. (1950) *The Perception of the Visual World,* Houghton-Mifflin, Boston.

Grimson, W.E.L. (1981) *From Images to Surfaces,* MIT Press, Cambridge, MA.

Grimson, W.E.L. and Lozano-Perez, T. (1983) 'Model-based Recognition and Localisation from Sparse Range or Tactile Data', *AI Memo 738,* AI Lab, MIT.

Haralick, R.M. (1979) 'Statistical and Structural Approaches to Texture', *Proc. IEEE 67,* 786–804.

Haralick, R.M. (1980) 'Edge and Region Analysis for Digital Image Data', *Comp. Graph. and Im. Proc.,* 12, 60–73.

Haralick, R., Watson, L. and Laffey, T. (1983) 'The Topographic Primal Sketch', *Int. J. of Rob. Res.,* **2**, No. 1, 50–72.

Herskovitz, A. and Binford, T.O. (1970) 'On Boundary Detection', *AI Memo,* AI Lab, MIT.

Hildreth, E.C. and Ullman, S. (1982) 'The Measurement of Visual Motion', *AI Memo 699,* AI Lab, MIT.

Hinton, G.E., Sejnowski, T.J. and Ackley, D.H. (1984) 'Boltzmann Machines: Constraint Satisfaction Networks that Learn', *Tech. Rep. CS–84–119,* CMU.

Hogg, D. (1983) 'Model-based Vision: a Program to See a Walking Person', *Im. and Vis. Comp.,* **1**, No. 1, 5–20.

Hopfield, J.J. (1982) 'Neural Networks and Physical Systems with Emergent Collective Computational Abilities', *Proc. Nat. Ac. of Sc.,* USA 79, 2554–2558.

Horn, B.K.P. (1973) 'The Binford-Horn Line-finder', *AI Memo 285,* AI Lab, MIT.

Horn, B.K.P. (1975) 'Obtaining Shape from Shading Information', *The Psychology of Computer Vision,* Winston, P.H. (ed.), McGraw-Hill.

Horn, B.K.P. (1977) 'Understanding Image Intensities', *Art. Intel.,* 8, 201–231.

Horn, B.K.P. and Schunk, B.G. (1980) 'Determining Optical Flow', *AI Memo 572,* AI Lab, MIT.

Horn, B.K.P. and Sjoberg, R.W. (1979) 'Calculating the Reflectance Map', *Applied Optics,* 18, 1770–1779.

Horn, B.K.P., Woodham, R.J. and Silver, W.M. (1978) 'Determining Shape and

Reflectance Using Multiple Images', *AI Memo 490,* AI Lab, MIT.
Hubel, D.H. and Wiesel, T.M. (1966) 'Receptive Fields, Binocular Interaction and Functional Architecture in the Cat's Visual Cortex', *J. of Physiol.,* 160, 106–154.
Ikeuchi, K. and Horn, B.K.P. (1981) 'Numerical Shape from Shading and Occluding Boundaries', *Art. Intel.,* 17, 141–184.
Jain, R. (1983) 'Direct Computation of the Focus of Expansion', *IEEE PAMI-5,* 58–64.
Jerain, C. and Jain, R. (1984) 'Determining Motion Parameters for Scenes with Translation and Rotation', *IEEE PAMI-6,* 523–530.
Julesz, B. (1960) 'Binocular Depth Perception of Computer Generated Patterns', *Bell Sys. Tech. J.,* 39, 1125–1162.
Julesz, B. (1975) 'Experiments in the Visual Perception of Texture', *Sc. Am.,* 232, 24–63.
Kanade, T. (1980) 'A Theory of Origami World', *Art. Intel.,* 13, 277–311.
Kelly, D.H. (1979) 'Motion and Vision II. Stabilised Spatio-temporal Threshold Surface', *J. of Opt. Soc. of Am.,* 69, 1340–1349.
Kender, J.R. (1976) 'Saturation, Hue and Normalised Colour: Calculation, Digitisation Effects and Use', *Tech. Rep.,* Department of Computer Science, CMU.
Kender, J.R. (1980) 'Shape from Texture', *PhD dissertation,* Department of Computer Science, CMU.
Kikpatrick, S. Gelatt, C.D. and Vecchi, M.P. (1983) 'Optimisation by Simulated Annealing', *Sc.,* 220, 671–680.
Kuan, D. and Drazovich, R. (1983) 'Model-based Interpretation of Range Imagery', *Proc. AAAI-83,* Washington, 210–215.
Land, E.H. (1977) 'The Retinex Theory of Colour Vision', *Sc. Am.,* 108–128.
Lawton, D.T. (1981) 'Optic Flow Field Structure and Processing Image Motion', *Proc. IJCAI-81,* Vancouver, 169–179.
Lee, D.N. (1980) 'The Optic Flow Field: The Foundation of Vision', *Phil. Trans. Roy. Soc. Lond.,* B290, 169–179.
Longuet-Higgins, H.C. and Prazdny, K.F. (1980) 'The Interpretation of Moving Retinal Images', *Proc. Roy. Soc. Lond.,* B208, 385–387.
Lowe, D. and Binford, T.O. (1981) 'The Interpretation of Three-dimensional Structure from Range Imagery', *Proc. IJCAI-81,* Vancouver, 613–618.
Lu, S. and Fu, K.S. (1978) 'A Syntactic Approach to Texture Analysis', *Comp. Graph. and Im. Proc.,* 7, 303–330.
Marr, D. (1976) 'Early Processing of Visual Information', *Phil. Trans. Roy. Soc. Lond., B275, 483–524.*
Marr, D. (1982) Vision, Freeman, San Fransisco.
Marr, D. and Hildreth, E.C. (1980) 'Theory of Edge Detection', *Proc. Roy. Soc. Lond.,* B207, 187–217.
Marr, D. and Nishihara, H.K. (1978) 'Representation and Recognition of the Spatial Organization of Three-dimensional Structure', *Proc. Roy. Soc. Lond.,* B200, 269–294.
Mayhew, J.E.W. (1983) 'Stereopsis', *Physical and Biological Processing of Images,* Braddick and Sleigh (eds.), Springer-Verlag, 204–216.
Nagel, H.H. (1983) 'Constraints for the Estimation of Displacement Vector Fields from Image Sequences', *Proc. IJCAI-83,* Karlsruhe, 945–951.
Ohlander, R.B. (1975) 'Analysis of Natural Scenes', *PhD dissertation,* Department of Computer Science, CMU.

Ohlander, R.B., Price, K. and Reddy, D.R. (1978) 'Picture Segmentation Using a Recursive Region Splitting Method', *Comp. Graph. and Im. Proc.*, 8, 313–333.

Okino, N. (1976) 'TIPS–1 / TIPS–2 Project: Relationally Structured Processor Base CAD/CAM System', *Proc. 4th NSF/RANN Conf. on Prod. Res. and Tech.*, Chicago.

Pentland, A.P. (1983) 'Fractal Based Description', *Proc. IJCAI–83,* Karlsruhe, 973–981.

Pollard, S., Mayhew, J. and Frisby, J. (1985) *Perception* (in press).

Popplestone, R., Ambler, A. and Bellos, I. (1978) 'RAPT – A Language for Describing Assemblies', *Ind. Rob.,* **5**, No. 3.

Prazdny, K.F. (1980) 'Egomotion and Relative Depth Map from Optical Flow', *Biol. Cyb.,* 36, 87–102.

Reiger, J.H. and Lawton, D.T. (1983) 'Sensor Motion and Relative Depth from Difference Fields of Optic Flows', *Proc. IJCAI–83,* Karlsruhe, 1027–1031.

Requicha, A. (1980) 'Representations for Rigid Solids: Theory, Methods and Systems', *Comp. Surv.,* **12**, No. 4, 437–464.

Selfridge, P.G., Sloan, D.R. (1981) 'Reasoning about Images: Application to Aerial Image Understanding', *Proc. Im. Understanding Workshop,* Baumann, L.S. (ed.), McLean, VA, 1–6.

Silver, W.M. (1980) 'Determining Shape and Reflectance Using Multiple Images', *MSc dissertation,* AI Lab, MIT.

Stiny, G. and Gips, J. (1972) *Algorithmic Aesthetics: Computer Models for Criticism and Design in the Arts,* Univ. of California Press, Berkeley, CA.

Stevens, K.A. (1980) 'Surface Perception from Local Analysis of Texture and Contour', *AI Tech. Rep. 512,* AI Lab, MIT.

Tennenbaum, J.M. and Barrow, H.G. (1977) 'Experiments in Interpretation Guided Segmentation', *Art. Intel.,* 8, 241–274.

Terzopoulos, D. (1982) 'Multi-level Reconstruction of Visual Surfaces', *AI Memo 671,* AI Lab, MIT.

Touvetzky, D. and Hintar, G.E. (1985) 'Symbols Among the Neurons: Details of a Connectionist Inference Architecture', to appear in *Proc. IJCAI–85,* Los Angeles.

Ullman, S. (1979) *The Interpretation of Visual Motion,* MIT Press, Cambridge, MA.

Ullman, S. (1984) 'Maximising Rigidity: the Incremental Recovery of 3-D Structure from Rigid and Non-rigid Motion', *Perception,* 13, 255–274.

Waxman and Wohn (1984) 'Contour Evolution, Neighbourhood Deformation and Global Image Flow', *CAR–TR–58,* Univ. of Maryland.

Witkin, A.P. (1981) 'Recovering Surface Shape and Orientation from Texture', *Art. Intel.,* 17, 17–47.

Witkin, A.P. (1983) 'Scale-space Filtering', *Proc. IJCAI–83,* Karlsruhe, 1019–1022.

Woodham, R.J. (1981) 'Analysing Images of Curved Surfaces', *Art. Intel.,* 17, 117–141.

Yin, B. (1983) 'A Framework for Handling Vision Data in RAPT', *Proc. IJCAI–83,* Karlsruhe, 814–820.

Yuille, A.L. (1984) 'The Smoothest Velocity Field and Token Matching', *AI Memo 724,* AI Lab, MIT.

Zucker, S.W. (1976) 'Toward a Model of Texture', *Comp. Graph. and Im. Proc.,* 5, 190–202.

Chapter 6

Algorithms and Architectures for Digital Synthetic Aperture Radar (SAR) Processing

B. ARAMBEPOLA and S. R. BROOKS

6.1 INTRODUCTION

Over the last two decades man has grown accustomed to viewing his world from space, either through the photographic cameras held by orbiting astronauts, or through sophisticated imaging systems, operating within a number of spectral ranges, spanning the electromagnetic spectrum. Significant developments in the use of remote earth sensors were introduced by the TIROS series of orbiting meteorological satellites. TIROS–1, launched in 1962, was a precursor to the Improved TIROS Operational System (ITOS) of the early 1970s. More recently, the Geostationary Operational Environmental Satellite (GOES) forms part of a network of five such satellites, including the European Space Agency (ESA) METEOSAT, which provides a global weather observation system. The Earth Resources Technology Satellite (ERTS–1), launched in 1972, opened new developments in earth observation from space.

Trends towards imaging systems with higher performance, through increased spatial resolution and extending the part of the electromagnetic spectrum used for observation, were continued in the NASA oceanographic satellite SEASAT. In using the radar frequency band, SEASAT provided a notable departure from the techniques used by earlier satellite imaging sensors. Among six sensing instruments, SEASAT used a synthetic-aperture radar (SAR) operating at a frequency of 1.3 GHz, designated as L-Band, to image the earth's surface at a resolution of 25 m × 25 m over a continuous strip of width 100 km. In generating a new view of the earth, SEASAT was an outstanding success. The radar being an active system was able to image independently of the time of day, and had the additional advantage of operating successfully in all weather and cloud conditions since L-Band transmission is little affected by atmospheric propagation factors.

SEASAT generated spectacular images of the earth over both land and sea surfaces (Bernstein, 1982). An example from SEASAT is shown in Fig. 6.1. The image shows an area from the English Channel. A small section of coastline is visible to the top right of the image. The image clearly shows tidal currents, interspersed with well developed wave trains. Numerous small ships, with their associated wakes, can be detected within the image. This image graphically demonstrates the potential of radar imaging systems

Fig. 6.1 A SEASAT SAR image of a portion of the English channel, from a height of 800 km. (*Courtesy of ESA/EARTHNET.*)

operating from satellite platforms. Because of the high data rates associated with the SEASAT radar, imaging could only be achieved when the satellite was in line-of-sight contact with a ground receiver.

The SEASAT orbiting radar has been flown on space shuttle flights in the Shuttle Imaging Radar (SIR) series. The shuttle system included real-time on-board data recording, and hence was able to image a greater variety of terrain, limited only by orbit considerations and the flight duration. Projected missions, such as the European Earth Resources Satellite, ERS–1, will continue to develop and enhance planetary observation techniques at radar frequencies.

The high spatial resolution in these radar systems is achieved by a technique called *aperture synthesis*. The technique enables much finer resolution to be achieved than would be possible with a conventional side-looking radar. In a side-looking airborne radar (SLAR), an antenna which is fixed parallel to the track directs a radar beam broadside and downwards from the platform track as shown in Fig. 6.2. The beam is scanned by the

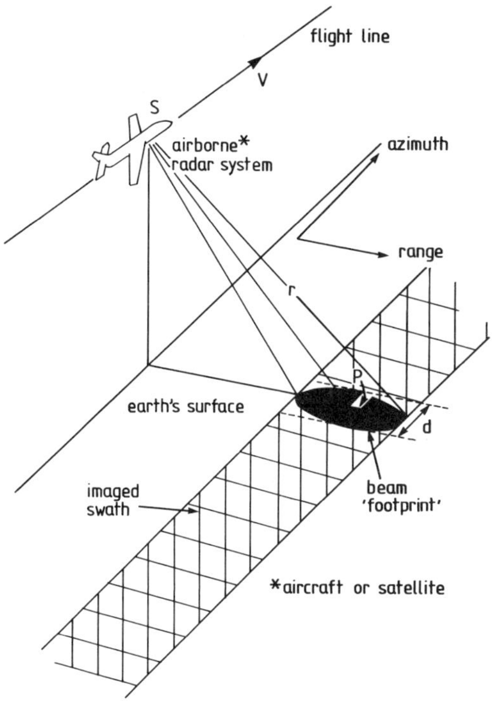

Fig. 6.2 SLAR imaging geometry.

motion of the platform so that the beam footprint is swept along a swath on the terrain surface. The radar receives and records the backscattered energy from this swath, generating an image of the surface reflectivity. The spatial resolution of the image is determined by the pulse width and the radar beamwidth in the range (or cross-track) direction. While the pulse width can be narrowed, and a finer range resolution achieved, improvement in the spatial resolution in the azimuth resolution can only be enhanced by extending the real radar aperture, assuming that the frequency of transmission is maintained constant.

In synthetic aperture radar (Hovanessian, 1980), Doppler processing is used to achieve high spatial resolution in the direction of motion of the radar platform. Each pixel in the SAR image is generated by coherently processing a large number of radar returns. Hence real-time SAR imaging requires very high processing throughputs and large memory requirements, well beyond the capability of current digital radar processors. Optical techniques have been widely used for processing SAR data. A good review of these techniques is found in Kozma *et al.* (1972). However, these techniques do not have the accuracy and adaptability properties, inherent in digital systems, which are essential for high precision SAR imaging. Hence with recent advances in VLSI technology there has been a growing interest in developing digital techniques and special purpose systems for processing SAR data.

In this chapter SAR signal processing operations are reviewed and methods of implementing these in digital systems are discussed.

A system for collecting SAR data is shown in Figs. 6.2 and 6.4(a). The radar platform travels at a velocity V relative to the Earth's surface. It transmits radar pulses at regular intervals and collects the returns. Consider the response of a point target P on the Earth's surface to these radar pulses. The transmitted waveform is usually a linear or hyperbolic frequency modulated waveform of duration τ, as shown in Fig. 6.3. Hence the radar return, which is delayed by $(2r/c)$, where c is the speed of light, also takes a similar form and is spread over a time duration τ. A waveform of this type is referred to as a 'chirp' waveform. The frequency modulation enables high range resolution to be achieved at low transmitter peak power. The high resolution is achieved in an analogous way to the azimuth compression process described below.

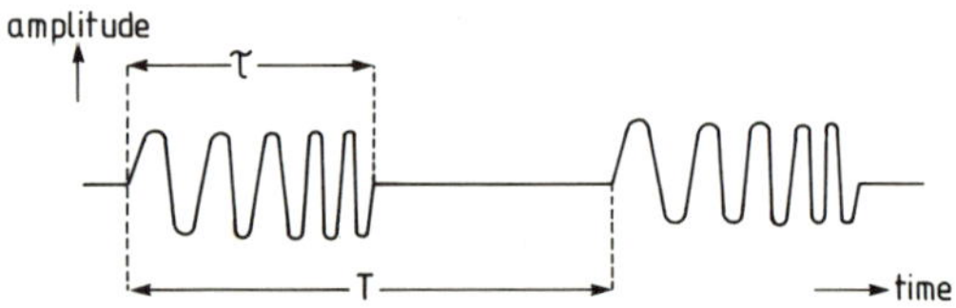

Fig. 6.3 Transmitted radar pulses.

The point target remains within the radar beam during the time it takes for the platform to travel a distance d. Since the interpulse period T is much smaller than (d/V), the point target responds to many successive radar pulses. Due to the relative velocity between the radar platform and the point target P, these radar returns are shifted in frequency. In order to obtain approximate expression for this Doppler frequency shift, refer to Figs. 6.2 and 6.4(a).

The distance r between the radar platform and P is

$$r = \sqrt{(r_0^2 + V^2t^2)}$$

A second order approximation to this is

$$r \simeq \left(r_0 + \frac{(Vt)^2}{2r_0}\right)$$

Hence the phase shift between transmitted and received signals is

$$\phi = \frac{-2}{\lambda}\left(r_0 + \frac{(Vt)^2}{2r_0}\right)$$

where λ is the radar wavelength, r_0 is the range perpendicular to the flight line and t is the elapsed time from when the platform passed its closest point of approach to the target.

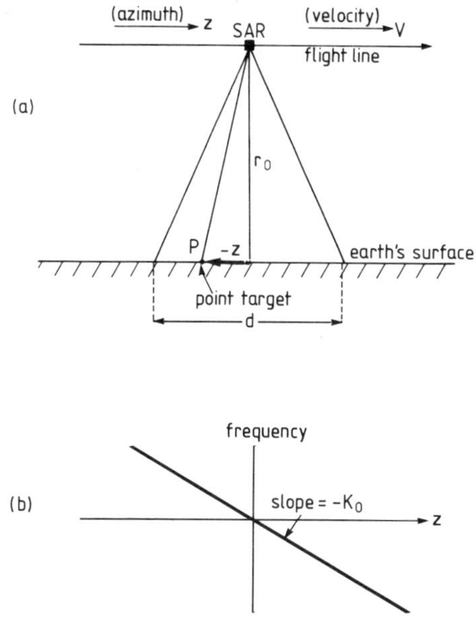

Fig. 6.4 (a) Configuration to obtain the azimuth point target response; (b) the point target response in the azimuth dimension.

The Doppler frequency shift between the transmitted and received pulses is given by

$$f_D = \frac{d\phi(t)}{dt} = -\frac{2V^2}{\lambda r_0}t$$

From the above expression it is observed that the azimuth spread of the point target response approximates a linear frequency modulated waveform. The Doppler frequency shift is highest when P enters the radar beam. This decreases linearly with time until it becomes negative and reaches a minimum before P moves out of the area of illumination. Assuming an antenna illumination pattern $A(t)$, the demodulated baseband representation of the azimuth spread function takes the form of Equation 6.1.

$$R\,(r_0, t) = A(t)\exp\left(j\,2\pi\frac{K_0}{2}t^2\right) \tag{6.1}$$

where the linear FM rate K_0 is given by Equation 6.2.

$$K_0 = -\frac{2V^2}{\lambda r_0} \tag{6.2}$$

This linear frequency modulated response is illustrated in Fig. 6.4 (b).

Although the expression for the azimuth response has been derived as a continuous function in Equation 6.1, due to the pulsed method of collecting

SAR data, only a sampled version of this response is obtained. The sampling frequency of the discrete function is equal to $(1/T)$. This yields.

$$R_a(r_0, t) = \sum_{i=-\alpha}^{\alpha} A(t) \exp\left(j\, 2\pi \frac{K_0}{2} t^2 \right) . \delta(t - iT) \qquad (6.3)$$

In designing SAR systems it is very important to select the pulse repetition frequency to ensure no significant aliasing of the Doppler frequency shifts. It is important to note that the linear FM rate K_0 of the azimuth response given by Equation 6.3 is a function of the range of the target.

Combining the appropriate forms for the range and azimuth responses it can be shown that the response due to a point target is spread over a region of the range-azimuth plane as depicted in Fig. 6.5. The curvature in this region is caused by an effect known as range cell migration or range walk, which results mainly from the change in range with azimuth variable z. The object of processing the SAR signals is to compress this spread into a well focused point in the image plane. In the presence of additive uncorrelated noise this compression can optimally be carried out by correlating the radar returns with a replica of the ideal point target response. This technique is known as *matched filtering* (Rabiner and Gold, 1975, Chapter 13).

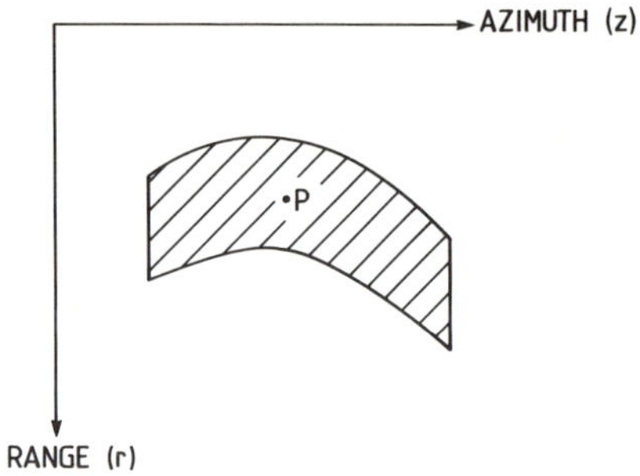

Fig. 6.5 Response to a point target in the range azimuth plane.

Since the point target response is two-dimensional (see Fig. 6.5), the pulse compression is essentially a two-dimensional correlation. However, the cross-coupling term in the two-dimensional operation is usually negligible and hence it can be decoupled into two one-dimensional operations, one in the range dimension and the other in the azimuth dimension.

The return of each radar pulse is correlated with a replica of the transmitted waveform to compress the signal spread in the range dimension. The correlation reference is often windowed to reduce the sidelobe effects. As the reference remains unchanged over relatively long periods this correlation can be implemented using surface acoustic wave devices. However, it is often necessary to adapt this reference function to changes in the transmitted waveform. These changes may be caused by changes in waveform coding or by variations in the parameters of the signal generator. In such circumstances it is necessary to take advantage of the flexibility of digital systems. In its sampled form this correlation can be expressed as in Equation 6.4.

$$y(k,l) = \sum_{i=0}^{N-1} h^*(i) \, . \, x(k+i,\ell) \tag{6.4}$$

where x is the radar return due to the ℓth pulse, $h(i)$ the transmitted waveform and y the range compressed signal. Note that the signals are sampled after demodulating to the baseband and that the sequences x, h and y are typically complex. The sequence $h(i)$ usually takes the linear frequency modulated form given in Equation 6.5:

$$h(i) = \omega(i) \exp\left(- \mathrm{j}2\pi \frac{K_r}{2} T_r^2 i^2\right) \tag{6.5}$$

where K_r is the linear FM rate, and T_r the sampling period. $\omega(i)$ is a weighting function which is zero outside the pulse duration τ. The spatial resolution that can be achieved in the range dimension can be increased by increasing the bandwidth of the transmitted pulse.

Range correlation removes the signal spread in the range dimension and compresses the point target spread into a curved trajectory shown in Fig. 6.6. This curvature is caused by the change in distance of the point P from the radar platform as P traverses the region of illumination. This is referred to as range cell migration or range walk, and in the case of an orbiting satellite radar, range walk caused by the earth's rotation can produce a significant effect.

Due to the range cell migration effect it is not possible to carry out azimuth compression simply by correlating the radar signal with a one-dimensional reference in the azimuth dimension. Instead, azimuth processing has to be treated as a two-dimensional operation. However, implementing azimuth compression using a two-dimensional correlation leads to increased memory and computational requirements. A further problem is caused by

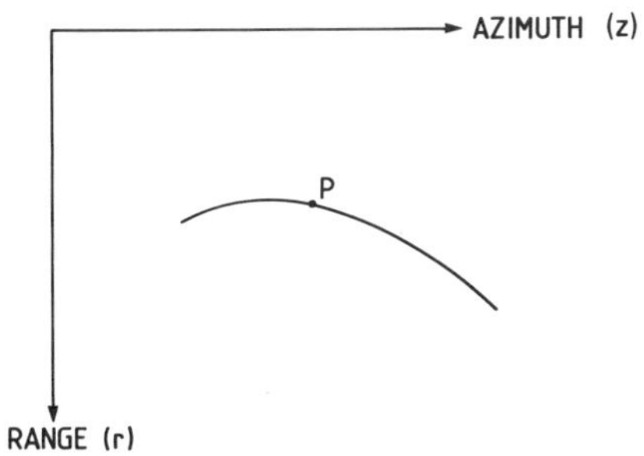

Fig. 6.6 Point target spread after range compression, showing the range cell migration effect.

the change in the reference function with the range variable r. Due to this it is necessary to make the data block size in the range dimension relatively small, which results in inefficient block processing. Hence it is common to implement the azimuth compression as a one-dimensional correlation, with corrections for the range cell migration effects.

Consider the range cell migration effect due to two targets, P and P_1, separated only in the azimuth dimension, shown in Fig. 6.7(a). In order to correct the range cell migration it is necessary to straighten out each of these curves into straight lines parallel to the azimuth axis. However, as the curves are different for each azimuth point it is inefficient to implement this in the time domain, especially with block processing. It can be shown that all these

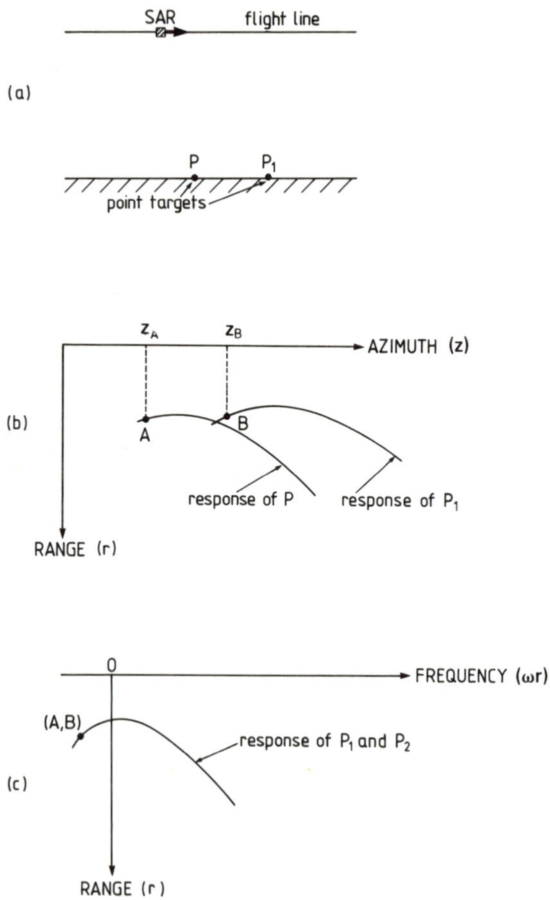

Fig. 6.7 The range walk effect. (a) Two point targets at constant range. (b) The range azimuth plane. (c) The azimuth-frequency plane.

curves map into a single curve in the azimuth frequency domain. In order to illustrate this, consider two points A and B on a r = constant line in the two curved trajectories in Fig. 6.7(a).

The Doppler frequency shifts at $z = z_A$ and $z = z_B$ due to targets P and P_1, respectively, are identical and these two points map onto a single point in the azimuth Doppler frequency domain. This argument applies to all the points in both curves and hence they map onto a single curve in the frequency domain as depicted in Fig. 6.7(c). Hence straightening this trajectory in the frequency domain will straighten the trajectories due to all targets at a distance r_0 from the path of the radar platform. The technique is therefore very useful in block processing. The straightening is carried out using interpolation and moving samples so that the curve in Fig. 6.7(b) represents a straight line $r = r_0$ in the r–z plane. In practice it is more accurate to carry out this range cell migration correction in two stages by splitting the range walk into two components, a linear one and a residual quadratic one. The correction for the linear range walk effects is applied in the time domain and the residual quadratic component is corrected in the frequency domain.

After correcting for the range cell migration effect, azimuth compression is carried out. As in range compression, this is achieved by correlating the signal in each range cell (i.e. r = constant line) with the reference given by Equation 6.3. This reference has to be updated fairly frequently with the range variable r.

It should be noted that this range parameter r is the distance between the path of the platform and the point target measured along a slanted plane (see Fig. 6.2). Hence after processing it is necessary to convert this range variable into a range variable measured along the surface of the earth.

The SAR signal processing operations described above can be implemented in the time domain, or in the frequency domain by Fourier transforming the signals. The main advantage of frequency domain techniques is that they involve much fewer computations than time domain techniques. However, the computations in the time domain processing are usually well structured and hence map very well into regular VLSI architectures. Furthermore, frequency domain techniques rely on processing large data blocks in order to achieve their computational advantages. Hence frequency domain SAR processing entails much larger system latencies and memory requirements than time domain processing. However, both methods require a considerable amount of memory to store many radar returns in order to process these coherently in the azimuth dimension.

Both these methods of processing SAR signals are described in this chapter. Frequency domain techniques are considered in more detail for two reasons. First, VLSI techniques for time domain signal and image processing have already been discussed in previous chapters. Secondly, most current fast SAR processors are based on Fourier techniques, mainly because Fourier processing involves much fewer arithmetic operations. Another reason is that it is much simpler to correct for range cell migration effects in the frequency domain.

6.2 SAR PROCESSING ALGORITHMS

6.2.1 Range processing

Let the signal reflected due to the i_zth radar pulse be denoted by $x(i_r, i_z)$. It is assumed that the received signal has been demodulated into baseband in-phase (I) and quadrature (Q) channels, and sampled at regular intervals to result in a complex data sequence. The signal is available in a sampled form in the azimuth dimension owing to the pulsed form of the radar imaging process. The first stage of SAR processing involves a pulse compression in the range dimension. This is achieved by correlating the i th reflection, namely $x(i_r, i_z)$ for $i = 0,1, \ldots ,N_r - 1$ and i_z = constant, with a replica of the transmitted waveform $h(i)$, as given in Equation 6.6.

$$y(i_R, i_z) = \sum_{i_z=0}^{N_r-1} h^*(i_r) \cdot x(i_R + i_r, i_z) \qquad (6.6)$$

In a digital system the correlation of Equation 6.6 can be carried out in the time or frequency domain. These techniques are studied in detail in the next section. Flexibility of digital systems allows easy and rapid updating of the reference $h(i)$. This updating may be required owing to changes in waveform coding or parametric changes in the signal generator.

6.2.2 Azimuth processing

Azimuth processing also involves a correlation operation:

$$Y(i_R, i_Z) = \sum_{i_z=0}^{N_z-1} g^*(i_z)\, y(i_R, i_z + i_Z) \qquad (6.7)$$

where $g(i_z)$ is the azimuth point target spread function. It has been shown earlier in this chapter that g can be approximated by a linear frequency modulated waveform.

The range and azimuth correlation operations are carried out in orthogonal dimensions. Hence a large amount of memory is required to store SAR data for range and azimuth processing. Fast semiconductor memories of current processors are usually incapable of holding the two-dimensional array required for efficient processing in range and azimuth dimensions. Hence large bulk storage devices such as disks are used to store SAR data between range and azimuth processing stages. The stages of processing involved in processing SAR data to result in a high resolution image are depicted in the flow diagram of Fig. 6.8.

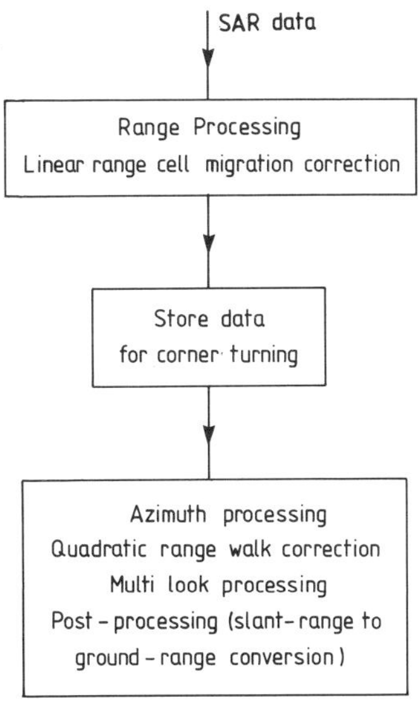

Fig. 6.8 SAR processing stages.

It has been shown in Section 6.1 that the point target spread functions approximate a linear frequency modulated waveform. Hence it is possible to implement azimuth compression by mixing the signal in each range cell with a linear chirp and then analysing its spectrum. Consider a target at an azimuth location $z = z_1$. After range compression the response due to this target is a chirp with the distance variable $(z-z_1)$. The effect of mixing this with a reference chirp function is shown in Equation 6.8.

$$K \exp(-k(z-z_1)^2)\exp(-kz^2) = K \exp(-2kzz_1-kz_1{}^2) \qquad (6.8)$$

The mixing operation converts the linear FM wave into a sinusoid with a frequency linearly proportional to the azimuth co-ordinate z_1. Hence this location can be obtained by analysing the spectrum of the mixed signal.

The coherent nature of the SAR image formation process gives rise to a considerable amount of noise in the image, known as *speckle noise.* This noise can be reduced by incoherently averaging several independent images of the same scene. Several independent images can be formed by segmenting the azimuth Doppler spread into several non-overlapping sections or 'looks' and processing each look independently to yield an image with reduced resolution. These can then be averaged incoherently to reduce speckle noise. The multi-look processing enhances the image signal-to-noise ratio at the expense of azimuth spatial resolution.

The final processing stage in the SAR image formation process is the post-processing stage. This involves processing operations such as slant-range to ground-range conversion, corrections for non-orthogonality of the range and azimuth axes, rotating the processed image to a specified orientation and multilook processing. The range parameter in the radar imaging process is the distance measured in an inclined plane containing the path of the platform and the point target. Hence the range=constant lines are usually not uniformly spaced in the ground plane. Hence an interpolation operation is needed to generate an image with samples spaced at uniform intervals. This is referred to as slant-to-ground range conversion. Both frequency domain and time domain techniques for SAR processing are described in this chapter. Since frequency domain techniques involve the computation of the discrete Fourier transform (DFT) (see Section 2.4.6), algorithms for computing the DFT are studied in the next section.

6.2.3 Algorithms for computing the DFT

The basic signal processing operations required for SAR processing are discrete Fourier transforms and convolutions. Algorithms for computing the DFT are described in this section and the convolution algorithms are studied in Section 6.2.4. These DFT algorithms are also often used in implementing digital convolutions. The N-point discrete Fourier transform (DFT) and the inverse discrete Fourier transform (IDFT) pair has been defined as in Equations 6.9.

$$X(k) = \sum_{i=0}^{N-1} x(i)\, W_N^{ik} \tag{6.9a}$$

$$x(i) = N^{-1} \sum_{k=0}^{N-1} X(k)\, W_N^{-ik} \tag{6.9b}$$

where $W_N = \exp(-j2\pi/N)$ and $x(i)$ and $X(k)$ are the N-point complex sequences in the time and frequency domains, respectively. We consider only the algorithms for computing the DFT since the IDFT can be computed using the same algorithms by conjugating the input and the output.

Algorithms for computing the DFT gain their computational advantages by converting a long DFT into a set of low order DFTs. This conversion is carried out by a set of mappings on the indices i and k in Equations 6.9. Two classes of mappings have been developed for this purpose and the corresponding algorithms have been referred to as the common and prime factor algorithms.

Common factor algorithms
Let $N=N_1.N_2$ be a factorisation of the transform length N in Equations 6.9. Let two index mappings be defined as in Equations 6.10.

$$i = i_2 + i_1.N_2 \tag{6.10a}$$
$$k = k_2.N_1 + k_1 \tag{6.10b}$$

Here i_1 and k_1 are in the range $[0,N_1)$ and i_2 and k_2 are in the range $[0,N_2)$. Substituting the expressions for i and k given in Equations 6.10 into Equations 6.9 yields Equation 6.11.

$$X(k_1,k_2) = \sum_{i_2=0}^{N_2-1} W_{N_2}^{\,i_2k_2}\,(W_N^{\,i_2k_1} \sum_{i_1=0}^{N_1-1} x(i_1,i_2)W_{N_1}^{\,i_2k_1}) \tag{6.11}$$

Here $x(i_1,i_2)$ is obtained by rearranging the sequence $x(i)$ as a two-dimensional array of size $(N_1 \times N_2)$ using Equation 6.10(a). Equation 6.9 can then be evaluated by evaluating the N_1-point and N_2-point row and column DFTs as in Equation 6.11. A set of point-wise multiplications by $W_N^{\,i_2k_1}$ has to be performed on the data between the two stages of DFTs. These have been referred to as the 'twiddle-factor' multiplications. The output is obtained as a two-dimensional array $X(k_1,k_2)$. This array can then be rearranged as a one-dimensional array using Equation 6.10(b). Owing to the difference in the index mappings of Equation 6.10(a) and Equation 6.10(b), the column-row ordering at the input is mapped into a row-column ordering at the output. This common factor algorithm has been more commonly referred to as the fast Fourier transform (FFT) algorithm. The data reordering effect mentioned above has been referred to as the 'bit-reversed' ordering effect of the FFT.

The choice of $N_1=2$ and $N_2=N/2$ in Equations 6.10 and 6.11 leads to the 'decimation in frequency' (DIF) radix-2 FFT algorithm. The first stage of this algorithm consists of $(N/2)$ 2-point DFTs and $(N/2)$ twiddle-factor multiplications. Since the 2-point DFT is simply an add/subtract operation, the basic computational kernel takes a form given in Equation 6.12.

$$\begin{aligned} x'(k) &= x(k) + x(k+N/2) \\ x'(k+N/2) &= (x(k) - x(k+N/2))\, W_N^{\,k} \end{aligned} \tag{6.12}$$

The computational kernel in Equation 6.12 is referred to as the DIF FFT butterfly operation. This equation, evaluated for $k=0,1, \ldots ,(N/2)-1$, completes the first stage of the FFT. The N-point DFT is then transformed into two $(N/2)$-point DFTs. These can further be factorised into four $(N/4)$-point DFTs using another similar stage of computations. Hence for $N=2^n$, the radix-2 DIF FFT algorithm contains n computational stages, $(N/2)(\log_2 N-1)$ complex multiplications and $N\log_2 N$ complex additions and subtractions. Higher radix algorithms as well as mixed-radix algorithms can be similarly developed. The factorisation $N_1=4$ and $N_2=N/4$ leads to the radix-4 FFT algorithm. The basic computational kernel of this contains a 4-point DFT and three complex multiplications. Since the 4-point DFT can be evaluated without any multiplications, this algorithm has 25% fewer multiplications than the radix-2 algorithm.

The choice of $N_1=N/2$ and $N_2=2$ leads to the other popular version of the FFT known as the radix-2 'decimation in time' (DIT) algorithm. The inner summation in Equation 6.11 then corresponds to a $(N/2)$-point DFT. Let us assume that the two $(N/2)$-point DFTs corresponding to $i_2=0$ and 1 have been computed and that the results of these computations are given by sequences $X_0(k)$ and $X_1(k)$, respectively.

These results can be combined as given in Equations 6.13 to yield the N-point DFT $X(k)$.

$$\begin{aligned} X(k) &= X_0(k) + W_N^{\,k}\, X_1(k) \\ X(k+N/2) &= X_0(k) - W_N^{\,k}\, X_1(k) \end{aligned} \tag{6.13}$$

Equations 6.13 are the basic computational kernel of the DIT FFT algorithm. The two $(N/2)$-point DFTs in the inner summation of Equation 6.11 can be similarly reduced to four $(N/4)$-point DFTs. This algorithm has the same number of arithmetic operations as the DIF FFT algorithm. However, the DIT algorithm is preferred for Fourier transform processors using multiplier accumulators.

Computational savings can be achieved by combining the decimation in time and decimation frequency operations within the same algorithm.

The differences in the input and output orderings in the FFT algorithms owing to the difference in the input and output index mappings have been observed earlier in this section. If the computations are carried out in-place the output is obtained in a bit-reverse order. However, if different memories are used for the inputs and outputs of each radix-2 stage it is possible to order the intermediate results of the computation process to obtain the output in the same order as the input. This has been referred to as the 'in-order FFT algorithm'. It is possible to configure the FFT algorithms in many ways, if it is not necessary to perform the computations in-place. One such algorithm, very useful in FFT processor implementation, is the constant geometry algorithm. In this algorithm the data addressing does not change from stage to stage of the FFT. Hence, simple addressing circuitry can be used to generate these data addresses. There is a technique for implementing the radix-2 FFT algorithm in-place and still avoid the bit-reverse ordering at the output (i.e. in-order). This is achieved by computing two butterfly operations, retaining their outputs in registers and interchanging these before storing them back in memory.

Prime factor algorithms

Consider the two index mappings given by Equations 6.14.

$$\begin{aligned} i &= (\alpha_1 N_2)\, i_1 + (\alpha_2 N_1)\, i_2 \\ k &= (\gamma_1 N_2)\, k_1 + (\gamma_2 N_1)\, k_2 \end{aligned} \tag{6.14}$$

Substituting for i and k in Equations 6.9 it can be shown that the N-point

DFT is transformed into a form given by Equation 6.15, provided $\alpha_1,\alpha_2,\gamma_1,\gamma_2$ satisfy the relations given by Equations 6.16. Equation 6.15 represents a $(N_1 \times N_2)$-point two-dimensional DFT.

$$X(k_1,k_2) = \sum_{i_2=0}^{N_2-1} \sum_{i_1=0}^{N_1-1} x(i_1,i_2)\, W_{N^2}{}^{i_1k_1}\, W_{N^2}{}^{i_2k_2} \tag{6.15}$$

$$\begin{aligned} \alpha_1\gamma_1 &= N_2{}^{-1} \bmod N_1 \\ \alpha_2\gamma_2 &= N_1{}^{-1} \bmod N_2 \end{aligned} \tag{6.16}$$

The relations in Equation 6.16 can be satisfied if, and only if, N_1 and N_2 are mutually prime numbers. Hence this algorithm has been referred to as the prime factor DFT algorithm. The two-dimensional DFT in Equation 6.15 can be evaluated using N_1 N_2-point DFTs and N_2 N_1-point DFTs, which is computationally simpler than computing the N-point DFT. The main difference between this algorithm and the CFA described in the earlier section is the absence of twiddle-factor multiplication in the PFA. Owing to this the PFA usually has fewer multiplications than the CFA. The main drawback of the PFA is the constraint that factors N_1 and N_2 of N have to be mutually prime. This makes the PFA more complex to implement than the CFA.

Equations 6.16 do not uniquely determine the coefficients $\alpha_1,\alpha_2,\gamma_1$ and γ_2 of the index mappings of Equations 6.14. Usually γ_1,γ_2 are chosen as unity and α_1,α_2 are chosen to satisfy Equations 6.16. This gives the well known Chinese Remainder Theorem (McClellan and Rader, 1979) index mapping for the index i, i.e. $i_1 = i \bmod N_1$ and $i_2 = i \bmod N_2$.

In general, if $N = N_1.N_2. \ldots .N_r$ and if $\gcd(N_i,N_j)=1$ for all $i \neq j$, the N-point DFT can be mapped into a $(N_1 \times N_2 \mathrm{x} \ldots \mathrm{x} N_r)$-point $r-$D DFT. Hence a long DFT can be mapped into a set of low-order ones of mutually prime order. This approach is efficient only if suitable algorithms can be derived for evaluating these low order DFTs.

Winograd has developed a set of efficient algorithms for evaluating low order DFTs. A brief review of these algorithms is given below. Let the length of the low order DFT be a prime number p. It can then be mapped into a $(p-1)$-point cyclic convolution using a technique originally developed by Rader (*ibid*). Let β be a generator of the multiplicative group modulo p. Then all integers from 1 to p can be expressed as distinct powers of the generator β. The p-point DFT can hence be transformed into $(p-1)$-point cyclic correlation given by Equation 6.17. The frequency component $X(0)$ is evaluated separately as the sum of all components $x(i)$.

$$X(\beta^v) = \sum_{u=0}^{p-2} x(\beta^u)\, W_N{}^{\beta^{(u+v)}} \qquad \text{for } v = 0,1,\ \ldots\ ,p-2 \tag{6.17}$$

A similar technique can be used to transform a p^r-point DFT into several low order cyclic correlations. These low order cyclic correlations can be expressed as polynomial products using the Z-transform. Winograd has developed an efficient technique for evaluating polynomial products using the Chinese Remainder Theorem for polynomials (*ibid*). A Winograd low order DFT algorithm for computing DFTs consists of the following computational stages:

Pre-weave stage. A stage of additions and subtractions which changes the length of the sequence from N to M.
Pointwise multiplication stage. M pointwise multiplications.
Postweave stage. A stage of additions and subtractions which results in the N-point DFT $X(k)$.

In matrix form these operations can be represented as in Equation 6.18.

$$\mathbf{X}(k) = \mathbf{C}\ \mathbf{B}\ \mathbf{A}\ \mathbf{x}(i) \qquad (6.18)$$

where **A** and **C** are preweave and postweave matrices of order $M \times N$ and $(N \times M)$, respectively. The elements of these matrices belong to the set $(-1,0,1)$. **B** is a $M \times M$ diagonal matrix containing the coefficients required for pointwise multiplication. A long DFT can be mapped into a set of low order DFTs using the PFA and these low order DFTs can then be computed using the Winograd algorithms. Winograd has achieved a further reduction in the number of multiplications by combining the pointwise multiplications in all low order DFTs into a single stage of pointwise multiplications. This algorithm has been referred to as the *nested algorithm* and has a form similar to the Winograd low order DFT algorithms.

The number of non-trivial multiplications in the nested Winograd algorithm for an $(N_1.N_2. \ldots N_t)$-point DFT, where $(N_i,N_j)=1$ for all $i \neq j$, is $(M_1.M_2. \ldots M_t)-(T_1.T_2. \ldots T_t)$. Here M_i and T_i are the total and trivial multiplications in an N_i-point Winograd low order DFT algorithm. (A trivial multiplication is defined as a multiplication by ± 1 or $\pm$j).

Winograd and prime factor algorithms have more complex computational structures than the FFT and hence are not very efficient for use in array processors. However, these algorithms involve fewer multiplications than the FFT and are expected to be more suitable for processor architectures based on digital signal processing (DSP) chips, where multiplication can be a relatively expensive operation.

6.2.4 Convolution algorithms

The simplest method of computing convolutions is the direct evaluation of the defining Equation 6.19.

$$y(l) = \sum_{i=0}^{N-1} h(i)\ x(l-i) \qquad (6.19)$$

A class of architectures known as systolic arrays (see Chapters 1 and 3) have been developed for implementing convolutions using the direct evaluation approach. These architectures are very regular and possess nearest neighbour connectivity. Hence these are well suited for VLSI implementation. The direct evaluation technique usually involves shorter latencies and requires less memory than the frequency domain block processing techniques.

The direct evaluation method requires N multiply-accumulate operations for each sample at the output, where N is the length of the reference function. The block processing techniques described below require only $O(\log_2 N)$ such operations per output sample. Hence for large values of N, block processing has fewer arithmetic operations than the direct evaluation technique. The use of these techniques therefore results in much smaller processing times, particularly in systems based programmable array processors.

Using the DFT

The DFT maps a cyclic convolution in the time domain into a set of pointwise multiplications in the frequency domain. An n-point cyclic convolution is obtained by defining $(l-i)$ modulo N in Equation 6.19. This is equivalent to periodically extending the sequences $h(i)$ and $x(i)$ outside the range $(0,N)$, with period N. This cyclic convolution can be computed using the DFT as follows.

N-point DFTs of the sequences $x(i)$ and $h(i)$
Pointwise multiplication of the two DFTs, i.e. $Y(k)=H(k).X(k)$
An N-point inverse DFT of the sequence $Y(k)$. In many applications the sequence $h(i)$ remains unchanged for many data sequences and hence can be precomputed and stored in the processor memory.

Frequency domain processing is very useful in SAR processing since it also simplifies the implementation of the range cell migration correction. As noted earlier in this chapter it is much simpler to correct for the range cell migration effect in the frequency domain.

The algorithms discussed in Section 6.2.3 can be used for computing the DFTs in the convolution algorithm. Although this technique involves fewer arithmetic operations than the direct evaluation technique, it requires more memory to store blocks of data and results in much longer processing latencies.

Most signal processing applications usually involve convolving a long data sequence with a relatively short reference sequence. In such cases the computation can be simplified using overlap-save or overlap-add techniques.

Using number theoretic and polynomial transforms

The polynomial representation of the N-point cyclic convolution in Equation 6.19 is given in Equation 6.20.

$$Y(z) = H(z).X(z) \bmod (z^N - 1) \qquad (6.20)$$

where $X(z)$ is defined by Equation 6.21.

$$X(z) = \sum_{i=0}^{N-1} x(i)\, z^i \tag{6.21}$$

$Y(z)$ and $H(z)$ are similarly defined. The polynomial $(z^N - 1)$ has the factorisation given by Equation 6.22.

$$(z^N - 1) = \prod_{i=0}^{N-1} (z - \alpha_N{}^k) \tag{6.22}$$

where α_N is a primitive Nth root of unity. The Chinese Remainder Theorem for polynomials can now be applied to transform the polynomial multiplication mod $(z^N - 1)$ in Equation 6.20 into a set of pointwise multiplications given by Equation 6.23.

$$Y_k = H_k . X_k \tag{6.23}$$

where

$$X_k = X(z) \bmod (z - \alpha_N{}^k) \tag{6.24}$$

Y_k and H_k are similarly defined. If the convolution is defined in the field of complex numbers, α_N is a complex exponential and the transformation in Equation 6.24 turns out to be the conventional DFT. However, the convolution in Equation 6.19 can be carried out by defining it in a finite 'ring' of integers, provided the ring is large enough to avoid overflows in the final results (intermediate overflows do not have an effect on the final result). Let this ring be chosen to contain an Nth primitive root of unity β. The transform in Equation 6.24 can now be defined using integer β as the transform kernel. This is referred to as a number theoretic transform (*ibid*). The number theoretic transform pair is given by

$$X(k) = \sum_{i=0}^{N-1} x(i)\, \beta^{ik} \text{ and}$$

$$x(i) = N^{-1} \sum_{k=0}^{N-1} X(k)\, \beta^{-ik}, \text{ in ring R} \tag{6.25}$$

where $\beta^i \neq 1$ for $N > i > 0$

and $\beta^N = 1$

An NTT exists in the ring of integers modulo L, only if the Nth primitive root of unity β is a factor of $(p_i - 1)$ for every prime p_i that divides L.

The definition of Equation 6.25 is very similar to that of the DFT. The only differences are:

the algebra is now defined in a finite ring
the transform kernel is an integer in this ring.

The algorithms described for computing DFTs can also be used for computing NTTs. The main advantages of NTTs over DFTs for computing convolutions are:

NTTs allow error-free evaluation of convolutions
NTTs can be made free of multiplications by selecting β to be a simple integer, say a power of 2.

One important class of NTTs is the class of Fermat number transforms (*ibid*). Here R is chosen as a ring of integers, modulo a Fermat number F defined as $F = 2^{2^{+}} - 1$. In this ring a transform of length 2^{t+1} can be defined by selecting β as 2. An FFT algorithm can be used to implement a Fermat number transform. The 'twiddle-factor' multiplications in this FFNT algorithm are by powers of 2 and hence can be implemented using simple binary shifts. Another useful class of NTTs have been developed based on Mersenne primes (*ibid*).

The main drawbacks in using NTTs for computing digital convolutions are listed below.

It is difficult to perform modulo arithmetic using conventional hardware and, with the advent of VLSI, fast parallel multiplication has become a relatively simple operation.
Simple NTTs impose considerable restrictions on the transform length and hence on the length of cyclic convolutions that can be computed using this technique. For example, the maximum length of the Fermat number transform that can be defined using a wordlength of 33 bits and β as a power of 2, is only 64. This length can be extended by utilising the 'pseudo' Fermat number transform (*ibid*).
NTTs can be used only as a tool for computing convolutions, whereas the conventional DFT produces a spectrum which is of physical and engineering significance.

Polynomial transforms have been defined by selecting the ring R in Equation 6.25 to be a ring of polynomials. These have been developed in a similar manner to NNTs and have the property of mapping a convolution of polynomials into a set pointwise multiplication of polynomials. Polynomial transforms are useful in computing two-dimensional convolutions.

6.3 SAR IMAGING USING A PARALLEL ARRAY PROCESSOR

A discussion on generating of high precision satellite SAR imagcs using a parallel array processor is presented in this section. The processor considered is a single instruction multiple data (SIMD) machine, comprising a regular array of processing elements. This class of processors has already been discussed in Chapters 1, 3 and 4. The SAR imaging specification considered here is that for the ERS–1 Satellite SAR System.

6.3.1 The processing system

The SAR processing system considered here consists of the following three components:

arithmetic unit and its data memory
control unit and its program memory and user interface
mass storage and input/output unit.

A schematic of the system configuration is depicted in Fig. 6.9. Brief descriptions of each of these components are given below:

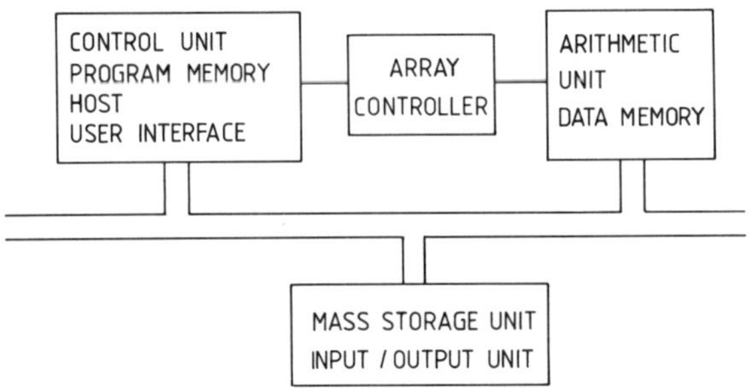

Fig. 6.9 A schematic diagram of the SAR processing system.

Arithmetic unit and data memory. The arithmetic unit consists of rectangular arrays of processing elements (PEs), as illustrated in Figs. 3.5 and 3.6. Each PE consists of a 1-bit arithmetic logic unit (ALU) and several 1-bit special purpose and general purpose registers. It can communicate with any of its eight nearest neighbours. The edges of the array can be connected in several ways, depending on the application. However, for this discussion it is assumed that the array is connected as a cylinder about both its orthogonal axes. The size of the array is assumed to be 64 × 64. Each PE in the array is connected to a column of external RAM below the array. The amount of external RAM available is assumed to be 8 kbits per PE. It is important to note here that the array size and memory size are not limitations of the processor architecture. However, the SAR processing technique used depends on these two system parameters, which makes such assumptions necessary. All PEs execute the same instruction, which is broadcast to the array by the array controller. This instruction may be an arithmetic logic instruction or a memory access instruction. In the latter case, the address of the bit plane is also provided by the controller. There is provision for activating only part of the array. This control is achieved by means of an activity register. The activity register plane may be set by loading a bit pattern from external memory or as a result of an operation. In addition to these parallel operations there are global indicators, e.g. to indicate whether the result of an operation is zero, throughout the active part of the array.

Since each PE has only a 1-bit ALU, arithmetic is carried out using bit-serial techniques. Since data is usually available in bit-parallel form it is first necessary to convert these into a bit-serial form before processing. This is carried out by an input/output device which resembles a two-dimensional shift register (see Section 3.4.5). Sixty-four samples of n-bit data are clocked asynchronously from the processor into this device. The processor array is then interrupted and the data is clocked into the processor memory in n-clock cycles. Data mappings and data routing are considered later.

Control unit, program memory and user interface. In SAR processing, the functions of this unit are broadcasting instructions to the arithmetic unit, performing the background computations to determine processing parameters and monitoring the activities in the complete system. These activities include data transfers for input/output and data transfers for processing. This unit will typically consist of a host computer with an array controller as the interface to the arithmetic unit. The host computer can be as simple as a small microcomputer. However, in order to carry out the complex background calculations for SAR processing and to monitor the high volume data transfers a more powerful minicomputer is required as the host. This will also contain, as peripheral components, an image display and an operator interface for process control and program development.

The programming of the array processor can be carried out at a high level or at assembly level. All instructions applying to the array processor are array instructions, which are to be executed of all active PEs. In addition to the instructions characteristic of a sequential machine it is necessary to introduce parallel constructs to the language, e.g. branch instructions which are executed only when a specified condition is satisfied at all active PEs or just one active PE. In order to get the best performance from the array it is necessary to code the computational kernels of frequently encountered algorithms as macros at the lowest convenient programming level. Examples of these computational kernels are the basic arithmetic operations and the data routing operations. Another frequently encountered operation is the fast Fourier transform. It is possible to derive a set of small computational kernels, which will ensure the optimum implementation of an FFT of any length on any number of sequences stored in any order in any size of PE array. In studying the application of this type of parallel processor to digital signal processing it is very important to identify such generic kernel operations. These kernels may then be combined for any processor configuration to implement the specified function.

Mass storage and input/output unit. Since this is not a real-time SAR processing system, raw data is assumed to reside in a large disk. The processed image is also stored in a second disk. In addition a third disk is needed to store intermediate results. The capacity required in each disk for this application is about 200 Mbytes. It is assumed that the system is pipelined. Hence, during the execution of the first processing stage involving disks 1 and 3, the processed image in disk 2 is transferred to an output device. Similarly, during the second processing stage involving disks 3 and 2, it is assumed that disk 1 is filled with raw data.

The limitations on processing speeds due to I/O bandwidths are not considered here. In practice, they could cause major bottlenecks. However, these speed limitations may be overcome by having parallel disk drives. Since these topics are not central to SAR processing they will not be considered any further.

6.3.2 Implementing signal processing algorithms and data routing operations in the parallel processor

In this section methods of implementing algorithms and data routing operations commonly encountered in SAR processing, using the parallel processor array, are considered.

Data mappings

Here we are concerned with methods of storing data under the processor plane and their significance for implementing signal processing operations.

Consider storing a 128 × 128 two-dimensional array in the memory of a 64 × 64 PE processor. Four samples have to be stored under each PE. There are two basic methods of selecting these four samples (see Section 3.4.3). In the first method, known as the 'pyramidal' mapping technique, a set of four nearest neighbours are stored under each PE. In the second technique, known as the 'windowed' mapping scheme, the array is folded under the PE array. For example, consider the samples stored under PE(0,0). In the pyramidal scheme these four samples are $x(0,0)$, $x(0,1)$, $x(1,0)$ and $x(1,1)$. In the windowed scheme they are $x(0,0)$, $x(0,64)$, $x(64,0)$ and $x(64,64)$.

Both mappings are useful for signal processing. Pyramidal mapping is more useful for computing low order convolutions, interpolations, decimations and other image processing operations such as median filtering. The windowed mapping is more appropriate for Fourier transformation and for the computation of very high order convolutions. For some SAR signal processing operations it is more useful to combine both these types of mapping. For example, a pyramidal mapping may be used in one dimension of the array, while in the other dimension the data can be mapped in a windowed form. Very often signal processing applications consist of a sequence of the basic signal processing operations mentioned above. For example, a radar signal processing application may consist of a Doppler analysis stage followed by a filter and detect operation. The Doppler spectrum analysis prefers a windowed data mapping of input and the filtering operation prefers a pyramidal data ordering. Hence it is very important to organise the algorithms so that appropriate data mappings are available at the input and the output. For this particular example it is very useful to organise the FFT algorithm so that it takes the input in windowed order and produces the spectrum in a pyramidal form. It is possible to achieve this by appropriately implementing the internal data routing operations in the FFT.

It is important to select data mappings which best suit the algorithms and which minimise the amount of data routing. For example, in processing many sequences in parallel it is more appropriate to store each sequence under a small set of neighbouring PEs. For example, in processing 1024 sequences using a (64 × 64) PE array it is obviously best to store each sequence under a (2 × 2) array of PEs.

However, it is in many cases not desirable or practical to minimise data routing using this approach. These cases are listed:

System incapable of tolerating long latencies. It is then not possible to block so many sequences before processing.

Memory limitations, in particular in small systems.

Processing required in both dimensions, for example in SAR. Isolation of sequences in the PE array is not very appropriate for these applications.

The digitised data from the imaging sensor cannot be mapped directly on the processor memory in the most favourable order. It is always possible to re-arrange the data in a relatively large external buffer by mapping it into processor memory. However, this involves additional hardware and increased latency. In real-time processing it is best to map the data directly into the processor memory.

Data routing

The memory of the parallel processor array is constructed using conventional random access memory (RAM) components. However, in this processor configuration the block of memory cannot be regarded as random access. Instead it has to be regarded as three-dimensional, with only one of the dimensions having the random access property. Only the 'vertical' dimension is random access, since any bit plane can be accessed in one cycle. If a PE of co-ordinates (x,y) wishes to access a bit in memory $(x\text{-}\Delta x, y\text{-}\Delta y)$, this has to be carried out using $(\Delta x+\Delta y)$ routing operations through the processor plane. It is possible to reduce the number of routing cycles using the diagonal connectivity property, min $(\Delta x, \Delta y) + |\Delta x - \Delta y|$. It is important to organise data mapping and algorithms to minimise the time wasted by data routing. Some of these aspects have been described earlier in this section.

Some of the common data routing operations are considered below. Consider a 4 × 4 matrix transposition in the x-z plane, where z is the random access dimension. This transformation can be implemented using two block data transfers as depicted in Fig. 6.10. These block data transfers can be conveniently implemented in parallel using the parallel processor, a method which can be extended to larger matrices. This type of operation is useful in matrix processing and Fourier transformation. The type of operations required for Fourier transformation may be somewhat different (see Fig. 6.11).

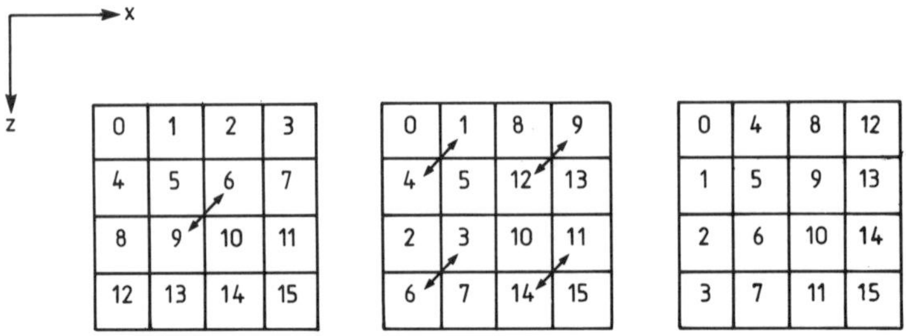

Fig. 6.10 4 × 4 matrix transpose in the x–z plane using two block data transfers.

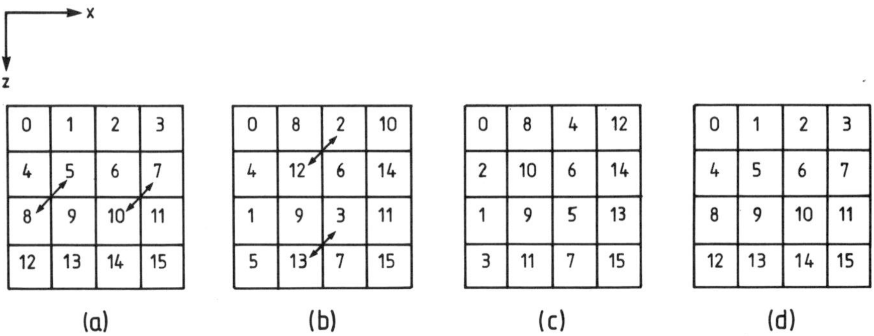

Fig. 6.11 Data routing for computing a 16-point FFT, using 4 PEs. (a) Initial windowed data ordering. (b) After the first routing operation (c) After the second routing operation. (d) Final ordering, allowing for bit reversal. (Note: each digit represents the index i of the sequence $\{ x(i) \}$.)

A second type of data routing that is useful in signal processing is linear data exchange in x or y dimensions. These are often required after computing FFTs to correct for bit-reversal effects.

Algorithms

Arithmetic operations are implemented in the processor using bit-serial techniques. These techniques are necessarily slow and the only way of achieving high throughput is using the high degree of parallelism inherent in the system. It is simple to consider operations such as pointwise addition or multiplication of two arrays and quote processor performance in Giga operations per second. However, signal processing operations are somewhat more complicated, and owing to reasons mentioned earlier in this section they cannot be carried out without a significant amount of data routing. Hence algorithm development is a very important stage in designing a system based on this type of parallel processor. This is considerably more difficult than developing algorithms for a conventional sequential pipelined processor. In developing algorithms the following factors have to be considered:

Minimise data routing. Due to reasons given earlier in this section, it is impossible to avoid routing and in many practical applications a considerable proportion of the processing time will be taken by data routing. Optimal I/O data mappings considered earlier in this section have to be considered here.

Distribute arithmetic operations. Unlike a sequential machine, minimising the number of arithmetic operations is no longer sufficient in this type of parallel processor. It is more important to distribute these operations, carrying out everything, as far as possible, in parallel and thereby minimising the total processing time.

Multiplications. Multiplications are very expensive in time in this type of processor. Hence minimal multiplication algorithms have to be developed.

Multiplication of an array. Multiplication of an array, or a plane of data, by a single value is much cheaper than multiplying two arrays. For example, the time for $(A(i).B(i)) >$ the time for $(k.B(i))$. This is because the coefficient k can be recorded using the canonical signed–digit code (i.e. using 0, 1, -1, for example 01101110 = 100 -1 00 -1) and carry out the shift and add in the multiplication only when the digit in coded k is non-zero. Since the coefficient k is held in the controller, this has the added benefit of saving coefficient memory. Hence algorithms have to be developed to take advantage of this property.

Computational kernels. In order to get the maximum parallel processing efficiency and to simplify high level programming it is useful to have a set of computational kernels for every algorithm. These should be designed to be independent of the array size, sequence lengths, data orderings and word lengths. These small kernels can be programmed using a low level language.

Two classes of algorithms, one for Fourier transformation and the other for filtering and interpolation, are described below. Both these algorithms are very useful in implementing SAR processors.

Fourier transformation. Algorithms for computing Fourier transforms in digital computers have been described in Section 6.2.3. Here we are concerned with their implementation in parallel processors.

Consider implementing a 16-point fast Fourier transform using 4 PEs. These 4 PEs form a linear chain in the x dimension. There are four samples for each PE and a windowed data mapping is assumed (Fig. 6.11(a)). The windowed mapping is used since the FFT first combines samples further apart and then combines the nearer ones. From Fig. 6.11 it can be seen that the data is stored in a very favourable order for computing the FFT. This computation is carried out in four stages:

A 4-point FFT computed at each PE. All 4-point FFTs are computed in parallel.

Three complex multiplications at each PE. (The fourth multiplication is by unity.) The coefficient for the three multiplications can be stored under each PE. However, analysis of the algorithm shows that for this case all the complex multiplications can be implemented more efficiently as scalar multiplications, with no memory at all used for coefficient storage. It is important to analyse algorithms carefully since these observations can make a factor of two improvement in speed and a similar saving in memory.

Data routing. This operation is very similar to the matrix transposition operation shown in Fig. 6.10, the objective of which is to get the four samples in each row under each PE in order to carry out a 4-point DFT in the horizontal dimension. However, it is important to note that the FFT results in an output in the bit-reversed order. Hence these data routing operations should be carried out to give the output in natural order. The required routing operations are illustrated in Fig. 6.11. The total time required for these operations is the same as that for a 4 × 4 matrix transposition.

Four-point FFT at each PE. The ordering of the spectral components is shown in Fig. 6.11(d). The data reordering due to bit reversal has been eliminated by the appropriate ordering of the data routing operations. In a 64 × 64 PE array 1024 of these 16-point FFTs can be carried out simultaneously.

These ideas can be extended for computing long FFTs. However, in order to simplify programming and to improve efficiency a set of low order computational kernels of the type FFT$mn(d)$ should be defined. Here m refers to the length of the FFT, n the number of PEs used in computing FFT and d the separation between these PEs. For the particular machine under consideration kernel routines for $m = 2, 4, 8$ and $n = 1, 2, 4$ appear to be adequate. The radix–2 and radix–4 FFTs are not the best algorithms for computing the DFT in most applications. Hence higher radix and mixed-radix algorithms should be considered in order to obtain the optimal algorithm for a particular situation. The best algorithm would depend on the size of the PE array, the length of the FFT and the number of FFTs being computed simultaneously. These algorithms can then be conveniently programmed using the kernel routines discussed above.

Filtering and interpolation. Consider the implementation of a convolution of the type given in Equation 6.19, for a small value of n. Let $n = 3$ and the samples x_i be stored under a linear array of PEs. The stages in implementing this convolution are:

Multiply the sequence $\{x_i\}$ by h_0. This is a scalar multiplication.
Store the product in a partial product sum array.
Shift the sequence $\{x_i\}$ one place to the right relative the PE chain. This involves data routing.

Multiply the shifted sequence by h_1 and add to the partial product sum array.
Shift the shifted sequence $\{x_i\}$ another place to the right.
Multiply by h_2 and add to the partial product sum.

This completes the convolution and the partial product sum now contains the convolution result. This method can be extended to two-dimensional convolutions.

When there is more than one sample per PE, pyramidal data mapping is most appropriate to implement convolutions. The user can then program his algorithm, assuming the size of the PE array matches exactly with the size of the signal array, although the physical size of the PE array may be smaller. The controller takes care of the mismatch between the physical size of PE and signal arrays by assuming a pyramidal data mapping.

Interpolations and decimations can be implemented using a similar technique. However, these operations encountered in SAR processing are somewhat different. The interpolations in implementing operations such as range cell migration corrections and slant range to ground range conversions are not spatially invariant. These spatially variant operations can be implemented by replacing the scalar multiplication by pointwise array multiplications.

6.3.3 SAR processing requirements

The processing requirements implicit in the following discussion have been derived from the options available for SAR processing in the ERS–1 satellite project.

SAR processing stages
A sequence of operations needed for processing SAR data to generate an image with specifications appropriate for the ERS–1 satellite project is given below:

Range processing. This operation consists of correlating each return of length 4700 complex samples with a 700-sample windowed reference. A Hamming function is assumed as the window.

Linear range walk correction. The range walk correction is carried out in two stages. The linear correction is implemented in the time domain after range correlation.

Fourier transformation in the azimuth dimension. DFT of each range cell is computed.

Correction for the residual range walk effect.

Separation of the spectrum into six sections each corresponding to a look.

Azimuth correlation. Each look has to be correlated with a 200-sample complex reference function. These correlations are implemented as pointwise multiplications in the frequency domain.

Inverse transformation. The IDFT of each of the six looks is computed. Slant-to-ground range conversion and resampling then follow. The range parameter in the processed image is measured in a slanted plane containing the path of the radar platform and the object on the ground. This image has to be converted to one in which the distance measured along the ground is uniformly sampled, with the required resolution and sampling period. A similar resampling operation has to be carried out in the azimuth dimension to derive the image on a defined grid.

Summation of looks. The six looks are incoherently summed, to reduce speckle noise and improve the signal-to-noise ratio at the expense of resolution. This operation involves computation of the square of the modulus of the complex samples and summing the corresponding power values in the six looks.

The sequence of operations listed above involves processing the two-dimensional signal in the range and azimuth dimensions. This processing is carried out on SAR data read from a mass storage device, such as a disk. Hence it is very desirable to read the data from disk, carry out processing in both dimensions and transfer the processed image to another device. However, it is not possible to process data in a single stage, owing to processor memory limitations. SAR data collected from an 80 × 80 km area take the form of a 5000 × 20 000 sample complex array. However, it is not necessary to process the complete array simultaneously. Processing can be carried out on much smaller blocks of data. Since the reference sequence lengths in the ranged and azimuth dimensions are 700 and 200 samples per look, the minimum block size required for efficient processing is about 2000 × 500. In the full resolution single-look case this minimum block size for efficient processing is 2000 × 2000 samples. If a 32-bit processing wordlength is assumed for in-phase and quadrature components, the total memory requirement greatly exceeds the 4 Mbyte capacity of the parallel processor under consideration. Furthermore, memory is required to store coefficients and for double-buffering for processor pipelining.

It is therefore necessary to carry out processing in two stages. However, in smaller SAR processing systems or in processors with larger memory capacities it is important to consider the possibility of carrying out processing in both dimensions in a single stage.

The sequence of SAR processing operations are implemented in the parallel processor in two stages:

Stage 1

Range correlation
Linear range-walk correction

SAR data is read in blocks of 64 range lines from disk 1 and the processed blocks are stored in disk 2. The data should be stored in disk 2 so that it can be conveniently accessed for processing in the azimuth dimension in

Stage 2. Since Stage 2 processing is carried out in blocks of 128 range cells, the 64 × 5000 sample range lines are stored in disk in blocks of 64 × 128 samples.

Stage 2

Azimuth Fourier transform
Residual range walk correction
Look-separation
Reference function multiplication
Inverse Fourier transform
Slant-to-ground range conversion and resampling
Look summation

Stage 2 processing is carried out using the overlap-store technique. Data is read from disk 2 in relatively small blocks and processed image blocks are stored in disk 3. Successive blocks are overlapped in the azimuth dimension by about 200 samples. This value is approximately equal to the sum of the length of the azimuth reference and the window length of the azimuth resampler. A small overlap is provided in the range dimension to allow for edge effects in range resampling. The azimuth reference has to be updated very frequently in the range dimension since it is a function of r. However, for a given range cell the reference can be retained constant for about 1000 samples. Hence the block size in the range dimension is limited to about 1024 samples. The processor memory limits the number of range cells, each of length 1024, that can be simultaneously processed to 128. Sufficient memory capacity has to be allowed for storing the next data block, for overlaps and for processing parameters and reference functions.

6.3.4 Stage-1 processing

As described in Section 6.2, the range correlation can be implemented using time or frequency domain techniques. The time domain correlation of 64 5000-point sequences with a 700-sample reference is estimated to take about 4s in this parallel processor, assuming (16 bit I + 16 bit Q) complex arithmetic. The same data block can be processed using frequency domain techniques in about 0.2 s, with (24 bit I + 24 bit Q) block floating-point arithmetic. The main reason for the inefficiency in the time domain technique is the large amount of time taken to perform the very large number of multiplications using bit-serial techniques. A brief description of the implementation of the frequency domain correlation in this processor is given below.

Each correlation is computed using a 5120-point DFT with $N = 5120$. It is assumed that the DFT of the range reference has been precomputed and stored in the processor memory. This has to be updated regularly, for example every 1000 range lines, during processing. All 64 5120-point DFTs in the block are computed simultaneously.

Each 5120-point sequence is stored under a row of processing elements as depicted in Fig. 6.12. Although this is not the ordering that minimises data routing, it is chosen for the following reasons. First, it is simple to achieve using the available I/O devices and does not need a large data buffer for data re-ordering. Second, this scheme permits the minimisation of coefficient memory requirements. Third, it considerably simplifies the linear range walk correlation that follows the range correlation.

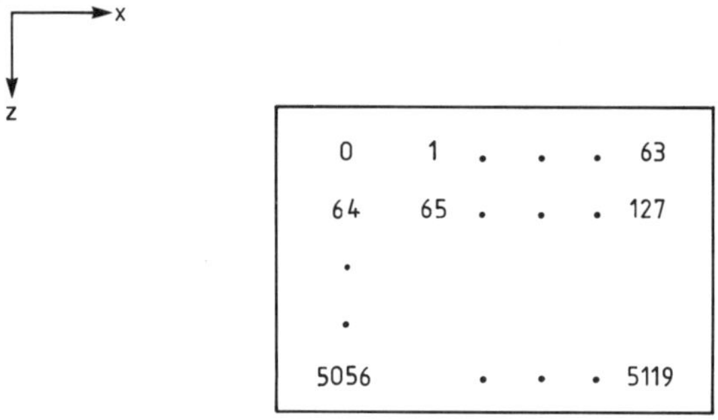

Fig. 6.12 Mapping a single range line under a row of PEs for Stage 1 processing.

Since the sequence length is not near to a power of 2, the conventional FFT technique is not very efficient for implementing this DFT. Hence a method comprising prime factor and common factor algorithms is used. Each 5120-point DFT is first reduced into a set of 80-point DFTs and 64-point DFTs using the method described in Section 6.2.3. These are implemented in the parallel processor in four stages:

an 80-point DFT at each PE
a set of complex multiplications
data routing operations in the x-z plane
set of 64-point FFTs.

Each 80-point DFT is computed using the prime factor algorithm. An in-place and in-order PFA is used to minimise the amount of memory overhead. This algorithm uses the 16-point and 5-point low order DFT algorithms developed by Winograd (see Section 6.2.3). The 64-point DFTs are computed using the FFT algorithm. The nature of the final data ordering in this DFT is not very important since all these steps are implemented in the reverse order in the inverse transform. However, this ordering should be known since it is needed to perform correctly the frequency domain pointwise multiplications. In order to save memory, the transform of the reference is not stored under every row of PEs. A set of 4 or 8 rows can share a transform of the reference function. It is then necessary to carry out data routing among these rows before and after the frequency domain pointwise multiplications. After these pointwise multiplications a 5120-point inverse DFT is computed in each row by inverting all the four steps and performing them in reverse order. The correlated output is then obtained in the same order as the input. There are several other options for implementing this frequency domain correlation. For example, it can be implemented using four 2048-point FFTs and the outputs combined, using overlap-save or overlap-add techniques. However, it can be shown that these techniques are not as efficient as the single DFT method described above. The time for correlating 64 range lines, using 24-bit complex arithmetic, is estimated as 200 ms.

The range correlation is followed by a correction for the linear component of range walk. This operation involves a linear shearing operation such as that depicted in Fig. 6.13. This is implemented by a set of interpolations. The interpolation differs from range line to range line and hence cannot be implemented using a spatially-invariant convolution. However, within each range line the interpolation remains unchanged for relatively long periods. Hence within each range line this interpolation can be implemented by a shift-invariant convolution. The shift-multiply-accumulate technique for computing low-order convolutions using this processor has been described in Section 6.3.2. However, unlike the convolution algorithm described there, the reference coefficients are stored under each PE. A set of reference coefficients is calculated for each range line in the host processor and stored under the corresponding row of PEs using the array controller. A windowed sinc function is usually used to generate the interpolation coefficients.

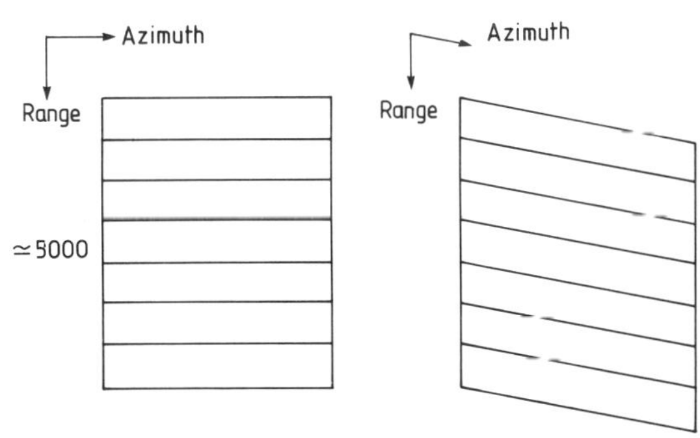

Fig. 6.13 Correction for the linear component of range walk.

The linear range walk correction can be efficiently implemented by storing a range line under a single row of PEs. As mentioned above, the interpolation coefficients are maintained constant for each range line over relatively long periods. However, it is necessary to update these coefficients several times across the swath. The time for linear range walk correction in this block of 64 range lines is estimated to be 140 ms. An 11-point real interpolator with 12 bits per coefficient is assumed. This time also includes the time needed to store the coefficients in the processor memory and for 16 coefficient updates across the swath. These coefficients may be evaluated in the host processor using a look-up table during the range correlation.

In order to transfer data from I/O buffer to processor memory and from processor memory to I/O buffer, the processor has to be interrupted. The processor time taken for these data transfers has been estimated as 20 ms per block of 64 range lines. This estimation assumes an input wordlength of (5 bit I + 5 bit Q) and an output wordlength of 12 bit I + 12 bit Q). Hence the total time for Stage 1 processing of a block of 64 range lines is about 360 ms. The time taken to complete Stage 1 on the whole image is approximately 120 s.

6.3.5 Stage-2 processing

The size of a data block used for Stage 2 processing is (128 × 1024) samples. The azimuth reference is updated every 32 range cells. Hence four reference functions have to be evaluated for processing this data block. Each reference function is approximated by a windowed linear FM waveform. The 4096 complex exponentials of the four reference functions can be computed in parallel in the 64 × 64 PE array, one at each PE. This is followed by four 1024-point FFTs of these four sequences. The frequency domain sequences are then separated into six non-overlapping segments, each corresponding to one of the looks. The time taken for reference function processing is very small compared to the time needed for data processing.

The (128 × 1024)-sample data block is stored in the parallel processor memory as depicted in Fig. 6.14(a). Each 1024-point range cell is stored in a windowed form under 32 PEs in one row.

A windowed data map is chosen since the first operation in Stage 2 is a Fourier transform. This data mapping does not result in the minimum number of data routing operations in computing DFTs. To minimise data routing each sequence has to be stored under 8 × 4 array of PEs. However, the latter mapping is not suitable for carrying out the operations following the Fourier transform.

The FFT algorithm is used for computing the DFTs of the 128, 1024-point, sequences. The method of implementing FFTs in this parallel processor has already been described. The data routing operations in the algorithm are designed to correct for any bit-reversal effects and to produce the output in the same order as the input. After the FFT a set of data routing operations are implemented to bring the output into the order shown in Fig. 6.14(b). This ordering is needed to carry out the signal processing operations following the FFT. The time taken to compute the 128 FFTs in 24-bit block floating point arithmetic is estimated to be about 30 ms.

The correction for the quadratic component of range walk is implemented using an algorithm very similar to that used for linear range walk correction

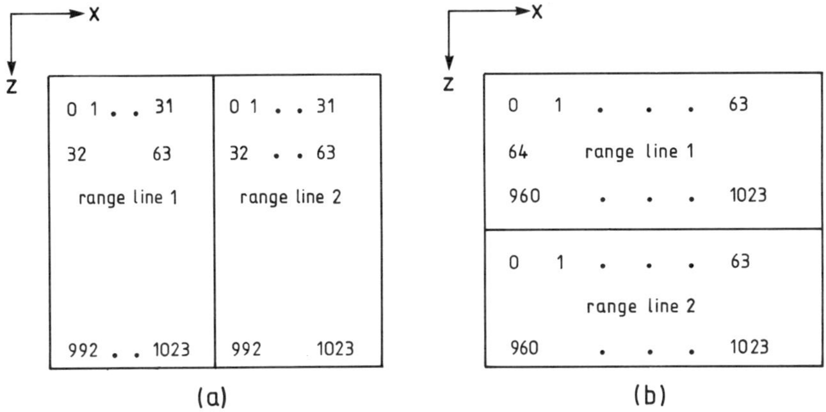

Fig. 6.14 Data mapping for Stage 2 processing. (a) Data mapping before FFT computation. (b) Data order after FFT computation.

in Stage 1. This requires interpolations in the range dimension. These computations are simplified by storing the 128 range cells in a pyramidal order. The time taken for residual range walk correction in each block is estimated to be 7 ms, assuming a 5-point interpolation reference and 8-bit real weights.

The 1024-point sequence in the frequency domain is separated into six non-overlapping segments, each having about 180 samples. Each segment of look is multiplied by the corresponding reference function spectrum and is inverse Fourier transformed. A 512-point inverse FFT is used in each look to achieve the required azimuth sample spacing. Each look is augmented by zero-valued samples up to a length of 512 before performing the IFFT. Hence in each range cell six 512-point IFFTs have to be computed. The time estimate for implementing pointwise multiplications and inverse FFTs in all six looks of a block of data is 120 ms. This estimate has been derived assuming (24 bit I + 24 bit Q) block floating point arithmetic.

The multi-look azimuth compression is followed by a resampling operation to convert slant range into ground range. The object of this processing stage is to obtain a set of equally spaced samples in the ground range in each range line. The sample spacing may be a specified by the user.

6.3.6 Summary

SAR signal processing using a processor array has been studied in detail in this section. It has been shown that all operations required for generating high precision SAR images can be very efficiently implemented using a processor of this type. There are two reasons for this high processing efficiency. First, the processor configuration is well suited for implementing the basic operations in SAR processing. Second, this type of parallel processor is very efficient in processing large data blocks relative to the size of the PE array. SAR processing is one area where processing is carried out on large data blocks, since it is necessary to combine many radar returns to achieve the required azimuth resolution.

The other main advantage of this processor is its regularity. This makes it well suited for implementing in VLSI. The GRID processor array, discussed in Section 3.4, is an important case in point. The processing power of the system can easily be increased by increasing the size of the array, and the data processing wordlength is not constrained by the machine word length. This allows users to select any word length for processing depending on the throughput and accuracy requirements.

The ERS–1 SAR processor application presented in this section is not a real-time application. However, this processor can be used in real-time SAR processing with smaller swath widths. There is also considerable scope for optimising the processor architecture for SAR and radar processing operations. Certainly, the signs are that VLSI technology and imaginative algorithm development are going to facilitate, perhaps dramatically, the development of compact, high performance real-time SAR imaging processors.

REFERENCES

Bernstein, R.L. (1982) Special issue on SEASAT, *J. Geophys, Res.*, **87**, C5.

Hovanessian, S.A. (1980) *Introduction to Synthetic Array and Image Radars,* Hughes and University of California.

Kozma, A., Leith, E.N. and Massey, N.G. (1972) 'A Tilted-Plane Optical Processor', *Applied Optics,* 1, 1766–1777.

McClellan, J.H. and Rader, C.M. (1979) *Number Theory in Digital Signal Processing,* Prentice-Hall.

Rabiner, L.R. and Gold, B. (1975) *Theory and Application of Digital Signal Processing,* Prentice-Hall.

Chapter 7
Whither Image Processing?

R. J. OFFEN

7.1 INTRODUCTION

This chapter is concerned with drawing together, in the context of both enabling technologies and representative future applications, some of the ideas discussed in preceding chapters of the book. We will start by looking at potential developments in three key areas of enabling technology – VLSI, computer architectures and Artificial Intelligence – and then proceed to look at a number of representative application areas involving image processing in some form or other. Rather than exploring specific instances in detail we will take a fairly broad generic view of those areas where the availability of sophisticated high performance image processing systems is likely to make a significant contribution. How the future is actually going to turn out must be discernible only in terms of a composite of insight and inspired guesswork. After all, who just over ten years ago, when the first Altair microcomputer kitsets appeared on the market, would have predicted the pervasive and central position that the personal computer now occupies. No one did at the time. Consequently, the choice of topics covered here must be regarded as being significantly influenced as much by the particular enthusiasms of the author as by any insight.

7.2 EXPANDING FRONTIERS

Progress in image processing is, and will continue for some time to be, contingent on advances in three key technological areas – VLSI technology, computer architectures and Artificial Intelligence. The following three sections will explore some of the technological potential inherent in these areas, primarily with a view as to how they might impinge in the future on the evolution of advanced image processing systems. There will be no discussions here of exotic implementation technologies, such as Josephson microcircuits and optical switches. This is because of both lack of space and the very long development times to be expected before there is any significant commercial use of these devices.

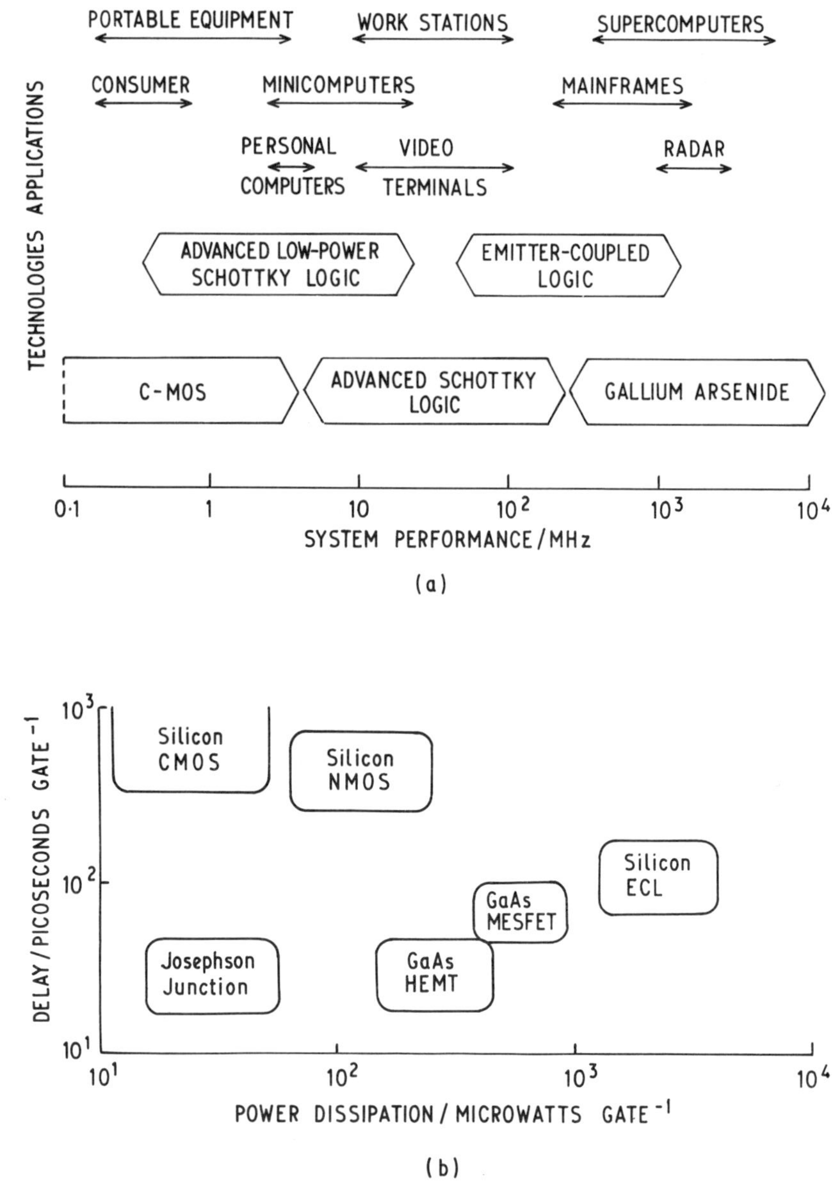

Fig. 7.1 Some typical (a) system and (b) performance trends for several different technologies. Adapted from *Electronics*.

7.2.1 VLSI technology

It is difficult to detect a clear, unambiguous motivating thread running through the extraordinary developments in microelectronic technology over the last couple of decades. This is mainly because of the tension which has always existed between systems driven advances and technology driven advances. Figure 7.1(a) shows schematically the linkages which exist between certain technologies and applications which reflect the appropriate performance. As formidable as the problems relating to reducing feature size, design checking, fabrication, controlling power dissipation, packaging and testing are, it is the question of what to do with a million or so transistors compacted onto a few square millimetres of silicon that is perhaps the most fundamental of all. It is to this last question that the systems designers have made significant contributions and helped provide a *raison d'être* for the technological feats that VLSI has wrought.

In this section we seek to outline some of the more significant factors which we expect to play a role in continuing still further the development of increasingly smaller, faster and more versatile computing elements. *Much of the material in this section is based on a review article by Seitz and Matisoo (1984), to which the reader is referred for a more detailed discussion.*

Four basic operations underlie all of computing. These are switching, of a transistor or a similar switching element, storage of information, as stored energy, the communication of information and input and/or output of information. Improvements in all four of these areas are being sought, but these improvements cannot be derived independently of each other. All existing digital technologies tend to have switching speeds and communication speeds which are comparable with each other, rendering global communication an increasing problem. Hence the interest in highly concurrent machines emphasising local communication in order to overcome this global communication problem as well as the very basic 'von Neumann bottleneck' which is intrinsic to conventional sequential machines. We will return to this point in the next section, though it should be borne in mind that many of the architectures which we will discuss have been motivated primarily by the limitations of the technology. It is worth noting that many of the potential technological improvements to be discussed in this section are going to do more to reduce the cost of a given computational functionality than produce revolutionary improvements in performance. It is going to take significant advances in the area of computer architectures, potentially involving massive concurrency, to provide this performance. This is especially true in the context of image processing.

Throughout the sixties and into the early seventies the number of components integrated onto the most complex chips existing at the time doubled approximately annually. From the mid seventies this rate slowed to a doubling about every two years. Currently, the most complex chips tend almost invariably to be memory devices, with two-megabit RAMs expected to be commercially available within the next couple of years, and Mitsubishi having already announced the availability of a four-megabit DRAM to be sampled in 1987/8.

The really fundamental question is, will this growth in chip complexity continue unabated? All the signs that are currently assessable indicate that this will indeed be the case. Two key factors will be significant reductions in feature size and defect density. The number of components which can be packed onto a chip of a given size varies approximately quadratically with feature size. Currently, a typical commercial feature size is about 2 or 3 μm, with devices of circa 0.1 μm feature size having been demonstrated in various research laboratories. This means that we can look forward to at least two order of magnitude increase in packing density as devices are scaled down to the smaller geometries.

The mainstream implementation pathways are still those concerned primarily with MOS and bipolar device technologies. The active devices of MOS technologies, in which we implicitly include CMOS, are field effect transistors, which were initially fabricated as a capacitative metal control electrode, called the gate, over a thin oxide layer on the surface of a semiconductor. Polycrystalline silicon doped to be a conductor has replaced the metal in modern processes as the next level of wiring above the silicon, with one or two levels of metal wiring above this polysilicon level. It is of interest to examine the consequences of shrinking such a device and the wires attached to it by a factor of 10, say from a contemporary feature size of 2 μm to some future feature size of 0.2 μm. The same scaling factor is applied in all three dimensions, retaining the same relative flatness of the surface, and ensuring that the electric fields within the device scale simply. A simple analysis (Mead and Conway, 1980a) shows that the voltage and maximum current both scale linearly. This result is important, because then the power per device scales down quadratically while the number of devices per unit area scales up quadratically, and the power per unit area remains approximately constant. The gate transit time scales linearly, which means that the smaller switch is also faster. The fundamental figure of merit for switching devices, the switching energy, is the energy required to make a single device change state. This quantity is clearly the same as the product of the power dissipated per device, when switching at maximum speed, and its intrinsic delay, and hence is also called the power delay product (see Fig. 7.1(b)). Total power is an excellent predictor of system cost, while switching delay contributes inversely to system performance. In this sense the switching energy of the technology used to build a computing system is a useful predictor of the cost-to-performance ratio of the system in which the switches are used.

The underlying reason for the remarkable advances in the cost-to-performance ratio of systems based on MOS technologies, and the promise inherent in the next factor of ten reduction in feature size, is now easily appreciated. The switching energy, the product of a quadratic scaling of the power per device and linear scaling of the delay, scales as the cube of feature size, assuming the electric fields remain constant in the scaling. The proposed example of a future MOS technology with 0.25 μm feature size can be regarded as very much a replica, scaled in size, energy and speed, of today's MOS technology. However, this technology is close enough to fundamental limits that a further tenfold decrease in feature size and voltage is probably questionable. This does not in any sense imply that a future 0.25

μm MOS technology will allow one to achieve a three orders of magnitude gain in performance at constant cost with the same computer design. The cubic scaling of the switching energy is the product of a quadratic scaling of area and power per function, and only a linear scaling of transit time. One finds in reality that this linear transit time scaling is very difficult to achieve in larger systems because of delays imposed by the interconnections, unless the system elements can operate concurrently. With conventional sequential systems it is accordingly much easier to exploit advances in MOS technology to reduce cost than to enhance performance.

On the basis of equivalent feature size, high performance bipolar circuits have an advantage in speed of about a factor of five over MOS technology. However, the integration levels of bipolar chips historically have lagged behind those of MOS chips by about an order of magnitude. Some of this lag arises from problems in cooling. The remainder of the lag results from the differing sensitivities the two technologies have to defects and to defect types. Consequently, bipolar devices and circuits are used in those portions of computers in which performance is a prime requirement. Generally, these are the processor, the cache memory and that part of the communication system in which high data rates are required. Bipolar technology has been responsible for the evolution of the very fast A–D and D–A converter chips which are essential for real-time image processing applications. Another more general high speed device where bipolar has held sway in the gate array. A typical state of the art bipolar gate array contains several thousand gates on a chip with logic delays under loaded conditions of about 500 ps.

Bipolar technologies can be scaled to much narrower linewidths than those currently implemented, much in the same way as MOS technology. However, the details of the scaling differ. The major difference results from a fundamental difference in operation of MOS and bipolar devices. The bipolar voltage levels are effectively set by the silicon band gap, so that bipolar scaling leaves the circuit voltage levels essentially unchanged. The scaling procedure begins with an optimised high performance bipolar device. As the lithographically determined horizontal dimensions are shrunk, the vertical dimensions of the emitter and base are correspondingly shrunk. To avoid problems with device operation it is necessary to increase the base doping substantially and to adjust the collector doping in proportion to the emitter current density. As the linewidth is shrunk from a typical current value of 2 μm to 0.2 μm, the relative delay contribution of various components remain approximately constant and the overall circuit delay reduces by approximately one order of magnitude, going from about 200 ps to 20 ps. Thus, from the point of view of performance, there is still considerable potential left in silicon bipolar technology.

The silicon technologies, bipolar and MOS, have been and will continue to be the most significant IC technologies. They both still possess the potential for considerable extensions of performance and level of integration. Another semiconductor material which appears to offer a significant increase in performance over silicon is gallium arsenide (GaAs), which exhibits electron mobilities, at low to moderate electric fields, which are considerably superior to those in silicon. Gallium arsenide is an example of the so-called generic III–V compounds, formed from elements located in groups III and V of the periodic table of elements, silicon and germanium being in group IV. An unloaded delay as low as 10 ps has been demonstrated in a high electron mobility transistor (HEMT) GaAs device. HEMT devices, which must be cooled to 77 K for best results, are still in the research and development stages and are difficult to fabricate. The effects of circuit loading will do much to dissipate this intrinsic speed advantage, to the point where, in general, room temperature GaAs devices will probably have an edge over their silicon counterparts by only a factor of two or so. At the present time a number of materials-related problems stand in the way of achieving a level of GaAs integration comparable with that achievable with silicon devices, though there is the potential to fabricate structures, such as 'superlattices', which have dimensions controllable on an atomic scale. Areas where GaAs is soon going to make its presence felt, if it is not already doing so, are monolithic microwave and millimetre-wave circuit fabrication, very high speed digital signal processing and optoelectronics.

Whatever the technology, as chips become larger, and wafer scale integration looks increasingly achievable, the question of how to cope with an irreducible density of processing defects becomes highly relevant. An important technique for coping with defects is the use of fault tolerant designs. The only devices to date to make extensive use of fault tolerant techniques are memory chips. This is because memory devices are very densely packed, hence increasingly vulnerable to defect problems, and usually display extremely regular structure. There is no general agreement amongst vendors on this topic, with some claiming that with good technology there is no need to incorporate fault tolerant features. The most popular approach is to add redundancy in terms of spare rows or columns of storage elements, with electrically or laser blown fuses used to alter the chip connectivity if faulty elements are discovered during testing. Yield improvements of better than an order of magnitude have been claimed, for a concomitant increase in chip area of typically less than 10%.

The success of fault tolerant techniques in manufacturing large memory chips has encouraged designers to think more positively about wafer scale integration. Applications which would gain most from such an approach are those where regular one- or two-dimensional structures, with regular interconnect, are required. These obviously include image processing where the basic repeated element is a pixel processor or pixel group processor and the connectivity requirements are predominantly local.

For successful wafer scale integration it will be necessary to know more about the detailed physics and statistics of the defect distribution – an average figure will just not be adequate – and how these relate, *inter alia,* to feature size and scaling parameters. It is already well known that defects are more likely to be found near the edge of a wafer rather than near the centre. A detailed discussion of these questions, including an extensive bibliography, can be found in Moore (1984). In the long term, the combination of structured, regular architectures and fault tolerant design techniques holds the greatest potential for increasing yield and making the design process tractable in the context of coping with complexity of function and interconnect.

7.2.2 Computer architectures

As we implied in the previous section, a basic and topical question is: 'having got all this technology, what do we do with it?'. Obviously, there will not be a unique answer to this question because the various technologies capable of large scale integration will each predicate their own solutions to our insatiable demands for ever increasing computing power. One of the paradoxes of the linkage between improvements in VLSI technology and raw processing performance is the fact that these improvements map most readily into reductions of cost rather than significant advances in computational performance. Really significant advances in computational performance, as we have implied previously, are going to require the utilisation of radically new machine architectures. The application of technological advances to the conventional von Neumann type of computer, which embodies a control flow architecture, is starting to reach the point of diminishing returns in performance for design effort and cost, with the rate of improvement in high performance machines decreasing smoothly towards an asymptote of circa 3 GFLOPS (billion floating-point operations per second). To reach this goal some formidable design and engineering obstacles will need to be overcome, especially those relating to communication delays and packaging. Even so, this expensively derived performance will not provide the performance required for a lot of the image processing tasks anticipated in various sections of this book, certainly not cost effectively.

The only viable route to the computational power required for image processing in particular and many other computationally intensive tasks in general is to employ the concurrent operation of a number, which could be large, of processing elements. This then raises the question of 'what architecture to use?'. The sheer diversity of possible answers is a theme that has at one level motivated this book. Do we connect together a few very fast, heavily pipelined, but otherwise conventional von Neumann machines sharing storage and communicating via a high speed switching network, or do we develop new exotic and intrinsically non-von Neumann machines? The former solution, employed by most existing multicomputers, does not scale attractively. Communication bottlenecks limit the number of linked computers to a fairly modest number, of the order of 16 to 64. Much larger machines are possible if each processor incorporates enough local memory or a cache to allow the switching network a latency exceeding the processor's instruction cycle. The Bolt, Beranek and Newman 'Butterfly Machine' is a case in point. Machines designed on the basis of a message passing model of computation tend to have a loose enough linkage between individual processing elements to be potentially very large, embodying perhaps millions of processing elements. The 'Connection Memory', discussed later in this section, is an example of this type of machine. It is fair to say that there are no foreseeable fundamental obstacles to prevent the development of very large concurrent machines. If they are to be useful, in that they are to be compact, versatile, easy to use, reliable and cost effective, then all the skills of the VLSI designer and technologist and the systems designer will need to be utilised.

In Chapter 4 we explored some of the data and control flow concepts underlying these and some other fairly abstract machine architectures. In this section we will look at some of the actual machines, built, under construction or proposed, which seem to be pointing the way towards the next generation of high performance machines and try to assess their potential. In the space available we will only be able to cover those machines which have a strong architectural and applications bias towards image processing, and inherently support the use of VLSI design and fabrication techniques in their implementation.

Since Unger (1958) proposed an architecture consisting of a two-dimensional array of processors for use in pattern recognition the main thrust in multiprocessor image processing computers has centred around the development of SIMD, or cellular, array architectures. Ideally, with this approach each processor has a single image pixel mapped onto it and is connected locally to a group of neighbouring processors. Initially, there was not the technology available successfully to implement such a machine, the ill fated Illiac machines (Baer, 1980) being a case in point, and it was not until the mid seventies that machines such as the CLIP, the DAP and the MPP, employing primarily MSI and LSI techniques, saw light of day. Of these only the DAP can be regarded as having been released in a commercial version. Subsequently, further machines such as the NTT array processor, the GEC Rectangular Image and Data (GRID) processor and the Hughes 3-D computer, all involving true VLSI techniques, have appeared or have been announced. For a general discussion of such image processing motivated machines, including several not discussed here, see Duff (1983).

As has been pointed out in previous chapters, there are some real and simplifying advantages to be gained by using SIMD architectures for image processing. A generalised SIMD array machine is shown in schematic form in Fig. 1.6. Local communication is implicit, with no need for indexing or special addressing techniques in accessing nearest neighbours. Local storage in the processing elements provides a very useful cache. Parallelism is inherent and easily accessible. The fairly standard choice of a bit-serial processor provides considerable flexibility of precision and data representations. Most of the claimed drawbacks of the earlier machines, such as the limited range of the local connectivity, the difficulty of mapping arbitrarily sized problems onto the processor array and the problem of achieving high speed data input and output, have been overcome by more recent systems such as the GEC GRID.

One of the first, and probably most thoroughly researched, machines to be actually implemented was the Cellular Logic Image Processor (CLIP). A series of CLIP prototypes has been developed by Duff and his co-workers at University College, London, with the 96 × 96 CLIP–4 array machine being perhaps the best known. CLIP–5 is an enhanced redesign of CLIP-4, but CLIP–6 represents a radical departure from the simple pixel per processor philosophy of CLIP–4 in that it consists of an array of 16-bit processors optimised to perform typical grey scale operations. The CLIP–6 array maintains the eight-way nearest neighbour connectivity of its antecedents. CLIP–7 is reputedly now on the drawing board!

The Massively Parallel Processor (MPP), developed by the Goodyear Aerospace Corporation, is a 128 × 128 SIMD array machine designed for numerical computations but also able to perform logical operations. It is well able to cope with a wide range of image processing operations, such as the statistical classification of pixels based on multi-spectral data and feature extraction. The array unit, consisting of 16 384 identical processing elements, 8 per chip, is linked to a staging memory, acting as a data input-output interface, and a sequential array control unit which generates instructions for the array and operates on scalar variables. A companion processor provides a link between the MPP and a host machine. It is capable of a raw processing performance of approximately 50 GBOPS (BOP = Binary Operations per Second) and a peak input-output rate of 320 Mbyte/s in its current 100 ns cycle time implementation.

The International Computers Limited (ICL) Distributed Array Processor (DAP) consists of a 64 × 64 SIMD bit-serial processor array embedded in the memory space of an ILC 2900 series mainframe. The processor is designed around a bit-serial full adder and incorporates X–Y buses for fast data shifting. The pioneering production machine is physically large, being implemented primarily in TTL, occupying a large air-cooled cabinet. Very much more compact gate array and VLSI versions are under development.

The UK General Electric Company (GEC) has developed the Rectangular Image and Data (GRID) processor in its Research Laboratories in Wembley, UK, which employs state of the art VLSI design and fabrication technology in its five-component chip set: array, input-output and three controller chips (see Section 3.4). A single GRID array chip, implemented in bulk CMOS VLSI technology, consists of an 8 × 8 array of bit-serial processing elements, each with 64 bits of on-chip dual-port RAM and capable of supporting up to 64 Kbits of off-chip RAM. The controller is a microprogrammed unit with the ability automatically to support sophisticated array-mapping algorithms. Applications programming is usually performed via a high level extended 'parallel C', which provides both single and multiple data stream control constructs. One unusual aspect of the GRID's sophisticated parallelism is its ability to provide the basic computing structure for a 'blackboard driven' system. Blackboard models are expert systems with a modular architecture consisting of multiple knowledge sources, represented as collections of production rules grouped into stimulus and response frames, a global database or blackboard, and a scheduler or controller. One use for GRID is as a precondition tester. All the preconditions of the production rules can be stored in the array memory and tested simultaneously, thus minimising the time required to identify the rules applicable to data newly available to the system. Another use is to store digitised images in the GRID array and encode the rules as subroutines for the GRID controller. These subroutines perform feature extraction and recognition tasks. These recognised features are then available to more abstract knowledge sources resident on the host machine.

The Nippon Telephone and Telegraph (NTT) Adaptive Array Processor (AAP) is built around an LSI chip comprised of an 8 × 8 array of one-bit processing elements, each of which consists of a RALU and two data transfer units, as well as 6 Kbits of external RAM. The chip has been fabricated in *n*MOS and contains about 81 000 transistors packed onto an area of 10.0 × 9.4 mm^2. This high packing density is the direct result of the geometrical regularity of the processor array.

The Hughes 3-D machine (Grinberg *et al.*, 1984) is a brave attempt to implement a SIMD array machine using wafer scale technology to provide a true 3-D machine organisation. The 3-D organisation has the advantage of reducing the average interconnect length from O(N) to O($\sqrt{N}$), where N is the number of circuit nodes. The resulting highly modular structure of the architecture provides for easy programmability and instruction rates of circa 10 GOPS, for a 128 × 128 implementation. The basic machine will consist of an array of 128 × 128 identical elements contained on a stack of silicon wafers. The layout consists of an array of processors arranged hortizontally, with the elements of each processor in the array assembled vertically. Consequently, if one looks down on the stack of wafers one sees an array of primitive computers, with the individual elements of each computer arranged vertically through the stack. Signals are passed vertically through the stack along a wire-like bus composed of wafer feed-throughs and 'microbridge' interconnections between wafers. Each wafer in the stack will contain a complete 128 × 128 array of a particular type of processing element, such as registers, comparators or accumulators. All the elements of any particular wafer array will be identical, enabling the incorporation of only those elements required for a particular application, with the attendant savings in cost, reliability and size. Five basic wafer types – memory, accumulator, replicator, counter and comparator – have been defined, of which only three are needed for most implementations. A prototype two-wafer partial machine has been constructed, with a full 128 × 128 machine some three or four years away. An unanswered question at present seems to be that of how to cope efficiently with wafer defects.

A very effective method of coping with some low level data driven image processing tasks involves the use of systolic arrays (Kung, 1980), especially for such operations as the inner products used in statistical pattern recognition (see Section 1.4.2). Their usefulness for higher levels of processing is dubious, though Shapiro (1984) has proposed a scheme for constructing general purpose parallel computers which incorporate the systolic approach as a software oriented methodology for algorithm design and implementation. His abstract machine is an 'infinite processing surface' of rectangularly disposed processors, which can be approximated by various torus based architectures.

MIMD machines, with their far greater generality, are obvious candidates for high performance image processing systems. Very different approaches are exemplified by the following examples.

The 'Connection Memory' developed by Hillis (1981) at MIT is a memory module for concurrently manipulating knowledge stored in semantic networks. Each node and link in the n-cube connected network has its own simple processor consisting of a few 32-bit registers for indexing and temporary storage, a state vector and a rule table. The rule table is identical for all processors. These connect to form a uniform locally connected network of processor-memory cells. All communication takes place by packet switched message passing, in the spirit of Hewitt's 'actors'. The communication connections are dynamically configured in such a way as to mimic the structure of the specific problem being solved. For example, the relational structure of a complex image could be captured by such a network. A 128 Kword version is currently under construction at Thinking Machines Inc., Boston.

Another highly concurrent experimental machine which makes use of a message passing model is the 'Cosmic Cube' (Seitz, 1985), consisting of 64 single-board 16-bit computers interconnected by a communications network with a six-dimensional hypercube connectivity. At present it can perform demanding scientific and engineering computations some five to ten times faster than a DEC VAX/780.

The Inmos Transputer, due for release sometime in 1985, is a high performance process oriented microprocessor specially designed for concurrent operation. It features a 32-bit 10 MIP RISC processor, 4 Kbytes of static RAM and high speed input-output interfaces on a single 250 000 device CMOS chip. Its native programming language is the 'process' and 'channel' oriented Occam, which is ideally suited for programming concurrent operations.

Already image processing researchers are looking forward towards even more exotic machines on which to carry out their computations. Much of this interest is motivated by the conflicting requirements imposed by low level processing, such as feature extraction, as opposed to high level operations like feature and object recognition. Most approaches involve a gradation of processing functionality from the computationally intensive data driven lower levels to the more abstract hypothesis driven higher levels. This gradation of processing functionality is commonly called the 'processing cone' (discussed in Section 2.2). Uhr (1984) has proposed that machine architectures for image understanding should reflect this pyramid structure.

Such a machine would consist of arrays of SIMD processors interleaved with arrays of memory, converging from a base, or low level layer, to a small number of processors at the vertex of the implicit structural pyramid. Each layer would execute the same instruction on different data fetched by each processor operating in the SIMD mode. Such machines have the advantages of having good global message passing characteristics, good local connectivity and the ability successively to transform, abstract, combine and converge fairly naturally.

An even more radical approach would be to mix processing modalities so that the lower level processing is performed by a high computational throughput SIMD array whilst the higher level processing is performed by an AI oriented MIMD architecture, with a sufficiently large interleaved layer of memory to provide an intermediate 'blackboard'. These architectures could provide the best of two worlds for complex image understanding tasks and perhaps make possible some of the more exotic applications to be discussed later in this chapter.

Taking a long term view, there is a growing interest in attempting to establish the actual physical limits which ultimately constrain the speed, size and dissipation of computing systems (Mead and Conway, 1980b). This is an abstruse subject area, more the realm of the theoretical physicist than the computer scientist. Early work was concerned primarily with logical, as opposed to physical, structures, and assumed the computation process was inherently irreversible because of the discarding of information about the computer's history in performing such computational operations as erasure or overwriting of data. These information reduction processes require of the order of $kT\ell n2$ energy dissipation per operation (about 10^{-21} joules at typical room temperatures). Models of general purpose computers, both classical and quantum mechanical, have been proposed, which are explicitly thermodynamically reversible; computation is carried out in logically reversible steps at the cost of no energy dissipation. Whether these idealised systems are relevant to physical computing structures is controversial. A further source of energy loss is in performing the read operation of the result of the calculation, which is analogous to a measurement. Furthermore, in a realistic system, the transmission and storage of information within the computer requires a signal energy of at least order kT per bit transmitted in order to prevent the signal being swamped against the background of thermal noise. A similar amount of energy is dissipated in order to maintain the structural integrity of what is essentially a non-equilibrium system, and so combat the spontaneous decay of any stored data. The interested reader is referred to Mead and Conway (1980b) and to some of the relevant papers listed in the Bibliography at the end of this chapter.

7.2.3 Artificial Intelligence

A topic that is increasingly going to impinge on the processing of images, especially at the higher interpretive levels, is Artificial Intelligence (AI). The study of AI is now a large, diverse and burgeoning field of endeavour, and one which is growing in relevance and vigour by the day. If anything, the danger in this sudden stardom which AI has assumed in the last couple of years is that far too much is being expected far too soon, both in terms of seminal developments and the concomitant profits, which the various AI start-ups seem to regard as self-evident. There are currently far too many examples of weak theories of knowledge chasing hard problems, and uncontrolled extrapolations from limited successes, for one to be too sanguine over the longer term commercial prospects. Many new venture activities seem more concerned with fuelling fads than satisfying evident needs.

The history of AI to date is brief, diverse and fascinating. An excellent and very readable coverage can be found in McCorduck (1979). As far as the overall field of AI is concerned, the three-volume *Handbook of Artificial Intelligence* (Feigenbaum *et al.*, 1981–2) is probably the key initial source of reference. The emerging commercial uses of AI are covered, albeit somewhat superficially, by Winston and Prendergast (1984).

In this section we shall attempt only to elucidate, in general terms, the nature and content of some facets of AI and then go on to explore its potential for supporting and facilitating sophisticated image processing systems.

What is AI and how does one make use of it? These are not easy questions to give hard and fast answers to. Defining 'intelligence', in the first place, is difficult enough, let alone deciding what we mean by 'artificial'. It has been suggested that intelligence is that attribute of human behaviour that we admire but do not understand! In general terms, AI is best regarded as that part of Computer Science concerned with creating and studying computer programs or computer systems that exhibit performance characteristics which we would normally identify as 'intelligent' in the context of human behaviour. Some such characteristics are knowing, reasoning, learning, problem solving, reasoning by analogy, language understanding and coping with uncertainty. At the back of many an AI researcher's mind is the notion that AI may help unravel some of the deeper questions relating to human cognition. In this context computation is being explicitly regarded as an *ad hoc* metaphor for cognition, with the implied goal of being able eventually to construct an information processing theory of intelligence. This metaphor has its own particular danger, in that a deep understanding of the human mind may still not enable us to develop truly intelligent machines. For example, the detailed study of birds has not provided much of an insight into how to build aeroplanes. Real understanding of how to do this came from the study of flight; not birds. One must conclude that deep problems are manifest in attempting to model nature explicitly rather than in terms of its underlying physical principles. After all, man has been spectacularly unsuccessful in constructing flying machines which flap their wings. We must conclude that intelligent behaviour on the part of a machine no more implies a close functional equivalence between machine and brain than there is between an aeroplane and a bird. It is essential in the long term that the computation-cognition metaphor be regarded as a bit of a red herring; this is an important point, because otherwise it will be difficult to produce intelligent systems which significantly outperform humans.

This deep yearning on the part of some researchers for understanding in AI to effectively unravel the intricacies of the human mind may or may not bear fruit, but there is no doubt that AI is steadily moving towards the development of a technology capable of producing intelligent systems. This will certainly have a profound effect on the lives of our descendants, if not on ours. Already computer programs exist which perform complex tasks or solve hard problems, which we would normally regard as requiring considerable intelligence, better than most humans could. In a very real sense it can be said that much of the current activity in AI is aimed at transforming complex, ill defined problems into tractable and suitably defined problems.

Mathematical logic and symbolic processing have been and still are central and energising concepts for much of AI. Indeed it was very early in the history of computing when Turing and Church, in the process of establishing a link between the notion of formalisation of reasoning and abstract computational machines, developed the concept of computation as symbol processing. In fact, these early pioneers of computing had already seen that numbers were an unessential element of computation – just one, but by no means unique, way of interpreting the internal states of a computer. After all, a 'Turing machine' is primarily symbolic rather than numeric in terms of its internal data states.

The first significant successes revolved around the development of 'problem solvers', especially programs which could play games such as chess, backgammon or checkers. In the process, the basic problems of search, 'pruning' of the search space, planning and choice of an efficient representation were all established as important enabling concepts. Other seminal developments have involved learning, reasoning, automatic programming, robotics, machine vision, expert systems and natural language understanding. Of these, machine vision and expert systems, the latter sometimes described as knowledge engineering, are the topics most germane to our discussion here.

Contemporary machine vision research is concerned with some aspect of enabling computers to understand the three-dimensional structure of their surroundings. There are quite a number of different views as to which is the best approach to this problem, such as binocular stereo, structure from motion, shape from shading, photometric stereo, top-down *vs.* bottom-up, connectionist, etc., etc. These have already been discussed in Chapter 5.

Expert systems are a means of embodying specific human skills within a computer program, though it should be pointed out there is no generally agreed definition. Hayes-Roth *et al.* (1983) regard expert systems as a means to investigate and utilise methods and techniques for constructing man-machine systems with specialised problem solving expertise, consisting of knowledge about a particular domain, understanding of domain problems, and skill at solving some of these problems. Early applications of expert systems have shown considerable commercial and industrial potential, allied with some enthusiastic investment, in areas as diverse as medicine, mineral prospecting, biotechnology, computer system configuring and chemistry. These existing systems are clearly restricted to applications with limited or highly structured domains, and much research is still needed if expert systems capable of coping with fairly general domains involving several different kinds of knowledge are to be at all successful. Most expert systems developed to date exhibit a tendency to trade off the degree of expertness with the conceptual depth of the underlying domain model. Intrinsically deep models tend, in general, to produce only a fairly basic common sense level of competence whereas a more superficial, but detailed, model will support a level of competence closer to that which we might regard as expert.

Expert systems are an obvious candidate for facilitating the mechanisation of high level vision. Much of the work to date in this area has grown out of the US DARPA-sponsored Image Understanding Programme, started in 1975. Its purpose was to investigate means of using *a priori* knowledge to establish the interrelationships between objects in a scene, especially those which have the potential to be automated. Efforts within the programme covered a broad spectrum, from elaborating cohesive theories of vision to hardware issues affecting actual implementations. Applications ranged from working within the confines of stereotyped 'blocks worlds' to the understanding of aerial photographs of complex terrain features. A general coverage of this work can be found in Cohen and Feigenbaum (1982). Unfortunately, there are really no eloquent experts in vision. People know that they see, but they do not know how they do it. As a result, vision expert systems usually end up using a collection of arbitrary rules satisfying the programmer's personal tastes and prejudices rather than embodying the knowledge of some real human expert. The limitations of these existing image understanding systems are painfully obvious, even to an optimist.

The intrinsically more complex and sophisticated intelligent knowledge based systems (IKBS) are presumed to be truly knowledge based rather than narrowly and pragmatically rule based as is the case with expert systems, which can in a very real sense be regarded as proto-IKBS. IKBS can be defined as semi-intelligent systems for carrying out complex tasks. This implies working with a large, incomplete, uncertain and rapidly changing knowledge store, use of inferential procedures for applying this knowledge in reacting to variegated and unreliable inputs in a changing environment, and the use of sophisticated and flexible control mechanisms. It is in a sense both applied AI and the central energising activity of AI. IKBS are the future intelligent systems of information technology, of which current expert systems can give us only a veiled glimpse. The whole concept of intelligent systems is bound up not only with what can be done, but also with how it is done. Learning is an important but hard problem in this context. Taken to its limit this implies some measure of self-awareness or self-referentiality.

Some of the key generic characteristics of IKBS are listed in Table 7.1 (Sloman, 1983).

Table 7.1. Possible characteristics of generic IKBS

Possess rich and flexible domain-specific knowledge bases.
Contain powerful and varied descriptive resources, formal and representational.
Use rigorous inference procedures; general and specific, logical and non-logical.
Able to reason under uncertainty.
Self-monitoring and able to learn.
Utilise meta-principles and other higher order knowledge.
Incorporate strategies for controlling search.
Use matching and concomitant match descriptors.
Support communication between subsystems, especially via 'demons' acting on data bases such as 'blackboards'.
Embody very large, very fast, flexible, relational content-addressable memory systems.
Facilitate rapid reorganisation of memory to reflect nature of intelligent processes.
Embody parallelism, fine grain, and course grain, for problems that decompose naturally into a number of modules or subtasks.
Possess versatile interrupt mechanisms to cope with complex and unpredictable situations.

Perhaps the most basic IKBS consists of an appropriate data base of facts and rules, together with an 'inference engine' which acts on the data base by selecting rules and then making any concomitant changes on the stored data. This processing can be either data driven or goal driven and can be made more sophisticated by incorporating likelihood values, uncertainty propagation and meta-rules.

An attractive alternative strategy is to have collections of active modules, possibly quite different from each other, communicating via a global 'blackboard', which is just another name for a globally accessible data base. This approach lends itself very naturally to concurrent organisation and has a lot of appeal for image processing. Another alternative is to broadcast messages to named classes of modules, establishing in essence an 'actor' architecture.

To look even further ahead, Advanced Information Processing Systems (AIPS) are, at least conceptually, the true IKBS of the future. (*Some of the notions underlying this discussion of AIPS were elucidated during the course of a study, supported by the CEC, undertaken during the ESPRIT 'preparatory phase' by the GEC Research Laboratories and Logica (UK)*

Ltd.) They will emerge from the combining of advances in advanced computer architectures, knowledge bases, inference techniques, programming languages, high bandwidth communications, user interfaces and micro-electronics. One important aspect of their functionality will be their ability to handle and process images.

Certain general characteristics of AIPS can be stated quite simply. They will obtain information directly from their environment by means of sensors and activators, and through dialogue with human users and experts. They will relate this information to knowledge already stored within their systems and will create new knowledge using rules and inference. Typically they will have to be able to cope with information inputs which may be unreliable or even in conflict; with large, incomplete, inconsistent and rapidly changing data bases; and with inference procedures which may be tentative, incomplete and perhaps of limited applicability. Two particularly desirable aspects of AIPS will be the ability to explain results and describe known patterns of inference as well as having the ability to learn new information, new rules and new patterns of inference directly from experience.

Such systems should be contrasted with the majority of contemporary computer systems which tackle particular problems by applying predetermined solutions in the shape of narrowly focused algorithms with highly circumscribed control regimes. Figure 7.2 schematically illustrates a simple data base oriented information system and its environment, based on a model proposed by an ISO working group (ISO, 1982). The 'universe of discourse' is the collection of all entities that have ever been, are, or ever will be in a selected portion of the 'real' world or some 'abstract' world. The 'conceptual schema' is the description of the relevant possible states of the universe of discourse, including the classifications, rules, laws, etc., of the universe of discourse. The 'information base' is the description of the specific entities that in a given instant, or period of time, are perceived to exist in the universe of discourse and their actual states that are of interest. The 'information processor' operates to produce changes in the information base or conceptual schema, but only on receipt of a 'message' which contains legitimate information and commands. The 'environment' is where these messages originate from. It is a part of the real world, which may or may not intersect with the universe of discourse. On receipt of an appropriate command, the information processor may itself provide details, as a message, about information contained in the conceptual schema and information base.

There are some important points, well worth noting, implicit in this formulation. The information system is a formal system, whereas the environment is not. The behaviour of the information system is completely defined by behaviour rules and constraints which are established, directly or indirectly, by the environment. The information system on its own initiative never establishes rules for the environment. An information system, being fully predictable, is unable to deviate from its rules and constraints whereas the environment can deviate from its rules. An information processor has no

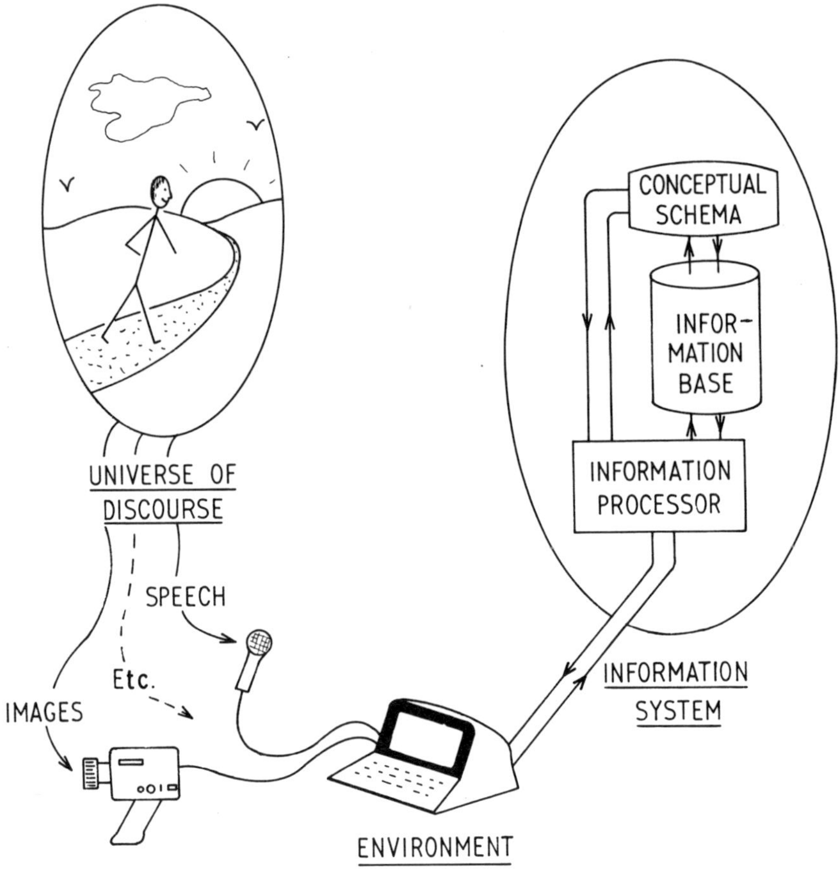

Fig. 7.2 An information system and its environment.

initiative of its own. It simply follows specified rules, transferring messages between the environment and the rest of the system. A 'user' is anybody or anything that engages in a dialogue with the information system via messages. A user is thus part of the environment. Some users, the group of experts, have authority to establish behaviour rules or constraints for the information system.

There are both static and dynamic aspects of the conceptual schema and the information base. The conceptual schema may change in response to changes in the universe of discourse. Dynamic aspects are also involved in the description of manipulations which are needed in order to make known parts of the conceptual schema and/or information base.

How does the generic concept of an AIPS differ from this rather general model of what is essentially a data base system? Perhaps the most crucial distinction is the more adaptable, less predictable, nature of the AIPS, which generally not only employs a variety of inference techniques but may also adapt those techniques in the light of experience, thus operating at a higher (meta) level of inference. In the most basic of terms the fundamental distinction is that an AIPS is a part of its own universe of discourse and environment. This essentially recursive definition enables generalised AIPS to recursively embed further sub-AIPS and so on. Any model of an AIPS should reflect this situation by making no distinctions between environment and domain of discourse. Figure 7.3 schematically shows the consequent generic model of an AIPS.

The IKBS has two main roles. It relates input signals to internally stored knowledge by interpreting them in terms of 'knowledge abstractions' compatible with the appropriate conceptual schema, and it applies inference to obtain further knowledge. Care has to be taken to distinguish between the two external interfaces, the EIF and the HIF, giving particular prominence to the pre- and post-processing of communications with human users. The HIF involves significant human factors aspects, with potential social, political and general cultural dimensions. A human user can also provide external input to the EIF; however, this is a far more restricted, non-interactive role for the human as a surrogate peripheral processor. By contrast, the HIF involves dialogue between the human user, attempting to find a solution to a problem, and the IKBS as well as between a human expert and the IKBS for the purpose of supplying or obtaining new expertise

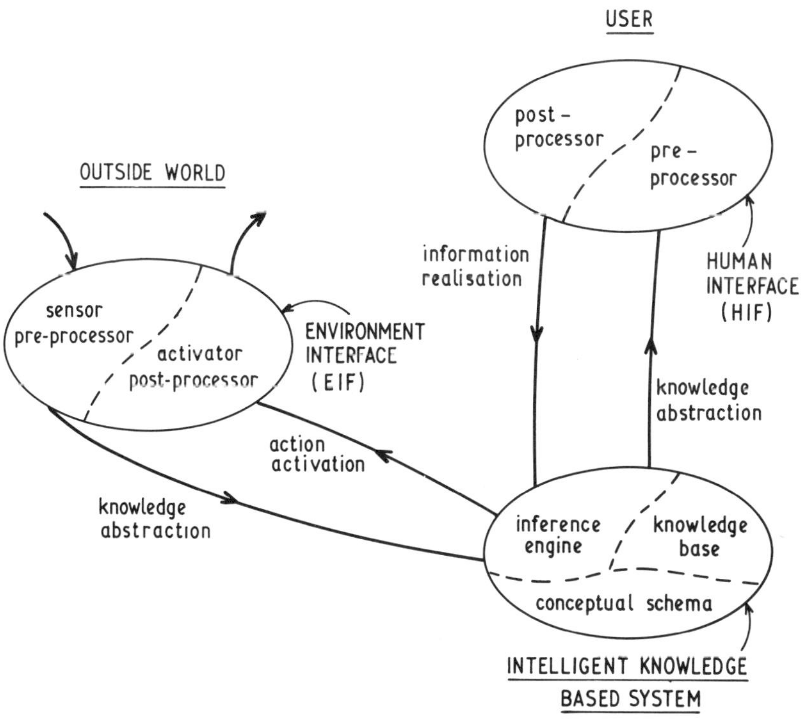

Fig. 7.3 Generic model of an AIPS.

at a meta-level of interaction. Other aspects of the EIF include sensors, activators, preprocessors and postprocessors. It is conceivable, but somewhat problematic, that the first two items are present in the HIF.

It is worth noting that a particular implementation of an AIPS should itself be a continually evolving system. To be more precise, included amongst the experiences to which an AIPS should adapt is its own implementation. This clearly is an ideal which may be difficult to realise. It introduces yet another recursive element for AIPS. Simply as a first step towards this goal, the design of an AIPS must take account of the extensive process of knowledge and expertise acquisition which is required before the system can function effectively. This is the process of 'knowledge engineering', providing an evolutionary first step in the evolution of AIPS.

The development of AIPS will depend critically on three of the central themes of AI research: the development of new languages, the development of new inference supporting computer architectures and the developing predominance of a knowledge based approach to AI over a generalist approach. Two of the most difficult problems arising within these fields are how to represent facts about the world, and how to programme a machine to learn from its experience. The facts about the world, in the present context, should include facts about the AIPS itself. This means that there are going to have to be techniques available for supporting clear semantic models of real-world knowledge, both in general and for specific domains in the real world.

These AI techniques will take a long time to appear in a versatile, robust and useable form. The emergence of true AIPS will be a gradual process. They will evolve slowly and incrementally from the information technology infrastructure which today involves predominantly VLSI devices, advanced computer architectures, data storage, displays and communications. These areas are becoming increasingly interdependent and focused on distributed systems. Soon it will be impossible to talk of one area in isolation. Even today it is difficult to discuss computer technology without mentioning, *inter alia,* device technology, data storage, communication and systems integration.

A related and potentially very important role for AI technology is in the arena of education. This applies to both explicit computing technology oriented training and more ubiquitously to general purpose 'computer aided instruction' (CAI). Here the notion of a picture, especially a coloured and dynamic one, being worth a thousand words is fundamentally true and worth taking advantage of. Education and computers seem to be destined for each other. As the amount of information increases, so does the need for effective, self-paced learning in a flexible instructional environment. As is the case elsewhere, much of the UK's education system is becoming increasingly bogged down by high student-to-teacher ratios, limited flexibility and an inability properly to assist the highly motivated student. Computers offer a potential remedy for this situation. The fundamental strengths of the ubiquitous personal computer as a purveyor of CAI are its immediate availability, high level graphics and real-time response.

Some interesting work, recently undertaken at MIT, has taken this a stage further and looked at the fusion of video games and CAI. Video games have the power to focus the player's attention for hours at a time, despite the often frustrating and difficult goals intrinsic to the games. If the same concentration could be focused on educational tasks, the learning rate and sense of achievement of students would grow by leaps and bounds. The basis of this fascination with video games, whether it comes only from motor-sensory stimulation or some additional combination of aggression and goal achievement, is only now being studied in any detail. Perhaps the most important feature of most video games is that the player never wins. The stakes just get higher. This factor, coupled the basic video game elements of fantasy, curiosity, challenge, uncertainty, revenge, evasion, progression and goal attainment should, in principle, be equally applicable to education. It may well turn out to be the case that the quality and the cost effectiveness of the accompanying graphics, image generation and animation facilities will be the deciding factor.

7.3 THE FUTURE

One of the most significant characteristics of most modern societies is the centrality of the concept of information in its various and manifest guises. The central meaning of any dictionary definition of information is that it refers to structured data in some manifestation or context. (We imply the existence of definite *ad hoc* operational distinctions between data (symbols), information (structured data) and knowledge (information plus rules for its utilisation.) Perceptive readers will not wholeheartedly agree with these trite but useful distinctions (Machlup and Mansfield, 1983).) It is in this very general sense that we shall regard it, because it ties in naturally with one's innate perceptions of its role in our own lives and in society as a whole. On the surface, the role of information in pluralist, democratic, libertarian societies appears to be very different from that in authoritarian regimes, but the underlying notion that 'information begets power' is, in the end, common to both. It is just the superficial aspects which are genuinely different. It is the fundamental and ubiquitous nature of information in our lives, combined with the rapidly increasing computerisation of its generation, control, transmission, dissemination and storage which makes it relevant to our present discussion. Increasingly, in the future, much information is going to be represented, and subsequently processed, as digitised images, be they X-ray scans, Landsat images, video film or whatever. This is no more than a reflection of the fact that our information processing channel with the highest bandwidth, by a long way, is the visual one. It is this primacy of images in information representations that renders a digression into the possible social impact of information processing germane at this stage of our discussion.

A fundamental deficiency of today's computers stems very much from the input-output mechanisms. Computers still cannot communicate with human beings in natural forms, such as spoken or written languages, pictures or images, documents and illustrations. Existing computers are far from satisfactory in their input-output speeds, especially when it comes to real-time speech, vision and translation. To develop 'human oriented' interactive computers we must first upgrade their capability to understand natural information representations and then respond to them intelligently. There are three major areas of information processing where technological advances are going to have a major significance. These are natural language understanding, speech processing, involving both analysis and synthesis, and what is the subject of this book, image processing and understanding.

In much of what has been discussed so far we have been concerned primarily with technology, its advancement and its application. To balance our discussion it is more than appropriate to look briefly at how these advances will impinge on our lives and those of our fellow citizens. We will look briefly at some of the more significant of the possible consequences of the increasing sophistication and ubiquity of information processing, with the implicit understanding that images or image metaphors, in a large number of different guises, will constitute much of the underlying information base.

A concept that is increasingly invoked in discussing the adoption and penetration of the results of information technology is the eventual emergence of an 'information centred society'. To some this evokes the threat of some Orwellian nightmare, while to others it heralds a new order of civilisation comparable to that ushered in by the industrial revolution. No doubt the truth will lie somewhat between these two extremes.

What characteristics can we predict for such a society? Table 7.2 lists some of the factors which will certainly be relevant in the establishment of such a future.

What is certain is that there are going to be some fundamental transformations in life, work and social structures. Some basic potential problems have to be recognised and adequately dealt with. The more important of these are listed in Table 7.3.

One country where there is a better internal consensus than elsewhere on the advantages inherent in an information centred society is Japan. There are a number of internal factors which facilitate the achievement of such a consensus. These include a relatively homogeneous, well educated (especially in science and engineering) population, a committed social ethos, a large consistent internal market, positive governmental attitudes and a generally better internal consensus than exists elsewhere. The Japanese are not significantly better than others at creating and developing new ideas but they do have the social ethos to support rapid discussion, dissemination and consequent action as well as an education system that supports a consistent intellectual diffusion in areas such as information technology (Karatsu, 1984). The Japanese 'New Generation Computer Programme', discussed in Section 7.3.1, is a case in point. There is a widespread view in Japan that the

Table 7.2. Characteristic themes underlying the emergence of 'information societies'.

1. An increasing role for information and knowledge in providing a driving impetus for society. They will become, in a sense, the new 'capital' underpinning future expansion.
2. The emergence of post industrial societies. As with the diminution of farming effort, now below 10% of the work force in many western countries, so the same will happen with industry as the information society emerges.
3. Shrinking of social space, with corresponding moves towards a world community and a universal language.
4. Expansion and diversification of information sources, with attendant 'democratisation' of information. The growth and specialisation of newspapers will be mirrored in similar media, such as cable services.
5. Centralisation and decentralisation will emerge as opposing, economically and politically motivated, themes. Centralised bureaucracies and the 'electronic cottage' represent two extreme manifestations.
6. Expansion and systematisation of knowledge, with its implicit threat to established specialist 'priesthoods'.
7. Control over production and distribution, accompanied by very much reduced numbers of workers involved in manufacturing. Information will become central to effective manufacturing automation.

economy is ready to move forward rapidly into a period of change that promises to be revolutionary in its overall impact and raises the possibility of catching a long term upswing in world wide economic growth.

Other advanced industrial countries, such as the UK and the USA, do not seem to be able to muster the single-minded enthusiasm displayed by the Japanese. Perhaps this is no more than a reflection of their intrinsic diversity of population and purpose.

However it arrives, quickly or slowly, easily or painfully, the information society is an almost inevitable consequence of the technical, commercial and political imperatives current in the world today. A significant part of the information that will be the new currency of our societies is certain to be in the form of images of some sort or other; video or graphical, synthetic or real, etc. The following four sections will seek to explore briefly some of the possible diverse applications, embedding some aspect of image processing, which we might expect in the near and more distant future, especially those facilitated by advances in computer design and fabrication technology. The various applications selected are intended to be representative and illustrative, rather than in any sense complete. In a sense they reflect no more than the interests and enthusiasms of the author.

Table 7.3. Some problems intrinsic to the development of an 'information society'.

1. How do we achieve equality of access to information? This is essential if one is to avoid alienating disadvantaged minorities, such as those with inadequate computer literacy. At least initially, the growth of information societies in the technically advanced nations will certainly exacerbate the disadvantaged positions of the third world countries.
2. How do we match job openings with available labour? Skill mismatches already exist in many western countries, with attendant high unemployment. How is education and retraining to be implemented? How do we protect communities reflecting established, but now decaying, industrial bases? It is less a problem of wealth than of social stability, so how do we achieve the latter?
3. How do we achieve a rapid and encouraging change in the public arena? By 'democratising' information? If so, will there be less of a need for governmental bureaucracies?
4. How do governments manage the increasing complexity of the underlying structure of an information society? This increasing complexity should not be mirrored in a mass of new and restrictive legislation. This will only result in perpetual deadlock over any controversial issues.
5. How do we get international cooperation to cope with the rapid growth of information societies? Such matters as spectrum use and satellite 'footprints' are key issues.

7.3.1 New generation computer systems

Image processing requires computers in a much more general context than the specialised high performance machines discussed so far. In practice most of these specialised machines are going to be seen as embedded engines which are attached to more general host machines. So what has the future to offer as far as these more general purpose machines are concerned? This section could just as legitimately have been titled 'Fifth Generation Computer Systems' (FGCS) but because of the lack of consensus on 'generation' terminology and recent speculation on sixth generation machines we will for generic purposes use the more general term 'new generation' and keep the term 'fifth generation' (or 5G) to describe the appropriate Japanese initiative. Speaking at an SRI International conference on advanced information technology, held in Cambridge, UK, during September 1984, Professor Hideo Aiso gave some brief details of putative Japanese plans for a 'sixth-generation' computer systems project. The main research targets would be the analysis of the structure of the human brain along with the application of biochemistry to the development of 'biological' chips, the development of optical computing, advanced communications and relevant aspects of human psychology. Perhaps the intended distinction is that these machines may actually be really intelligent as opposed to merely manipulating knowledge in fairly circumscribed ways.

What is new generation computing? As first envisaged by the Japanese, and now endorsed by much world wide activity, they are computer systems which are directed primarily at knowledge and information processing, based on the enabling concepts of innovative architectures, inference mechanisms and VLSI technologies, and developed with the express purpose of meeting the perceived requirements of the 1990s. These systems include both intelligent interaction between man and machine and inference using knowledge bases. The functionality of such machines falls fairly naturally into four categories: problem solving and inference, knowledge bases, intelligent interfaces and intelligent programming techniques.

The problem solving and inference functions are intended to enable the system to find solutions to problems by carrying out logical reasoning using data and knowledge stored in the system as well as information provided from the external world. This capability is intended to cover deductive inference, inductive inference, reasoning based on incomplete knowledge, and cooperative problem solving.

The knowledge function is aimed at providing systematic storage and retrieval of not only so-called data but also reasonable judgements and test results organised into consistent knowledge. Beside knowledge accumulation, it includes knowledge representation tailored to problem solving, knowledge acquisition and updating, and simultaneous utilisation of distributed knowledge sources.

The intelligent interface function is intended to enable computers to handle speech, graphics and images so that computers can interact with humans flexibly and smoothly. It might be regarded as giving computers the equivalents of human ears, mouth and eyes, but its primary objective is to provide computers with integrated communication abilities akin to that of man.

The intelligent programming function is intended to enhance the intelligence of computers so that they can take over the growing burden of programming from humans. While its ultimate goal is to achieve an ability automatically to convert functional specifications into efficient computer programs, it is aimed initially at achieving a modular programming system and a program verification system and at establishing appropriate specification and description languages.

In order to achieve these various functions some significant advances will be needed in such diverse areas of endeavour as parallel machine architectures, language semantics, formal techniques, knowledge based systems, speech processing, image processing, natural language understanding, and VLSI design and fabrication technologies. The full implementation and commercial exploitation of such machines is a long way off, but some interesting and useful 'fall out' can be expected along the way.

It should be mentioned at this point that new generation computing and the so-called supercomputer projects under way around the world are very different activities, certainly in Japan (Fernbach, 1984). The only significant common ground is the underlying need for advanced VLSI technology, capable of implementing considerable parallelism. Most supercomputer projects are concerned primarily with the development of super-fast machines aimed at complex computational and simulation problems in, *inter alia,* physics, meteorology and engineering. The raw power inherent in these machines will ensure their continued use in the future for such activities as the real-time or near real-time processing of large complex satellite-derived images. The fundamental requirement is for raw computing power rather than sophisticated inference or knowledge based support. A supercomputer is perhaps best defined as the most capable arithmetic processor available at any given time. Currently the emphasis is on achieving scalar performances of several hundred MFLOPS and vector performances of several GFLOPS by the early 1990s. Figure 7.4 summarises the trend in supercomputer performance since the sixties.

The emerging new generation computers will not suddenly emerge as gleaming sophisticated 'mega-machines'. Rather, they will start to appear as low cost, ultra-powerful, user oriented computer systems with massive versatile data bases, parallel processing, IKBS software and powerful interconnection facilities. In the final analysis computing is a continuum, with supercomputers and FGCS representing different and to some extent overlapping parts of this continuum. There is no reason to believe that in the longer term these two approaches to computation won't merge, as they do in the human brain. The Japanese new generation computing initiative, along with some of its earnest American and European imitators, will now be briefly outlined.

Information processing, based on emerging new generation computing techniques, is in some ways an ideal industry for Japan. It requires no material natural resources, is not energy intensive and is highly appropriate for a country which has one of the most highly trained labour forces in the world. Hence it is not surprising to find Japan establishing itself as the natural focus for the emergence of the next generation of computer systems.

The Japanese new generation computing initiative, sometimes called the Fifth Generation Computer Systems Project (Moto-Oka, 1982), is modelled on the very successful VLSI Technology Research Association, which was the central organisation of a four-year VLSI technology project (1976 – 1980). It was this initiative, supported by the Japanese Ministry of International Trade and Industry (MITI), which launched the Japanese semiconductor industry on a path which led to its now increasingly dominant position in world markets. In a very real sense this initiative and its very tangible success marked a turning point in Japanese attitudes to high technology collaborative ventures (Sakakibara, 1983) and, *de facto,* in the West's view of Japanese endeavours.

In April, 1982, MITI, in cooperation with eight leading Japanese

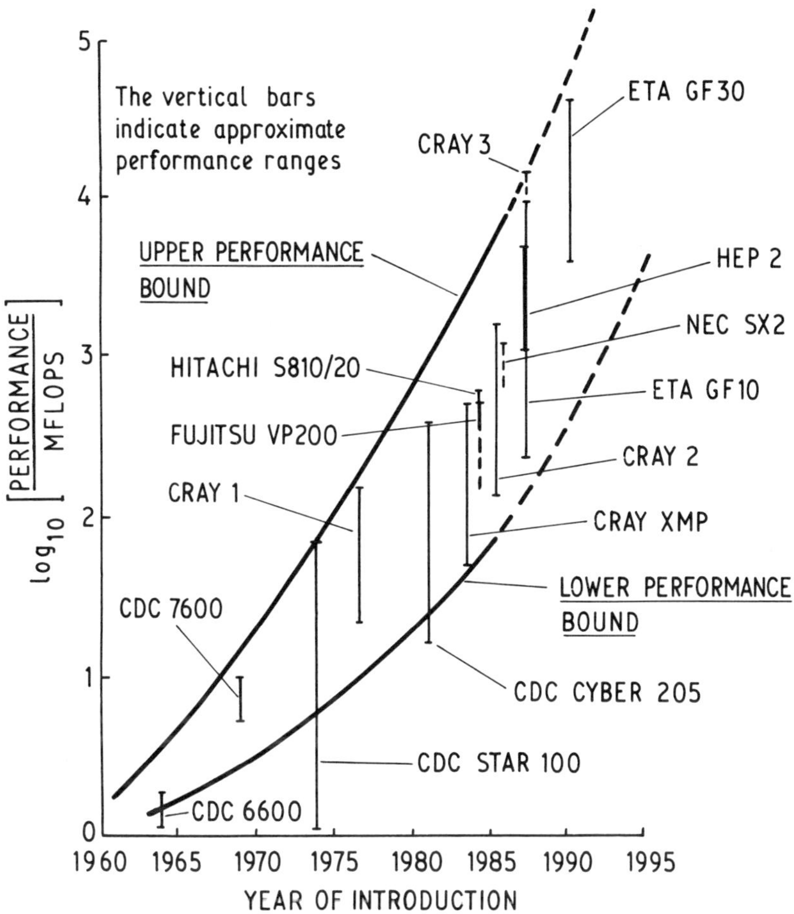

Fig. 7.4 The actual and predicted growth of supercomputer performance since 1960 (Fernbach, 1984; Hwang and Briggs, 1984).

electronics companies, launched the FGCS project. Spanning ten years, its aim is to establish the Japanese computer industry as a world leader, with all the concomitant prestige and commercial success. It is also intended to refute the increasingly hollow accusations, still heard in some parts of the world, that Japan tends to exploit only knowledge imported from abroad, without contributing any of its own to benefit the world at large. Hence, the project is aimed at original research and plans to make its results freely available to the international research community. A metaphor that repeatedly reappears in discussions with Japanese researchers is the comparison of their FGCS programme with the US space shuttle programme.

In the light of the above comments it is worth noting in passing that the traditional Japanese view of creativity is one that prefers to regard improvement as the most crucial element in its expression. In this context creativity implies a jump forward from a current context. It does not readily embrace the notion of an idea springing out from nowhere. Rather, it relies on incremental advances, dependent on a continuum of small ideas, which by themselves may seem insignificant.

The main focus for the project is the Institute for New-Generation Computer Technology (ICOT), situated in central Tokyo, currently with some forty research staff, the bulk of them seconded from industry. Stripped of its window dressing, the clear and consise aim of the ICOT researchers is to build parallel computers, whose machine (kernel) language is based on Horn clause predicate logic, and to interface them with data base machines whose data description and query languages are based on Horn clause logic. A Horn clause is of the general form:

$$A \text{ if } B1 \ \& \ B2 \ \& \ B3 \ \& \ B4 \ \& \ \ldots \ \& \ Bn$$

where *A* can be read as a procedure head and the *Bs* as its body. Many of the stated aims, such as giga-LIPS (LIPS = logical inferences per second) inference engines, have much more to do with consciousness raising than serious project planning. It is worth noting that despite the very coherent and single-minded external face presented by ICOT, there are internal misgivings over ICOT's almost obsessional prediction for logic programming in general, and the PROLOG programming language in particular. Logic programming can be summed up in the following two equations:

program = *set of axioms*
computation = *proof of a statement from the axioms.*

Some of the events leading up to this predilection for logic programming have been discussed by Shapiro (1983).

Figure 7.5 shows the now well known basic conceptual configuration of an FGCS. The system description and kernel language can be regarded as loosely defining the interface between the applications software and the machine hardware. The software unfolds upwards, and the hardware downwards, in a fairly self-explanatory manner. It is not intended to discuss this in detail, apart from pointing out that pictures, or images, constitute an important interface modality.

The ten-year project has been broken down into three distinct phases, as outlined in Fig. 7.6. The initial phase, which ended in April 1985, was concerned with establishing the fundamental technological elements required to build 5G systems. In the intermediate phase, the algorithms and basic architecture to be used in the subsystems that will constitute the fundamental building blocks of 5G systems will be determined and a small scale FGCS prototype developed. It is in the intermediate phase that the main work on image and graphics interfacing will take place, coupled with speech and natural language understanding. The final phase of research and development has as its goal the completion of a prototype 5G system, completely integrating all the results of the preceding two phases.

At a recent conference in Tokyo (ICOT, 1984) the ICOT researchers

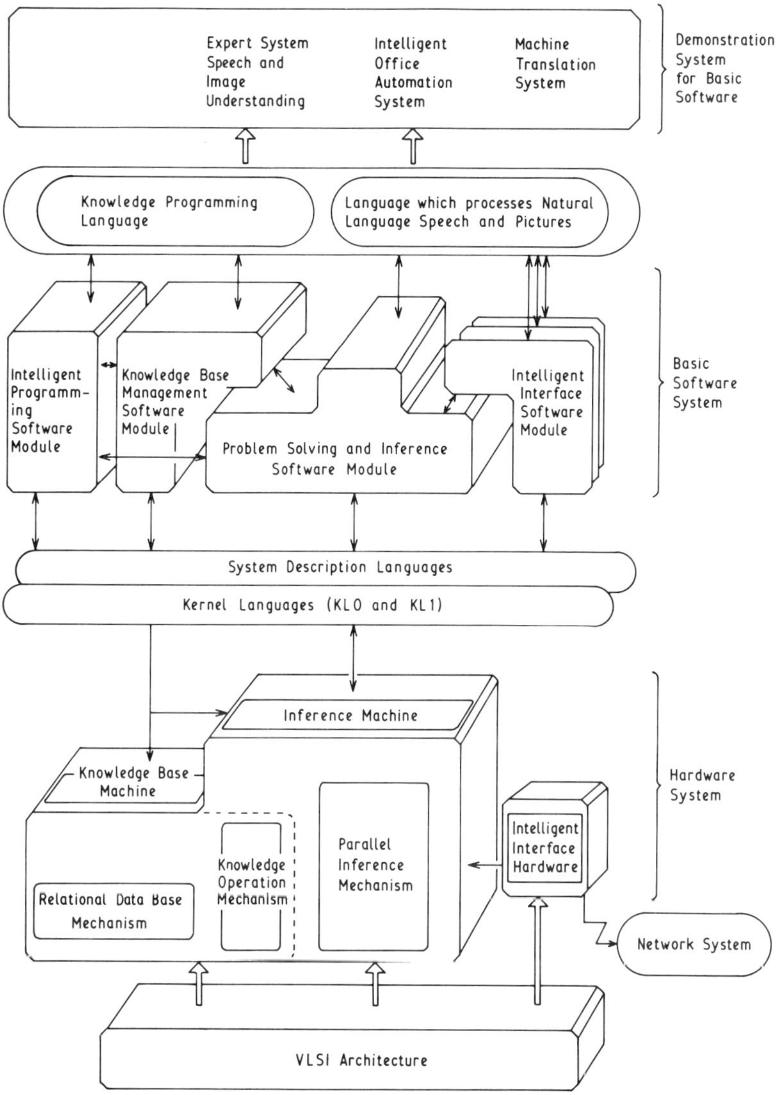

Fig. 7.5 The basic conceptual configuration of an FGCS.

announced, and demonstrated, some of the results of the first phase. These include a relational data base machine, 'DELTA', (Shibayama, 1984), a 30 kLIPS personal sequential inference machine (PSI) together with its 'SIMPOS' operating system, a knowledge acquisition support system (KAISER), a discourse understanding system (DUALS), a logic design support system, expert systems and parsers. The two most substantial items, seen as the workhorses of the intermediate phase, are PSI and DELTA. Neither of these prototype systems in themselves represent significant

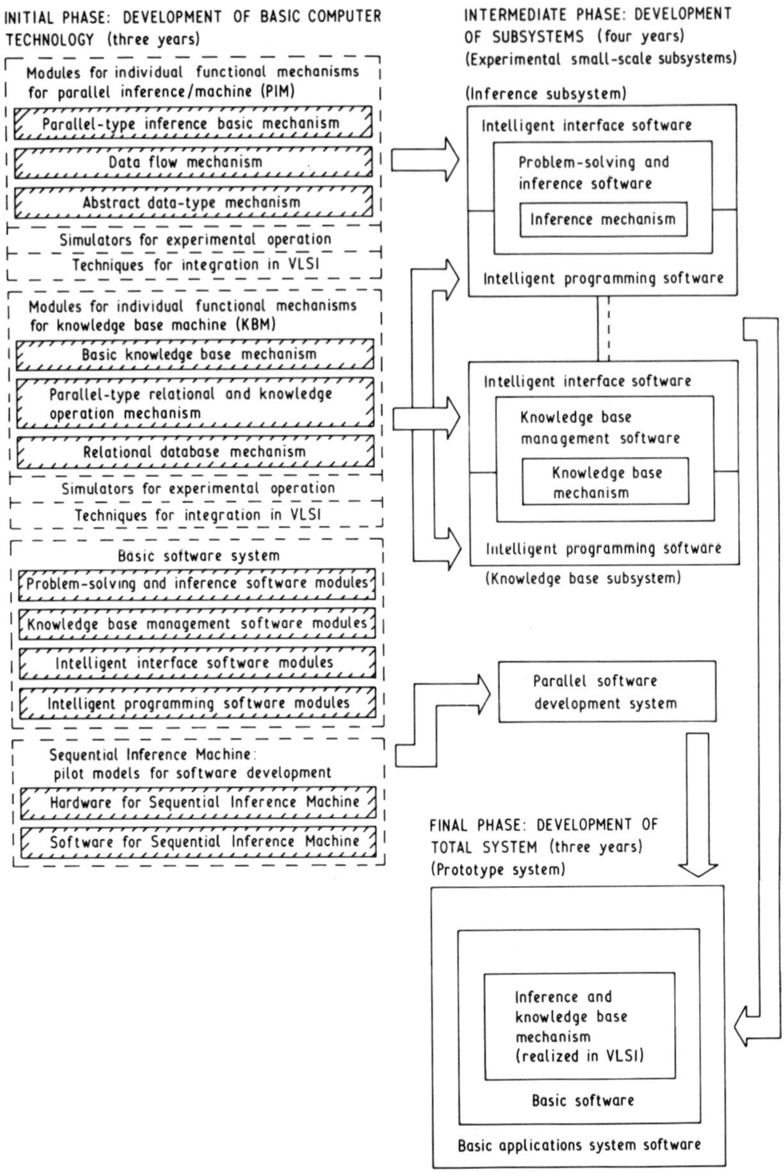

Fig. 7.6 The three phases of the ICOT FGCS research and development programme.

advances, e.g. PSI is very similar to existing US tagged architecture LISP machines, but they are 'home-grown' and in that context must be regarded as highly significant. Once again the Japanese have demonstrated their consummate skill and tenacity at catching up in a technological area where, until recently, they lagged considerably behind the USA and Europe. If the world wide effort to produce innovative new information processing parallel computers can be regarded as a race, then on present performance the Japanese deserve to win! Despite the programmes now established elsewhere (see below) the abiding ethos, especially the obsession with military systems, in the USA and Europe means that the rest of the world may well be reduced to becoming helpless bystanders as the Japanese forge ahead in the world's marketplaces. As it is we can expect to see the appearance of commercial products, such as cost effective, high performance personal LISP and PROLOG workstations, based on the initial phase deliverables in the very near future.

By all appearances the FGCS project is more or less on target. In a real sense, ICOT represents only the tip of an impressive iceberg, with the bulk of the really innovative research and development activities taking place in the big companies, the MITI Electrotechnical Laboratory and the Nippon Telephone and Telegraph Laboratories. We should look forward to the results of the intermediate phase and its concomitant activities with keen interest, if not trepidation.

The UK's Alvey initiative is an undisguised and direct response to the Japanese FGCS project. Named after the chairman of the *ad hoc* committee which first proposed the programme, it embraces the areas of VLSI design and technology, computer architectures, software technology, IKBS, man-machine interface and networks. A number of large scale 'demonstrators' have been set up which are aimed at facilitating the transition of new technologies into applications areas. The primary project mechanism is pre-competitive R&D involving companies and universities working collaboratively. The companies have to put up half their own project costs. A budget of some STG£350 million has been allocated for the first five years of the programme. There are significant long term programmes on both low level image processing and high level IKBS machine oriented vision within this initiative.

In addition to the UK's Alvey initiative, there are, within Europe as a whole, a number of somewhat similar national programmes as well as the pan-EEC ESPRIT programme. ESPRIT (European Strategic Programme of Research and Development in Information Technology), which was jointly conceived by twelve large European IT companies and the Commission of the European Community, formally started in 1984 after some one-year pilot projects had demonstrated the feasibility of pre-competitive research collaboration across national boundaries. The initial five-year, ECU1500 million programme emphasises three key enabling technologies, VLSI, advanced information processing and software technology, as well as two applications areas, office automation and computer integrated manufacturing. Various aspects of image processing are regarded as important components of a number of the research programmes covered by these generic topics. In the long term it is hoped that the ESPRIT initiative will redress a situation where at present Europe represents about one third of the world wide IT market but supplies only about one tenth of world wide production.

Within the USA both the Administration, through the Department of

Defence (see Section 7.3.4), and industry have responded to the perceived Japanese challenge. The US computer industry's most significant response has been the Microelectronics and Computer Technology Corporation (MCC), which is a collective US corporate response to the Japanese FGCS project. The MCC, based in Austin, Texas, is a cooperative long term research and development activity involving some nineteen US companies, including RCA, NCR, Boeing, Rockwell, Motorola, Sperry and Honeywell. At present the MCC is a private, non-profit corporation with no visible Federal support. The membership fee is currently US$1 million, with any project related costs being in addition to this. It is envisaged that when at full strength the MCC will employ some 450 research staff, about double the present number. The MCC R&D programme is aimed at producing significant technical advances over a five- to ten-year time scale. About half of the budgeted programme will relate to advanced computer architectures, including work on parallel processing, data base management systems, human-factors technology, AI, and knowledge based systems. This is very much in the image of the FGCS project. Other activities will relate to software development and to semiconductor device design, fabrication and packaging. In a US context the whole enterprise is highly experimental, and not without its critics. As for its long term viability, only time will tell. Not the least of the MCC's potential worries is the possibility of it inadvertently violating US anti-trust laws, with the consequent threat of treble damages being sought and awarded.

It might appear to the reader that the above discussion has little to do with image processing. To the contrary, we would strongly emphasise the point that in the future image processing, in a broad generic sense, will become as much an integral part of a computer's environment as the more basic features we are familiar with today. After all, at a very basic level it is an obvious and important means for increasing the user–machine bandwidth. Consequently the new generation computer systems which will embed this capability are certainly germane to our theme, and hence their inclusion here.

7.3.2 Manufacturing and automation

When the use of images and image processing are mentioned in the context of automating manufacturing processes, most people tend to think only of the use of vision systems for inspection and testing or perhaps, in the longer term, as an aid to robotic manipulation. Much has been written on this burgeoning topic (Gomersall, 1984; Miller, 1983), which we will not consider further here. In fact, the role of images, and image metaphors, is much more central and subtle, especially now that computers are increasingly involved in design, manufacturing, testing and distribution.

This growing centrality of computer technology in manufacturing technology has led to the coining of the generic descriptor 'computer integrated manufacturing' (CIM) as an all embracing term to describe this emergent technology. CIM as a generic concept embraces a large number of different design and manufacturing activities as illustrated in Fig. 7.7. This figure is based on a control orientated model developed by a technical panel, chaired by the author, which established the European Community ESPRIT CIM workplan (CEC, 1983) during 1982 and early 1983. This model portrays a fairly high level of decomposition, but will suffice for the purposes of this section. In no sense is the figure meant to imply that the model's various

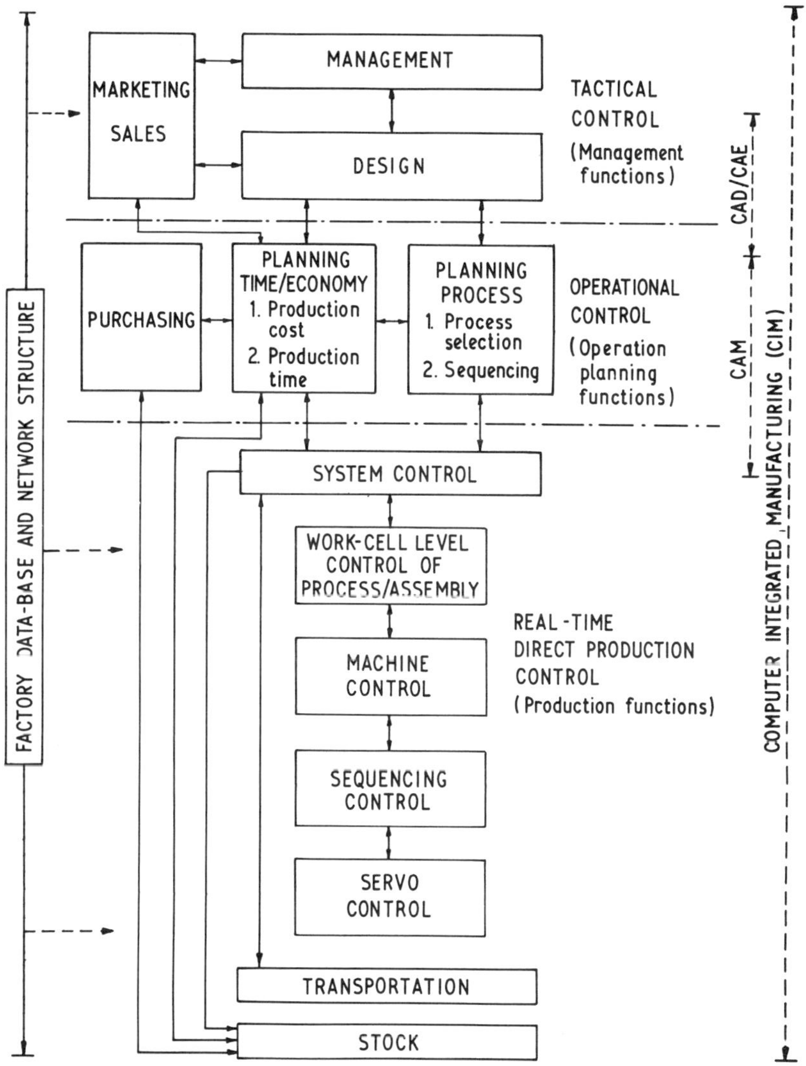

Fig. 7.7 Generic model of a manufacturing control system.

modules are contiguous or even on the same site. The implicit computer based control is divided into three main classes or levels: tactical, operational and direct. The first two tend to involve off-line functions, whilst the third is very much an on-line activity. The tactical level includes management, marketing, sales and design control. At the operational level, the production planning unit gets control information from the design and documentation unit, selects the production processes for the product described in the design documentation, plans and costs the process sequencing, establishes the parts inventory, and surveys the state of production equipment. At the direct control level, the system control unit controls the production distribution over the appropriate work cells, which may be involved with machining, processing, measuring, assembling, transportation or testing. This level receives product process plans and schedules from the operational level. The work cell control units decompose the received production programmes into subprogrammes for each individual machine or transporter within the work cell. These are decomposed, in turn, into real-time axis control vectors.

The penetration of computers into the various levels of manufacturing technology outlined above has led to the coining of a number of often ill-defined acronyms such as, *inter alia,* CAD (computer aided design), CAE (computer aided engineering), CAM (computer aided manufacturing) and CAT (computer aided testing). Any meaning we wish to give these, or similar terms, will arise naturally from the context of what follows. Until recently much effort has been expended in implementing increasingly sophisticated systems which separately perform these functions. Now the emphasis is very much more on overall integration.

Central to the model outlined above is the continuous flow of product and process specifying information from the design stages through to the manufacturing and testing activities, and of course in the reverse direction. It is the automating of this flow which is central to the whole concept of CIM, along with the notion of flexibility, versatility and cost effectiveness within this flow. This flow from design to product is the subject of much research and development activity in the USA, Japan and Europe. Most approaches involve, to a greater or lesser extent, the incorporation of IKBS techniques in order to cope with the intrinsic complexity of the design-to-product information and process flow. The IKBS techniques are being used primarily in order to capture the skills and knowledge of the human designers and operators involved in the current manual activities, and to facilitate the various transformations inherent in an automatic process.

The first steps in this design-to-product flow revolve around the computer aided design (CAD) workstation. Current CAD systems incorporate facilities for sophisticated colour graphics, the designing and draughting of mechanical parts, the simulation of kinetic behaviour, the analysis of stress-strain relationships, the formulation of three-dimensional solid geometry models, and the development of appropriate data bases. At the present time there is no general agreement on how CAD systems should be designed, implemented or used, especially in the context of integrating them into the broader CIM environment. This has led to considerable interworking incompatibility, especially between the older wire frame and the newer solid model representations, and the emergence of somewhat *ad hoc* standards.

Geometry and modelling are at the heart of any CAD system used for mechanical design. It should be mentioned in passing that this is also a truism for integrated circuit design, the differences being more in spirit than in kind. A solid modeller will normally contain a rich set of primitives, such as plane, sphere, cone, cylinder, torus, helix, ellipsoid, etc., and various facilities for combining these primitives, as well as introducing lighting and textural effects. The designer interactively creates a geometric model of the part or product to be machined, processed or assembled. It is important to realise that the geometric model represents an engineering design and not just a sophisticated picture drawn on a graphics display. The geometric model represents a formal structure stored in the computer on which the designer bases all investigations and analyses in order to create a product. The richer the geometric capability of the system, the greater the flexibility given to the designer.

The current trend in CAD is towards representing the product as an object, stored in one master three-dimensional geometric data base. Drawings are merely two-dimensional projections of this model which have been appropriately detailed. This so-called 'total product model' (TPM) not only maintains geometric connection between various three-dimensional objects, but also supports consistency in shapes, constraints and technical data derived from geometrical positions and dimensions. Design constraints, rules and assembly sequences are, to a greater or lesser extent, legitimate and important parts of the TPM. The importance of having good three-dimensional representations, the TPM, of potential products or part-products lies in the fact that it then becomes feasible to simulate both the product and its manufacture at the design stage. This leads naturally to the extreme case of off-line programming of the manufacturing process since such simulations must *de facto* generate, in some form, all the information necessary to specify fully the manufacturing process. Figure 7.8 schematically illustrates this point. This technique is already in fairly wide use for tightly circumscribed numerically controlled machining (CNC) operations, often within a flexible manufacturing system (FMS) cell. There are still some fascinating and fundamental problems which need to be solved before this approach can be used for complex assembly tasks. This is because assembly, as opposed to machining, is likely to be a 'sensor-rich' activity, if only because of the need to cope with design tolerances, uncertainties, errors, defects and choices. These sensors will incorporate, *inter alia,* both visual and tactile modalities. This is very different from the in principle fairly straightforward use of machine vision for industrial inspection and non-destructive testing. In the present context the sensors will be tightly linked into real-time servo loops and will need to be fast and precise in operation. The most difficult problem inherent in this approach is that of deciding how

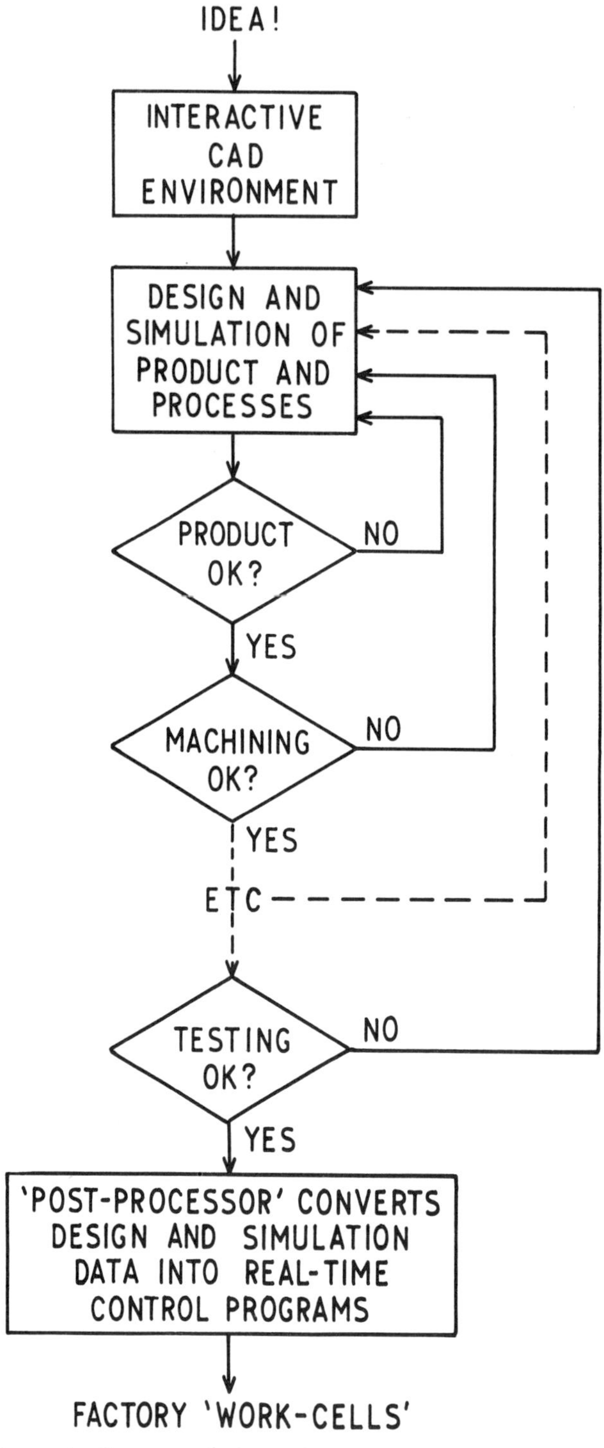

Fig. 7.8 Schematic diagram of the software 'flow' in a future CAD system incorporating an automatic off-line programming facility.

to model the uncertainties that the sensors are there to resolve. A knowledge based approach is almost certain to be mandatory. Answers to this and related problems are bound to have a significant effect on the manufacturing methods themselves.

It should now be clear that images and image processing have a lot to do with the concepts and processes outlined above. The data base containing the three-dimensional product models is in a sense a metaphor for the real world of products and manufacturing in exactly the same sense that an image is a metaphor for the imaged scene. The manipulation of the product models is an activity closely akin to image processing, especially in the context of incorporating visual servoing into assembly operations. Certainly a fundamental requirement for the future is going to be the ability to match models stored in the data base with those, perhaps in the form of a 2½-dimensional sketch, derived from a visual or tactile imaging system, and *vice versa.* The primitives, such as surface descriptors and their interrelationships, used for the data base solid model representations, and for the visual or tactile recognition systems, must have a large degree of commonality if they are to be made to interwork successfully. The ability to generate realistic sensor subsystem compatible 'scenes' from the model data base will be an important prerequisite for process and assembly simulation.

This centrality of models and images will emerge only slowly, with much of the impetus coming from global attempts to automate the design-to-product generative pathway completely. With this long term goal in mind some fairly basic two-dimensional industrial imaging systems are being supplied by a number of vendors (Winston and Prendergast, 1984) but it is only with the development of versatile and robust real-time three-dimensional imaging systems that significant progress will be made.

7.3.3 Medical image processing

Currently there are a number of imaging modalities to be found in medical diagnostic practice. These include conventional X-ray radiology, computed axial tomography (CAT) X-ray scanning, nuclear magnetic resonance (NMR) imaging (MRI), nuclear medicine, ultrasound techniques, positron emission tomography, cardiology and digital radiography.

Diagnostic imaging covers a wide range of technologies and techniques used to non-invasively image the body's internal anatomy and function. Similar techniques are also used for the non-destructive testing of inanimate objects in advanced manufacturing processes. The essence of a digitally based medical diagnostic imaging system is illustrated schematically in Fig. 7.9

Radiation is constrained to propagate through the object being imaged in such a way that an appropriate response function can be measured by a sensor or sensor array. Although the object has a three-dimensional

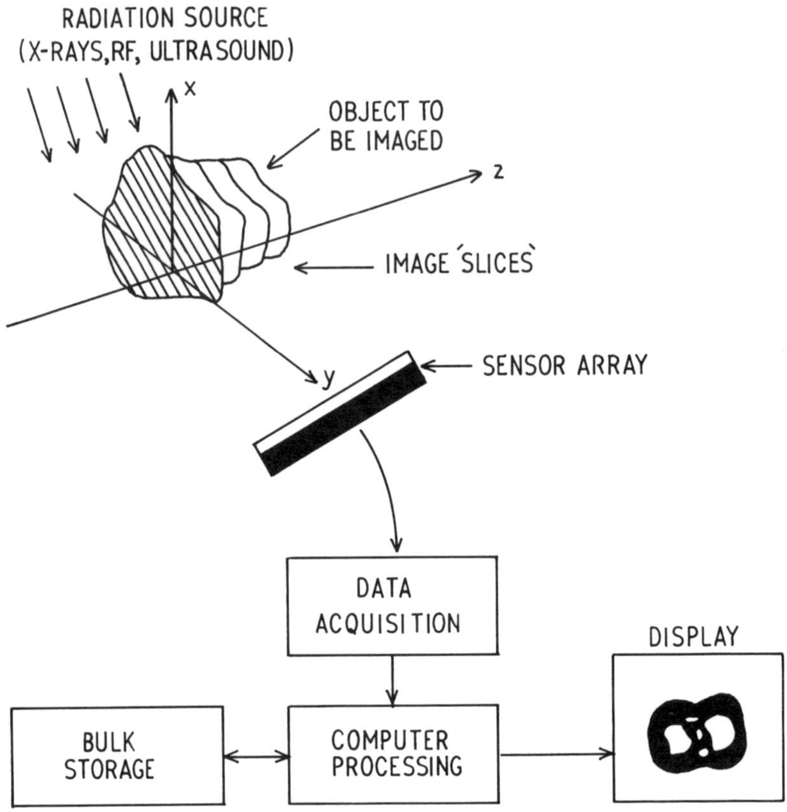

Fig. 7.9 Schematic diagram of a generic digitally based medical diagnostic imaging system.

structure, this dimensionality is lost in the irradiation and measurement process. Consequently, the output of any measurement sequence is sufficient only for a two-dimensional projected view of the organ being imaged. To overcome this deficiency, the radiation source is usually applied over a large number of angular positions, thereby generating a multiplicity of projections at the sensor array (see Section 1.2.8). Data aquisition is usually fairly conventional. The acquired data is passed either to a computer based processing system or directly to some form of storage and archiving system. Display of the computed image is performed by video processing subsystems, usually under operator control. Improved image quality is often achieved through the use of sophisticated mathematical algorithms, which can transform images in order to correct for patient movement, filter out noise, sharpen edges and enhance contrast (see Chapters 1 and 2). The gradual emergence of 1024 × 1024 image display formats and sophisticated reconstruction algorithms imposes some fairly stringent requirements on the processing system. Figure 7.10 shows the generic structure which these

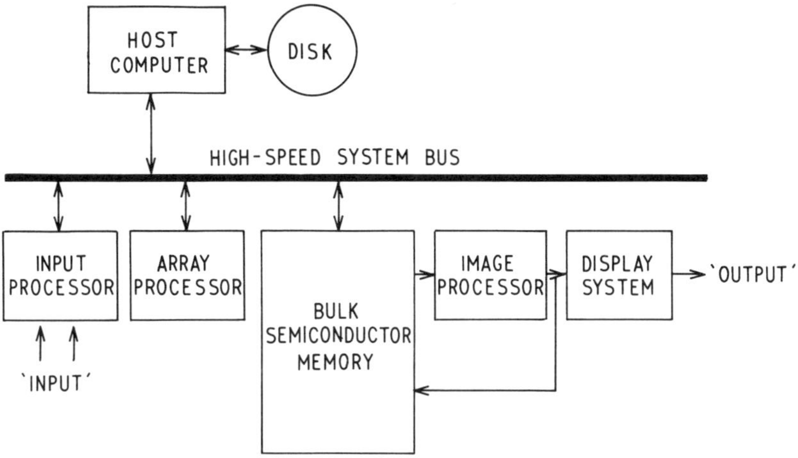

Fig. 7.10 Schematic diagram of a computer system suitable for processing medical images.

systems are tending to assume. Large amounts of high speed bulk memory, often heavily interleaved and multi-ported, are an essential ingredient if acceptable speeds of image generation and manipulation are to be achieved. This memory passes complete images to a pixel stream processor, the output of which is fed, in turn, to a display system. Data input to the system is usually transferred directly onto a high speed internal data bus, thus avoiding the necessity for transfer via a host computer link. In many applications an array processor has direct access to the bulk memory via the same high speed data bus. Emerging systems are starting to place an increasing emphasis on indirectly addressed data and table oriented operations, corresponding to the storage and processing of three-dimensional surface information.

Reconstruction techniques have been discussed in Section 1.2.8 and will not be discussed further here. Another diagnostic image generation and manipulation technique which is developing rapidly is 'digital radiography' in general and 'digital subtraction angiography' in particular. Digital radiography is an extension of the traditional technique of X-ray fluoroscopy. It involves detecting the X-rays transmitted through the patient with a scanning image intensifier and converting the output signals to a digital format for use by a computerised processing system. The image derived in this way is intrinsically two-dimensional. Digital radiography provides the opportunity to apply sophisticated image processing algorithms, including subtraction, edge enhancement and contrast enhancement (see Chapters 1 and 2) to the captured X-ray image. This has the important consequence of reducing the radiation exposure experienced by the patient, since lower dose levels are compensated for by improved sensitivity. Angiography is the X-ray examination of blood vessels, especially arteries, made more visible by the injection of a contrast medium. Digital subtraction angiography then involves the subtraction of two images, of the blood vessels under investigation, acquired before and after the injection of the contrast material. Some degree of enhancement is also involved. This application of digital processing techniques to angiography has a number of advantages: increased safety as a result of less stringent injection requirements and reduced X-ray dosage levels, and the ability to store the captured or processed images.

In this, and the various reconstruction-based imaging techniques, the increasing availability of cost effective VLSI based array processors and graphics engines will do much to reduce costs, at present high, and improve performance and reliability. The development of more efficient image processing algorithms will also play a part.

Having obtained some appropriate diagnostic images, the question of how to use and handle these images is a far from trivial one. A modern hospital is an excellent example of a situation where really formidable amounts of data, much of it in the form of diagnostic images, have to be captured, stored or accessed reliably and on a continuous basis. Ideally, a hospital information system must combine the provision of detailed patient specific records and services, along with easy access to a very large and diverse data base of patient records, text, data and images, some of which may stretch back over several decades. The establishment of such systems will not be an easy task. First, there is the sheer quantity of data, of which we shall say more later. Then there is the question of how economically and reliably to transfer an existing 'paper, film and filing cabinet' system onto a fully electronic one, or more appropriately how to combine the two for an interim period which may last for many years. The question of how to annotate image data with a clinician's comments is a difficult one. The best solution is to use voice annotation but this implies considerable storage overheads as well as predicating a full voice/data/image communications network, but there is no doubt that this is the way that such systems will move in the long term.

The aspect of such a system that is of interest to us here is the 'picture archiving and communications system' (PACS). In order to get some sort of feel for the scale of the system requirements we will look at the various hospital imaging modalities in a bit more detail. (*The discussion of PACS is based on material prepared by Brian Dalton and Ian Young of Picker International, Wembley, UK.*)

By far the largest source of patient data is the radiology department of any hospital, which tends to be regarded as a central resource for image based examinations. At the present time the bulk (circa 75%) of such a department's image data is produced and archived in the form of large high resolution photographic films, with most of the balance being produced in a digital form, stored on tape or disc, or on cine/video film. It is widely expected that there will be a steady rise in the use of all digital techniques, due in the main to the introduction of inherently digital imaging systems, such as CAT X-ray and MRI scanners and fully digital conventional X-ray systems. The latter technique, digital radiology and the allied digital subtraction radiology, are expected to have an explosive growth and may well capture as much as 50% of the total X-ray market by the end of the decade.

The radiology department of a typical large teaching hospital routinely performs some 150 000 radiographic examinations each year, producing some 450 000 radiographic images (Lehr, 1983) – roughly 2000 per working day. During a particular diagnostic session the radiographer will need immediate access to the current images, but will, in all probability, want equally easy access to particular images for that patient taken over a period of many years. For a single radiology department this is an imposing requirement but when one recognises that other hospital staff may well wish to have access to this data the difficulties become compounded many fold. It has been estimated (*ibid*) that in such a hospital some 10 000 images will be examined in a typical working day. If we assume that the conventional X-ray images, 75% of the total, are stored as 1024 × 1024 by 10-bit arrays, and the CAT and MRI images as 512 × 512 by 10-bit, we have a total data storage requirement of about 16 Gbits/day (4 Tbits/yr) and an access requirement of about 85 Gbits/day. Storing and archiving this amount of data in a reliable and accessible manner presents a formidable problem. Any image annotation will be on top of these figures.

A close relative of radiology is cardiology, which may be sufficiently specialised to warrant its own specialist equipment and facilities. The data captured is typically on cine film or video tape, which although at present mainly analogue is bound to become increasingly digital. A basic problem is how to store these relatively low quality moving sequences efficiently. The requirements for the inherently digital images arising in nuclear medicine are to be able to cope with low resolution and in many cases moving images. Ultrasound imaging, which also produces low quality moving images, is often found outside the radiology department, in obstetrics for example. At present ultrasound images are not usually archived, but rather tend to be used in a real-time interactive diagnostic mode. There is also a requirement to archive low grade image and graphic data such as EEGs, ECGs, thermography and charts as well as tables corresponding to patient drug and diet regimes, chemical, pathological and histological test results.

One important aspect of such a system will be the viewing facilities provided for the radiologists and clinicians. It is probably true that at this interface the system will be seen to be a success or a failure. The image–viewer interaction is a complex and sophisticated one, with much experience and practice having been built up around the relatively versatile cost effective light-box film viewer and sheet high resolution photographic film. Typical 350 × 430 mm X-ray film resolution would be matched, at typical viewing distances, by a 1024 × 1024 image display on a high resolution studio quality TV monitor, although there is a feeling that 2048 × 2048 would be preferable. Unless the digitised display can readily offer features not available with standard film, as is bound to be the case in practice, people are going to be reluctant to abandon the more familiar photographic film. There are considerable perceptual differences between viewing photographic hard copy and TV images. These perceptual differences are particularly complex in radiography because of the requirement to view simultaneously a large number of different images in order to build up a composite solid picture of the imaged object. This is an area where AI approaches to inference based image interpretation could prove useful, in both aiding and deskilling diagnostic procedures. The transition is going to have its difficulties, if only because human nature is what it is. After all, how is one to maintain the strong placebo action invoked solely by the doctor's close interaction with the patient? The ease of access provided by a distributed digital archiving system is bound to raise important issues of professional self-interest, security and confidentiality.

As we have indicated, the staff of our typical teaching hospital will access, on the average, some 85 Gbits of radiological data per day. If this is uniformly distributed over a working day, a very unlikely assumption, the average data transfer rate is 1 Mbit/s. This appears to be well within the raw bandwidth of a typical baseband LAN such as the 10 Mbit/s Ethernet. In practice the non-uniform distribution of useage, the overheads intrinsic in the LAN operating system service interfaces and the other networked services, mentioned above, that the LAN will be expected to provide, means that a broadband system with a bandwidth of as much as 100 Mbit/s will be required. This implies either VHF co-axial cable or optical fibre. At present there is little coherent consensus on the best solution for such a hospital network, though some companies are readying systems for what promises to be a large and profitable market.

The actual image archiving and the provision of ready access, if only because of its scale, is probably the most difficult aspect of a hospital information system. The circa 2000 images (16 Gbits) produced every day, by themselves imply a daily storage media requirement of four multi-platter discs, or forty standard magnetic tapes, or two high density tapes, or two optical discs. An image archive covering some twenty years would need approximately 14 000 optical discs, for example! The fact that a ten-year-old radiograph should be as accessible to a clinician as one taken the previous day poses some fundamental data base design problems. It is not clear at present how this is to be resolved in a cost effective manner. In the longer term a knowledge based system will no doubt be used.

7.3.4 Defence and aerospace systems

There are two compelling reasons why image processing will play an increasingly important role in future defence and aerospace systems. These are the need for autonomous operation and the need to make greater use of the outputs from a diverse range of sophisticated sensors. Consider for example the importance that satellite based remote sensing of the earth's resources has already assumed, and the beautiful and graphic planetary images sent back to earth during the various interplanetary space probe missions to date.

For obvious reasons, rapid response is becoming increasingly important in defence systems. This increasingly forces decisions to be made without explicit operator intervention. Indeed, in many cases operator intervention will not be physically feasible. Targeting, surveillance, command and control activities all need to rapidly make sense out of large amounts of disparate and possibly unreliable information. This is bound to require the application of advanced AI methods, especially those relating to image understanding. Subsequent control may be passed to an autonomous system which will attempt to select an appropriate target from the captured image data set, and initiate an appropriate response. The emergence of high performance autonomous systems has been inhibited by the inability of present day computers robustly to direct actions that fulfil complex objectives in unpredictable situations. One obvious area of concern is the role of autonomous systems during a transition from peace to hostilities, and *vice versa,* when the rules of engagement may alter very quickly. An extremely cogent example of such a dilemma is the case of any projected defence against strategic nuclear missiles, where systems act so rapidly that almost complete reliance may have to be placed on automated systems in circumstances where the complexity and unpredictability of factors affecting irreversible decisions is very great indeed.

The current trend in sensor development and utilisation is to produce far more information than a human can reasonably assimilate, especially in a saturation attack or when trying to cope with complex tasks such as flying a supersonic, terrain following aircraft. As equipment becomes more complex it is going to have to take more responsibility for its own adjustment and operation. It must retain its function and efficiency even when used by a tired or untrained operator. In the foreseeable future we should not look for significant improvement in the physical performance of the sensors themselves. Rather, it is going to be increasingly important to make much greater use of the information that is gathered and to give the sensor subsystems greater intelligence.

In imaging systems the display is often the weak link which can be significantly improved by such features as auto-cueing and menu selection. Transferring images from place to place is an expensive activity, both in terms of facilities and time, and there is bound to be considerable pressure to send coded information compressed into drastically reduced bandwidths. Context is important here. Sending the full picture of a stereotype of a pine tree can involve several megabytes of data. As an extreme example of compression, the two words 'pine tree' may well be quite adequate in some contexts.

Image processing plays an important role in the US 'Strategic Computing Programme' (DARPA, 1983), managed by the Department of Defense's Advanced Research Projects Agency. Having started in fiscal year 1984, it aims to spend US$600 million over a five-year period with the intention of creating a new generation of military oriented 'machine intelligence technology'. The hope is that the resulting technology will have unprecedented capabilities and hold the promise to increase greatly both national security and economic strength during the coming decade. The programme is very much applications driven. It starts by focusing on three demanding military applications that require machine intelligence technology. These are an autonomous vehicle, a pilot's associate and a carrier group battle management system. These applications are expected to generate diverse requirements for functions such as machine vision, speech recognition and synthesis, natural language understanding and expert systems technology as well as encourage synergistic interactions among the developers of the new technology. These intelligent functions are to be implemented on advanced machine architectures and designed and fabricated using VLSI microelectronic techniques which meet the application performance requirements. Table 7.4 lists the key areas of technical advance which DARPA expects to facilitate the production of high performance machine intelligence systems. Figure 7.11 is a schematic summary overview of the overall programme's structure and goals.

Table 7.4. Key areas of technical advance according to DARPA (1983) that can be expected to produce high performance machine intelligence systems.

Expert Systems: Codifying and mechanising practical knowledge, common sense, and expert knowledge.
Advances in AI: Mechanisation of speech recognition, vision and natural language understanding.
System Development Environments: Methods for simplifying and speeding system prototyping and experimental refinement.
New Theoretical Insights in Computer Science.
Computer Architectures: Methods for exploiting concurrency in parallel systems.
Microsystem Design Methods and Tools.
Microelectronic Fabrication Technology.

Of the three 'demonstrator' applications, the autonomous vehicle is the one which is pivotally dependent on very significant advances in image processing technology. The proposed vehicle will contain a small modular computer control system. Vision modules will be included that provide basic scene processing and object recognition capabilities. With vision modules as input devices, a symbolic processor module will then be able to directly process fragments of pictorial, graphic and three-dimensional scenic images. When further supported by rule based inferencing and image understanding in a compact but powerful symbol processor and interfaced with specialised motor controlled systems, these vision modules will enable the computer controlled autonomous vehicle to 'see', to move about and to interact

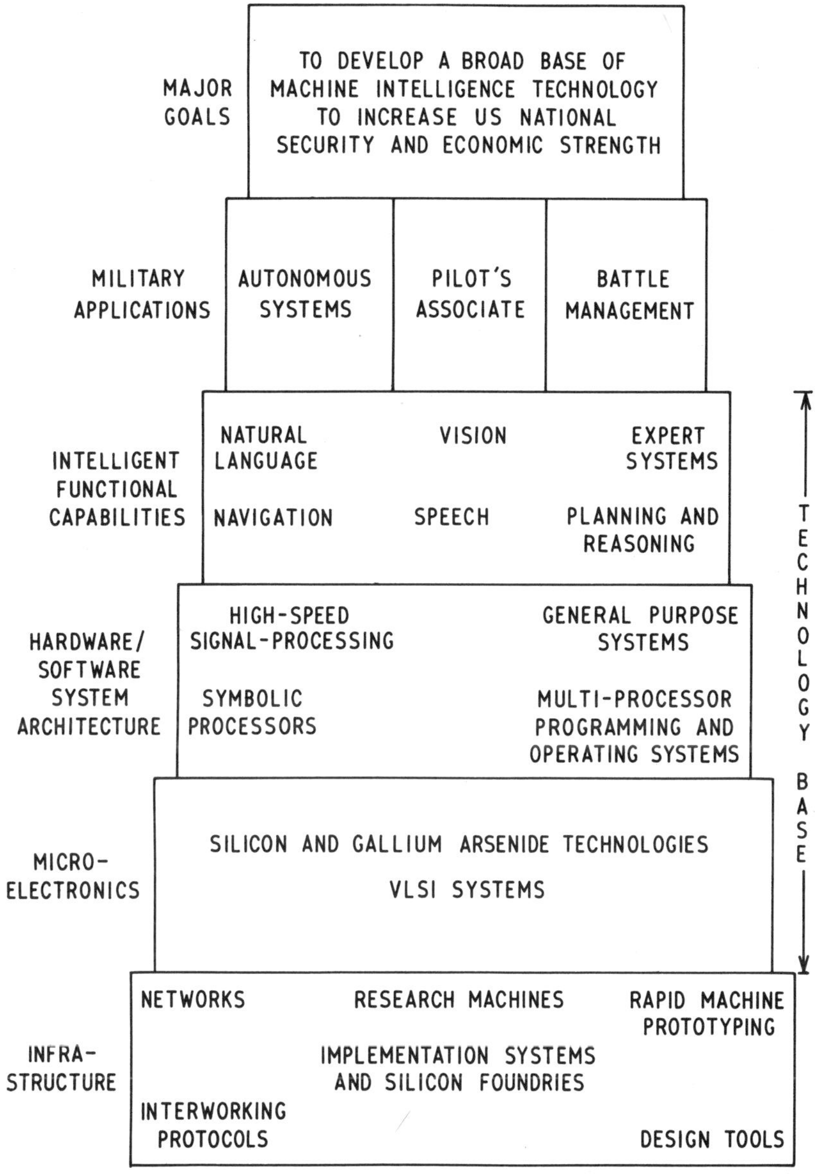

Fig. 7.11 Schematic representation of the goals and program structure of the DARPA Strategic Computing Programme.

intelligently with its environment. The resulting vision, scene interpretation, and motor control processes will be, it is anticipated, at the very least analogous to those found in lower animals.

To develop an autonomous land vehicle with the capabilities envisaged requires an expert system for navigation as well as a sophisticated vision system. The expert navigation system must plan routes using digital terrain and environmental data, devise strategies for avoiding unanticipated obstacles, estimate the vehicle's position from landmarks and other data, update the on-board digital terrain data base, generate moment-to-moment steering and speed commands, and monitor vehicle performance and on-board systems. All these functions must be accomplished in real-time to near real-time while the vehicle is moving at speeds up to 60 km/h. Scaling up from laboratory experiments indicates that such an expert system demonstration would require of the order of 6500 rules firing at a rate of 7000 rules/s. Current systems contain fewer rules, 2000 on average, and fire at a rate of 50–100 rules/s.

The vision system must take in data from imaging sensors and interpret these data in real-time to produce a symbolic description of the vechicle's environment. It must recognise roads and road boundaries, select, locate and dimension fixed and moving obstacles in the roadway; detect, locate and classify objects in open or forrested terrain, locate and identify man-made and natural landmarks, and produce thematic maps of the local environment, while moving at speeds up to 60 km/h. Scaling up computing capabilities used in laboratory vision experiments suggests an aggregate computing requirement of 10–100 GIPS to accomplish the above tasks. This compares with capabilities, for example, of 30–50 MIPS in today's most powerful von-Neumann type of computers. Of equal importance with these required computing capabilities is the weight, space and power required by the computing systems. For a land reconnaissance vehicle, for example, the computers should occupy no more than about a cubic metre, should weigh less than 250 kg, and should consume less than 1 kW of power, including environmental support. The requirements represent at least 1–4 orders of magnitude reduction in weight, space and power over today's computing systems. For aerospace and undersea autonomous vehicles, the constraints and requirements will be even tighter and need to include the capability to operate in high radiation environments.

As a foretaste of what is to come, note that in June of 1984 the dummy warhead of a Minuteman-One ICBM, launched from the Vandenberg Air Force Base in Southern California, was intercepted and destroyed high over the Pacific Ocean by an optically guided missile launched from the Kwajahein test range. A long wavelength infra-red sensor in the interceptor was activated after the second-stage booster rocket had been fired. With the aid of two computers, processing data at some 18 MIPS, the interceptor homed in on the dummy warhead and destroyed it while it was still some 160 km above the ground; despite a deliberate initial misalignment of some 32 km, which was intended to test further the interceptor's homing capabilities. Proponents of the US 'star wars' strategic defence initiative (SDI) hailed this success as solid proof of their convictions, while critics were quick to point out that the imaging sensor had no way of distinguishing between real targets and decoys and may well have completely failed in the nuclear environment of a full scale attack. Whichever view is right, the test was still a convincing demonstration of the power of a sophisticated imaging guidance system in a space environment. Even if the SDI systems are never installed, for whatever reason, the associated R & D programmes are bound to come up with some impressive advances in aerospace imaging technology.

7.4 EPILOGUE

In conclusion, we hope that a number of discernible themes have emerged from the preceding chapters. The following list summarises what we hope are the aspects of contemporary image processing that the reader has come to regard as especially significant.

(i) The ubiquitous nature of image processing and its centrality in advanced information processing.
(ii) The importance of advanced computer architectures to the realistic implementation of image processing algorithms.
(iii) The emerging need for formal tools which can link abstract algorithmic and computational aspects of a system design with potential implementations.
(iv) The strong enabling push provided by VLSI design and fabrication technology.
(v) The central and energising role of AI techniques.

In the end, we recognise that there is still much to be done, and whatever the future, it will be in reality different from what we have implied here. It is our sincere hope that from the various and diverse themes and topics discussed in this book you, the reader, have gleaned some sense of where image processing, both as a science and an engineering discipline, currently stands and the directions in which it is still evolving.

BIBLIOGRAPHY

Benioff, P. (1982) 'Quantum Mechanical Models of Turing Machines that Dissipate No Energy', *Phys. Rev. Letters,* **48**, 1581–1584.

Bennett, C.H. (1973) 'Logical Reversibility of Computation', *IBM J. Res. Develop.,* **17**, 525–532.

Bennett, C.H. (1984) 'Thermodynamically Reversible Computation', *Phys. Rev. Letters,* **53**, 1202.

Keyes, R.W. and Landauer, R. (1970) 'Minimal Energy Dissipation in Logic', *IBM J. Res. Develop.,* **14**, 152–157.

Machlup, F. and Mansfield, U. (eds.) (1983) *The Study of Information: Interdisciplinary Messages,* J. Wiley.

Seitz, C.L. (1984) 'Concurrent VLSI Architectures', *IEEE Trans. on Computers,* **C-33**, 1247–1265.

REFERENCES

Baer, J-L. (1980) *Computer Systems Architecture,* Computer Science Press.

CEC (1983) *ESPRIT Workplan,* IT–TF–0001–83.

Cohen, P.R. and Feigenbaum, E.A. (1982) *The Handbook of Artificial Intelligence,* Vol. 3, Pitman.

DARPA, (1983) *Strategic Computing. New-Generation Computing Technology: A Strategic Plan for its Development and Application to Critical Problems in Defence,* 28 Oct., DARPA.

Duff, M.B.J. (ed.) (1983) *Computing Structures For Image Processing,* Academic Press.

Fernbach, S. (1984) 'Supercomputers – Past, Present and Future', *FGCS,* **1**, 23–30.

Feigenbaum, E.A. *et al.* (1981–82) *The Handbook of Artificial Intelligence,* Vols. 1, 2 and 3, Pitman.

Gomersall, A. (1984) *Machine Intelligence: An International Bibliography with Abstracts on Sensors in Automated Manufacturing,* IFS and Springer-Verlag.

Grinberg, J. *et al.* (1984) 'A Cellular VLSI Architecture', *IEEE Comp.* Jan. 69–81.

Hayes-Roth, F., Waterman, D.A. and Lenat, D.B. (1983) *Building Expert Systems,* Addison-Wesley.

Hillis, W.D. (1981) 'The Connection Machine', *Report AIM–646,* MIT.

Hwang, K. and Briggs, F.A. (1984) *Computer Architecture and Parallel Processing,* McGraw-Hill.

ICOT (1984) *Proc. of the Int. Conf. on Fifth-Generation Comp. Sys.,* 6–9 Nov., ICOT.

ISO (1982) *Report on Concepts and Terminology for the Conceptual Schema and the Information Base,* ISO/TC79/SC5/WG3.

Karatsu, H. (1984) 'Applicable Fields and Social Impacts of Advanced Computer', *ICOT J.,* Oct., 2–12.

Kelly, M.J. *et al.* (1982) 'Submicrometre Device Physics', *GEC J. of Sc. and Tech.,* **48**, No. 2, 108–111.

Kowalski, R. (1984) 'The Relation Between Logic Programming and Logic Specification', *Phil. Trans. Roy. Soc. Lond.,* A, **312**, 345–361.

Kung, H.T. (1980) 'The Structure of Parallel Algorithms'. *Adv. Comput.,* **19**, 293–326.

Lehr, J.L. (1983) 'The Impact of Manual and Computer-Assisted PACS for Automated PACS', *SPIE,* **418**, PACS II, 6–13.

Machlup, F. and Mansfield, U. (eds.) (1983) *The Study of Information: Interdisciplinary Messages,* J. Wiley.

McCorduck, P. (1979) *Machines Who Think,* W. H. Freeman.

Mead, C.A. and Conway, L.A. (1980) *Introduction to VLSI Systems,* Addison-Wesley.

Mead, C.A. and Conway, L.A. (1980a) 'Physics of Computational Systems', in *Introduction to VLSI Systems,* Addison-Wesley, 333–371.

Miller, R.K. (1983b) *Machine Vision for Robotics and Automated Inspection* (3 Vols.), Technical Insights Inc.

Moore, W.R. (1984) 'Fault-tolerant Techniques for Yield Enhancement', *GEC J. of Res.,* **2**, No. 1, 1–15.

Moto-Oka, T. (ed.) (1982) *Fifth Generation Computer Systems,* North Holland.

Sakakibara, K. (1983) 'From Imitation to Innovation: the VLSI Semiconductor Project in Japan', MITI WP 1490–83.

Seitz, C.L. and Matisoo, J. (1984) 'Engineering Limits on Computer Performance', *Physics Today,* May, 38–45.

Seitz, C.L. (1985) 'The Cosmic Cube', *Comm. ACM.,* **28**, No. 1, 22–33.

Shapiro, E. (1983) 'The Fifth-Generation Project: A Trip Report', *Comm. ACM.,* **26**, No. 9, 637–641.

Shapiro, E. (1984) 'Systolic Programming: A Paradigm of Parallel Processing', *Proc. Int. Conf. on Fifth-Generation Comp. Sys.,* ICOT, 458–470.

Shibayama, S. *et al.* (1984) 'A Relational Data-base Machine with Large Semiconductor Disc and Hardware Relational Algebra Processor', *New Generation Comp.,* **2**, 131–155.

Sloman, A. (1983) 'Intelligent Systems: A Brief Overview', *Intelligent Knowledge-Based Systems,* **1**, Annexes, SERC-DoI, 1–20.

Uhr, L. (1984) *Algorithm-Structured Computer Arrays and Networks,* Academic Press.

Unger, S.H. (1958) 'A Computer Oriented Towards Spatial Problems', *Proc. IRE.,* **46**, 1744–1750.

Winston, P.H. and Prendergast, K.A. (eds.) (1984) *The AI Business: The Commercial Uses of Artificial Intelligence,* MIT Press.

Index